CONSTRUCTION MANAGEMENT

CONSTRUCTION MANAGEMENT

DANIEL W. HALPIN
Georgia Institute of Technology

RONALD W. WOODHEAD
University of New South Wales

JOHN WILEY & SONS
New York • Chichester • Brisbane • Toronto • Singapore

Library of Congress Cataloging in Publication Data:

Halpin, Daniel W
Construction management.

Includes index.
1. Construction industry—United States—Management.
2. Construction industry—Law and legislation—United
States. I. Woodhead, Ronald W., joint author.
II. Title.

HD9715.U52H324 624'.068 80-11540
ISBN 0-471-34566-0

Printed in the United States of America

20 19 18 17 16 15 14

Preface

The construction industry is the largest single sector in the U.S. economy. It is a fragmented and diffuse industry encompassing both very small and very large contracting groups as well as the in-house construction forces of government and semigovernment agencies. It also includes many professional groups such as architects, engineers, construction managers, and management consultants. In addition, materials suppliers and vendors as well as other support groups are part of this massive industry. It is not surprising then that it speaks with many voices, and that at times highly quantitative methods are appropriate while at other times the intuitive or empirical approach is all that is available. The management of construction is at one and the same time an art and a science. It contains elements of both. Therefore, construction managers must be masters of a wide range of qualitative and quantative subjects. In this sense, their success is founded not on extraordinarily high proficiency in one or two areas, but instead on a very high level of competency in a large number of areas. Construction managers are like decathalon athletes. It is their aggregate score on a number of events that determines whether they win or lose.

This textbook is written to help students gain a perspective regarding the industry and some cross-sectional understanding of the things to be mastered if they wish to be successful as construction managers. More than a certain amount of skill and knowledge is required to accomplish this task successfully. The student must be aware of the resources—money, machines, material, and men—that are basic to realizing a construction project. These basic resources are often referred to as the four M's of construction. They must be carefully and professionally committed and managed within a construction environment of contracts, through the communication of ideas and under the changing impact of weather, unforeseen events, and varying conditions.

The text material is built around a set of chapters that introduce the nature of the four basic resources (money, machines, material, and men) and the concepts for their management. After initial chapters on the construction environment and the traditional building process approach to project design and construction, the student is introduced to the legal and management structures within which the basic resources are managed. Finally, a chapter on the construction management approach to projects attempts both to pull together most of the concepts introduced thus far and to present to the student a fairly recent and innovative management form that has been developed by the industry.

The text presents unusual end-of-chapter Questions and Exercises. Some are designed to help the student grasp the material presented in each chapter. Others encourage the student to go out into the construction world and gain an understanding of the relevance of the basic material as well as first-hand knowledge of current practice and management procedures. These Questions and Exercises can form the basis for class assignments, term projects, or research efforts.

The material in this textbook has been successfully taught at the junior and senior level at a number of universities. The coverage is broad in scope and designed to provide an introduction to construction management to students from a wide range of academic backgrounds. It should be attractive to students with architectural, engineering, engineering technology, and urban planning backgrounds.

Many people have made valuable contributions to this book. Mr. Peter Forster made many comments and suggestions that led to the improvement of Chapters 7, 8, 10, and 11. Mr. Frank Spears of Henry C. Beck Co., Atlanta, contributed to Chapter 12, Materials Management. Information in Chapter 18 is based on interviews with Mr. Richard S. Hartline of J. A. Jones Co., Atlanta, conducted by Mr. Robert Graves. In addition, material presented throughout the text is based on research projects submitted by students over the past seven years in an undergraduate course taught at the University of Illinois and Georgia Institute of Technology. We would like to thank the many students who by their interest and dedication have motivated us to write this book. Finally, we are grateful to Miss E. Caterson for her dedicated typing of the final manuscript.

Daniel W. Halpin
Ronald W. Woodhead

Contents

CHAPTER ONE

The Construction Environment

1.1 INTRODUCTION

Construction can be described as the creative effort that converts the four M's of construction—materials, manpower, machines, and money—into a constructed facility. The construction process is concerned with the fabrication and field erection at specific locations of specially designed facilities defined by the customer-owner. Construction is therefore field and project oriented, producing unique products rarely suited to mass production or a controlled manufacturing environment.

The magnitude and scope of the construction effort involved in a project will depend on its size and complexity and many other factors. Items that must be considered include (1) the variety of technologies and the type of construction involved; (2) required quality control standards; (3) the geographic location of the construction site and the nature of the work environment; and (4) the planning and management skill of the constructor. At any time the dynamic movement of men and equipment on a construction site graphically reflects the level of construction activity. During the period of construction, the variety and sequence of the construction operations and the manner in which they are performed provide visual indication of the construction effort. Not so obvious perhaps to the casual observer are the planning, procurement, and management efforts expended on the project. Nevertheless, these supporting efforts are vital to the successful realization of a constructed facility.

Construction is a complex undertaking involving the interaction and co-ordination of the skills and efforts of a large number and variety of agents.

Some agents acting for the customer-owner are concerned with defining the project and ensuring that performance requirements are met in the constructed facility. These agents focus on quality control, general supervision, and monitoring functions during construction. Agents associated with the constructor are involved either with the planning, directing, and management of the construction effort or with the actual work effort itself. Other agents, however, may be concerned with supporting the construction effort through the acquisition and supply of the resources needed for the project. In traditional construction practice, these groups of agents can be readily identified with architect/engineers, contractors, labor, and material and equipment suppliers. In some cases, however, this traditional division and organization of the construction process may not be considered responsive to the needs of the owner, the contractor, or those concerned with the management of the project. In such cases, a variety of organizational forms can be utilized for the management of the construction process.

We can include management as the fifth "M" of construction. Construction management comprises the planning, initiating, and directing of the construction process required in a specific project. The nature and scope of the construction management effort will depend on the technological complexity and magnitude of the construction effort and on the business, organizational, and contractual environment established between the various agents involved in the project and construction process.

This chapter outlines the nature of the construction process and introduces the basic problems associated with the planning and management of construction projects. The scope of construction management is introduced and related to the content and layout of the material presented in this text.

1.2 THE TOTAL BUILDING PROCESS

Construction is engineering in action. It is performed by the skilled application of human effort to the basic operations that make up each construction method. Construction is the productive effort that produces the finished product. From this point of view, the construction process is a part of a much more comprehensive effort, which is triggered by the awareness of a need for a facility. It is terminated when the completed facility has outlived its usefulness and is scheduled for demolition or conversion to another use. This larger process, of which the construction process is a component, can be referred to as the *total-life building process*. Other components of the building process that reflect its total life focus are project formulation, planning, engineering design, use and maintenance, and disposal. The basically linear nature of the building process is shown in Figure 1.1. The manner in which the construction process is related to the engineering and design processes for a particular project establishes a characteristic framework and environment for the construction

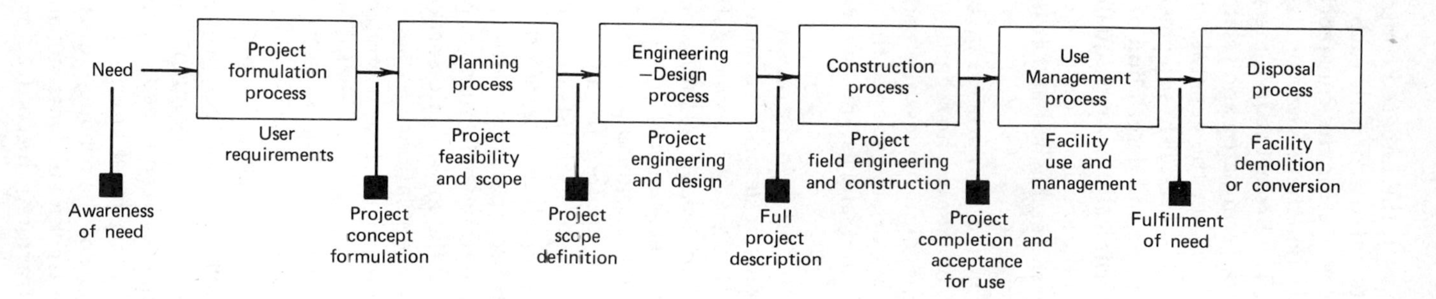

FIGURE 1.1 Linear nature of total building process.

effort. This framework together with the imposed contractual and working relationships between the agents entrusted with each component process establishes the organizational structure for the construction of the project.

The building process is initiated when there is a recognized need for a facility commitment to satisfy this need. The project scope emerges as user requirements for the conceived facility are formulated and related to available technological and financial resources. Within the general framework of what is attainable and authorized, engineering establishes the technological details of the project. Design teams develop the full description of the project. Project definition emerges in terms of technical drawings, quality control and performance standard criteria, and in the specifications and contract documents. Once full project details have been established and resource commitments authorized, construction can commence.

As mentioned previously the construction process involves the interaction and coordination of a large variety of agents who regard, and participate in, the construction process from different points of view and with different technical and professional responsibilities. The organization of the construction process for a particular project is built around, and establishes, the manner in which these agents interact.

The relationship of one agent to another in the construction environment is often very complex but can generally be described by one or more basic types. These basic relationships include (1) the *master-servant* relationship in which one agent hires the services of another for wages; (2) the *business-service* relationship that characterizes the freedom of interchange of goods in the market place; (3) the *contractual* or formal *legal obligation* relationship in which one agent or group of agents freely bind themselves to another for a consideration to perform some services under uniquely defined constraints or contract conditions; and (4) that of the *intimate cooperation of equals* in a team effort. The frankness of communications, or the interest in cooperation, between agents will be affected by the form of the relationship, and obligation, of one agent to another. In this respect, the manner in which the construction process is organized for a project establishes the nature of the relationships between large groups of agents and defines the construction management environment.

The traditional approach to construction basically follows the linear form of the building process, and is illustrated in Figure 1.2. Most construction is performed in this organizational and contractual relationship between owner, consultant, and construction agents. Usually a consulting architect/engineer (A/E) group, acting for the customer/owner, performs the detailed project analysis and design and provides overall management and supervision of the contract. The contractor, generally referred to as the principal or prime contractor, working under the specific terms of a contract assumes the detailed control, planning, and management of the productive effort required to build the facility. In many cases, the contractor subcontracts out specific portions of the project to specialized subcontractors. The contractor may find this action

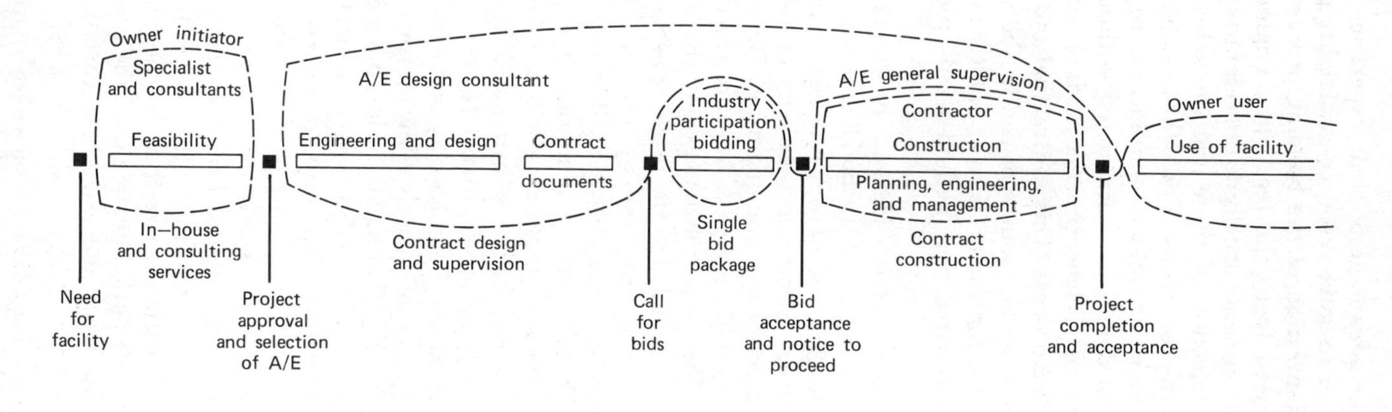

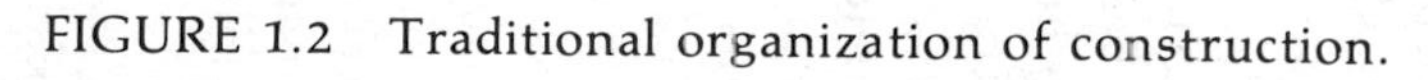
FIGURE 1.2 Traditional organization of construction.

desirable in order to take advantage of their expertise, to reduce financial commitments and risks, or to solve resource availability problems.

In this traditional organization of the building process, the construction process is totally separated from the feasibility, engineering, and design processes. The contract documents usually do not refer to specific construction methods and leave the selection of these and the solution of related field engineering problems to the contractor. Thus, the design basis and its implied construction rationale are not readily communicated to the contractor. In some cases project design details are formulated without consideration of construction methods and costs. Hence practice tends to enforce the segmental nature of the total building process. This contractual and construction management environment is often the source of conflict between the architect/engineer consultant and the contractor, if an attempt is made by the contract manager to enforce both construction method details and product performance on the contractor.

In the traditional process potential bidders can be supplied with a full project description, since the linear nature of the building process generally ensures that the preceding engineering and design processes have been completed. With this detailed knowledge of a project, bidders can eliminate many uncertainties relating to material types and quantities. Rarely in practice, however, is project definition totally complete, nor are the contract documents sufficient in every detail. Therefore, to some extent, the lack of details and the interpretation as to what is included in the project raise many conflicts in the management of the construction process. From the owner's (architect/engineer's) point of view the competitive bidding environment, in which prospective constructors bid the construction price of the project, enables the owner to select the best contractor (not necessarily that one with the cheapest bid). It also provides him* with a "ball park" estimate of what a segment of the construction industry expects the project to cost. However, the owner must be wary of those bids that appear low. Low bids may be the result of the contractor overlooking some detail or not adequately allowing for risks associated with the construction effort. On the other hand, the low bid may reflect an innovative solution to a construction method and correctly reflect the expertise of the bidder. In establishing the bid price, the contractor must first carefully estimate the material, equipment, and labor content of the project and the cost of managing the project. He must also establish a profit margin that reflects the risk level of the project, the competition and environment in which he operates, and the monetary return on his investment that he needs to stay in business. Finally, if the project is of sufficient size that the construction effort will be extended over a considerable period of time then the contractor must either consider potential inflation costs and wage-base variations in his initial bid

*From this section on, we usually use masculine pronouns when referring to the people involved in the construction process. We have done so for convenience and succinctness.

price or request rise-and-fall contract clauses to offset variations in resource costs from that known at the time of the bidding.

The traditional approach to project definition and construction management will be examined in detail in Chapters Two and Three. Other methods of developing the building cycle and the contractual basis for these approaches will be presented in Chapter Four. Finally the "Construction Management" approach will be discussed in Chapter Eighteen.

1.3 THE PROJECT ENVIRONMENT

The project concept as a convenient approach to work definition may arise in a number of ways. First it may emerge naturally as a single and unique package fulfilling the needs of a customer-owner who ventures into interaction with the construction industry on a one-time basis. Construction of a new addition to a hospital complex is a typical example of this type of one-time requirement for construction assistance. Alternatively, an industrial or commercial enterprise or a governmental agency may have a continuing-growth program or ongoing role to perform in society. In this environment, a portion of their works program may be conveniently packaged into a project definition and let out for construction. General Motors, for instance, is continually expanding its plant and, therefore, packages each capital plant expansion in the form of a project. Finally, local government organizations which perform in-house construction and maintenance programs, may find it convenient to identify a portion of their effort as a specific project if only for ease in procurement of resources and management control. For instance, the department of streets and public works identifies project work packages (e.g., patch pavement on Peachtree St. between locations A and B) as work units for control and management purposes.

The project concept comes naturally to the contractor because it represents the common interest he has with a specific customer. As such, it is defined in terms of special contractual, financial, and construction management relationships. The project format is convenient to the contractor's mode of operation. It provides the basis for effective project control in both the owner-contractor relationship that must be established and in the contractor's in-house construction management approach to the project.

Construction projects can be broadly classified as building, engineering, or industrial construction depending on whether they are associated with housing, social works, or manufacturing processes. The building construction category includes facilities commonly built for habitational, institutional, educational, light industrial, commercial, social, and recreational purposes. Typical building construction projects include office buildings, shopping centers, sports complexes, banks, and automobile dealerships. Building construction projects are usually designed by private architect/engineers and are constructed by a contractor for a private client. The materials required for the construction emphasize the architectural aspects of the construction.

Engineering construction usually involves structures that are planned and designed primarily by engineers (in contrast to architects). Normally, engineering construction projects provide facilities that have a public function and, therefore, public or semipublic (e.g., utilities) owners generate the requirements for such projects. This category of construction is commonly subdivided into two major subdivisions; thus engineering construction is also referred to as (1) highway construction and (2) heavy construction.

Highway projects are generally designed by state or local highway departments. These projects commonly require excavation, fill, paving, and the construction of bridges and drainage structures. Consequently, highway construction differs from building construction in the owner/designer/contractor relationship. In highway construction, owners normally prepare the design with their own in-house engineering staff so both the owner and the designer are public entities.

Heavy construction projects are also publically oriented and include sewage plants, utility projects, dams, transportation projects, pipelines, and waterways. The owner and design firm can be either public or private depending on the situation. In the United States, for instance, the U.S. Army Corps of Engineers (public body) uses its in-house design force to engineer public flood protection structures (dams, dikes) and waterway navigational strucures (river dams, locks, etc.). Public electrical power companies use private engineering firms to design their power plants. Public mass transit authorities also call on private design firms for assistance in the engineering of rapid transit projects.

Industrial construction usually involves highly technical projects in manufacturing and processing of products. Private clients retain private engineering firms to design such facilities. In some cases, specialty firms perform both design and construction under a single contract with the owner/client.

A slightly modified approach to project classification is shown in Figure 1.3. This figure shows the Construction Scoreboard Section of the *Engineering News Record* magazine, which reflects the weekly dynamics of the construction industry in the United States. This breakdown of construction identifies three major construction categories as follows:

1. Heavy and highway.
2. Nonresidential building.*
3. Multiunit housing.

These major categories are further dissected as shown, to retlect the major areas of specialization within the construction industry.

*This includes both building and industrial construction as defined above.

CONSTRUCTION SCOREBOARD

LATEST WEEK

COST INDEXES ENR 20-cities 1913=100	Feb. 1* index value	Change from last month %	Change from last year %
Construction Cost	2,877.5	+ 0.2	+ 7.3
Building Cost	1,740.4	0	+ 7.8
Common labor (CC)	5,474.0	+ 0.3	+ 6.4
Skilled labor (BC)	2,484.9	+ 0.3	+ 6.1
Materials	1,286.0	− 0.2	+ 9.8

*Official February Indexes

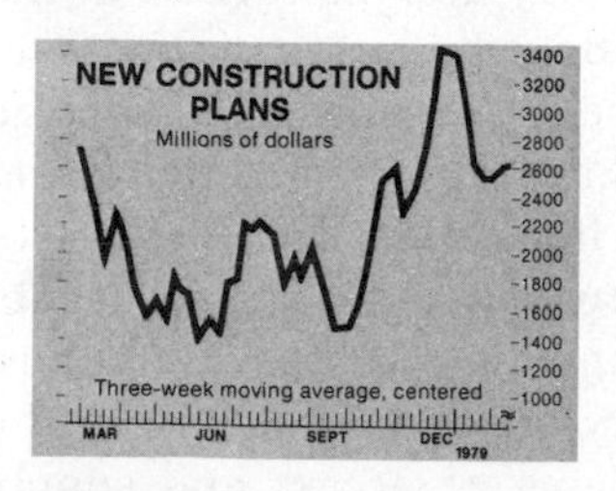

50-state totals ENR-reported	Week of Feb. 8 ($ millions)	1979 ($ millions)	Cum. 6 weeks change '78-'79 %
BIDDING VOLUME			
Total construction*	1,422.2	8,937.3	+ 75
Heavy & highway	603.7	4,645.6	+ 50
Nonresidential bldg	660.1	3,726.7	+118
Housing, multiunit	158.4	565.0	+ 89
NEW PLANS			
Total construction*	2,706.7	15,687.0	+ 30
Heavy & highway, total	879.0	7,396.0	+ 40
Water use & control	216.0	1,911.3	− 15
Waterworks	47.6	416.6	+ 20
Sewerage	137.5	1,232.6	− 26
Treatment plants	58.4	410.0	− 33
Earthwork, waterways	30.9	262.1	+ 3
Transportation	525.0	3,376.9	+ 34
Highways	276.9	2,158.6	+ 21
Bridges	19.2	320.9	− 30
Airports	39.0	157.7	+ 40
Terminals, hangars	34.5	81.6	+118
Elec, gas, comm	65.0	1,769.4	+419
Other heavy const	72.2	336.9	+ 99
Nonresidential bldg	1,468.1	6,271.9	+ 33
Manufacturing	331.4	1,128.7	+134
Commercial	470.0	2,138.7	+ 17
Offices	242.1	917.7	+ 17
Stores, shopping ctrs	174.1	923.5	+ 14
Educational	185.1	766.0	+ 29
College, university	19.0	258.0	+ 34
Medical	185.4	650.5	+ 14
Hospital	129.0	457.9	+ 1
Other	296.1	1,588.0	+ 29
Housing, multiunit*	359.6	2,019.1	+ 45
Apartments	288.1	1,254.0	+ 24

*Excludes 1-2 family houses. Minimum sizes included are: industrial plants, heavy and highway construction, $100,000; buildings, $500,000.

NEW CONSTRUCTION CAPITAL

	Week of Feb. 1 ($ millions)	1979 ($ millions)	Cum. 52 weeks change '78-'79 %
Total new capital	465.1	2,585.6	− 37
Corporate securities	35.0	396.5	− 77
State and municipal	430.1	2,189.1	− 8
Housing	56.3	1,021.9	+ 30
Other bldg and heavy	373.8	1,167.2	− 27

LATEST MONTH

WAGE RATE, 20 cities' average	Feb. 1979	%change from Jan. 1979	%change from Feb. 1978
Common	10.40	+0.3	+6.4
Skilled (average 3 trades)	13.79	+0.3	+6.1
Bricklayers	13.46	0	+5.0
Structural ironworkers	14.41	+0.4	+6.4
Carpenters	13.50	+0.4	+7.0

MATERIAL PRICES, 20-cities' averages	Feb. 1979	Jan. 1979	Feb. 1978
Structural steel (average 3 mills), base, per cwt	15.73	0	+8.1
Lumber, 2x4 fir, per Mbf, CL	336.28	−1.9	+15.1
Lumber, 2x4 pine, per Mbf, CL	282.33	−6.0	+1.2
Cement, bulk, per ton, TL†	52.63	+5.3	+15.4
Sand, per ton, CL†	5.12	+1.1	+7.2
Ready-mix concrete, 3,000 psi, per cu yd, 15 cu yd	36.47	+5.1	+19.2
Crushed stone, 1½'', per ton, CL†	5.72	+1.0	+20.4
Concrete blocks, sand/gravel, 8''x8''x16'', ea, TL*	0.60	+1.3	+21.0
	Jan. 1979	**Dec. 1978**	**Jan. 1978**
Gypsum sheathing, ½''x2'x8', per Msf, TL*	120.91	+0.7	+6.1
Plywood, plyform, ¾'', per Msf, CL†	560.48	−1.0	+12.8
Plywood, ⅝, per Msf, CL†	365.93	−0.6	+9.5
Brick, common, per M, TL*	121.59	+1.4	+9.2
Reinforced bars, whse base, per cwt	16.87	+3.7	+19.9
Structural clay tile, 3''x12''x12'', per M, 2,000 up*	423.22	0	+5.0
Vitrified clay sewer pipe, premium joint, 12'', per ft, CL*	4.62	0	+11.8
Concrete sewer pipe, 12'', premium joint, per ft, CL*	5.44	+2.5	+21.3

*Delivered †f.o.b. city CL-Carlots TL-Trucklots

ENR INDEX REVIEW

Base year = 100	Construction Cost 1913	Construction Cost 1967	Building Cost 1913	Building Cost 1967	Wage Rates Skilled 1913	Wage Rates Skilled 1967	Wage Rates Common 1913	Wage Rates Common 1967
1978								
Feb.	2681.43	249.63	1614.57	238.99	2342	235	5146	254
Mar.	2692.67	250.68	1617.96	239.49	2343	235	5169	255
Apr.	2698.11	251.18	1621.29	239.98	2344	235	5174	255
May	2733.19	254.45	1652.44	244.59	2356	236	5197	257
June	2753.46	256.34	1662.74	246.15	2374	238	5241	259
July	2820.65	262.59	1695.73	251.00	2442	245	5399	266
Aug.	2828.68	263.34	1704.81	252.34	2452	246	5406	266
Sept.	2850.58	265.38	1719.67	254.54	2461	247	5434	268
Oct.	2850.66	265.38	1721.13	254.76	2465	247	5434	268
Nov.	2861.45	266.39	1731.61	256.31	2472	248	5442	268
Dec.	2868.50	267.04	1734.26	256.70	2474	248	5455	269
1978 aver.	**2775.98**	**258.43**	**1673.53**	**247.72**	**2405**	**241**	**5303**	**262**
1979								
Jan.	2872.36	267.40	1739.93	257.54	2479	249	5455	269
Feb.	2877.47	267.88	1740.40	257.61	2485	249	5474	270

TRENDS TO WATCH

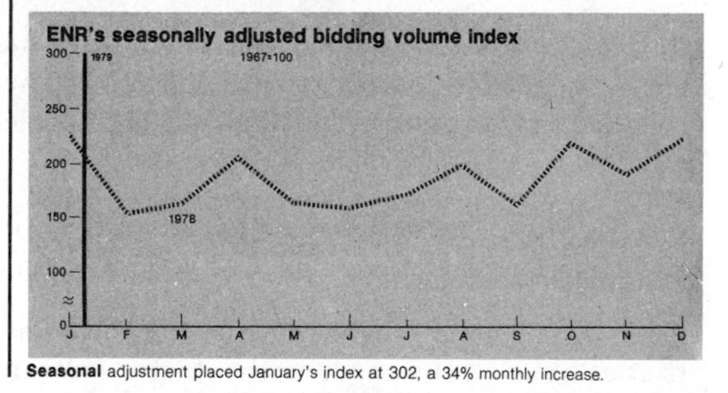

Seasonal adjustment placed January's index at 302, a 34% monthly increase.

February 8, 1979 ENR **27**

FIGURE 1.3 Construction classification: construction scoreboard, ENR (Courtesy of *Engineering News Record*).

The *Engineering News Record* (*ENR*) publishes an updating of information based on this set of construction categories each week. In addition to this information regarding individual project categories, the *ENR* indexes derived from a 20-city base are also reported. These indexes indicate industry trends and provide the construction manager with a nationwide view of the construction environment (see Figure 1.3).

Local geographic and environmental conditions play dominant roles in the work content and definition of a project and in the determination of construction methods and activity sequences. In many cases, and especially for large engineering projects such as dams, establishing lines of access, supply, and communication require major engineering efforts. It may even be necessary to develop and build site support facilities such as housing, quarries and aggregate crushing plants, concrete and bituminous mix plants, and cable ways before the major construction effort can commence. City building projects, on the other hand, require a different emphasis. Problems of access and supply imposed by city congestion and limited on-site storage facilities become critical. Project definition must include all preliminary and ancillary works if a realistic plan is to be developed.

1.4 PROJECT RESOURCE FLOWS

The planning and construction of a project require that a variety of resources be identified, mobilized, and applied effectively to work tasks throughout the project life. Six basic resource types are relevant to most management situations.

1. *Manpower.* The various work tasks must be worked with crews made up of personnel from various trades, and field organizations must be staffed.

2. *Machines.* Construction equipment and hand tools of all types are required dependent on the technologies to be used. The number and availability of equipment pieces constitute a major project resource affecting the rate of construction.

3. *Materials.* The delivery and site handling of materials comprises a dominant portion of the construction effort expended on a project.

4. *Money.* The pervasive influence of cash and finance is a project resource that establishes the speed with which work can be performed. It pays for the workers and machines and the purchasing of materials. It is also a means of compensating others for the completion of work performed.

5. *Information.* The continuous monitoring of project status and the communications required to initiate the flow of project resources such as materials and manpower constitute the information flow for a project.

6. *Management Decisions and Orders.* The various decisive acts of management in decision making constitute a project decision flow. Decisions represent the application of management and technical skill to the manipulation of project resources. Usually, decisions are manifested through an information flow such as labor directives or work order requests to initiate a work task or project activity.

The first three resources (men, machines, materials) are the actual physical resources needed to perform an activity or carry out a work task. The last three resources (money, information, decisions) are the management-oriented resources that establish the feasibility, required conditions, and directives necessary to initiate the work effort.

The availability and application of project resources over the life of a project can be regarded as project resource flow. This application of resources over time requires the effective utilization of management entities such as individual personnel, crews, and equipment. At the work task level, the basic resource flows involve the discrete movement of isolated entities. In bricklaying, for example, a certain number of trucks, scaffolding, hoists, masons, laborers, and brick pallets are involved. As all these discrete flows are viewed from higher levels of the decision hierarchy, they tend to integrate into continuous flows of men, material, and resources of various types.

For a particular project, activity, or work task, the interaction of the various resource flows can be modeled as shown in Figure 1.4. Initiating work on a work task or activity is conditional on technological prerequisite conditions that must be satisfied. Initiation of work is also dependent on the availability of certain minimum levels of materials, machines, men, and money. These minimum levels are called the *planning* levels and reflect the fact that in order to eliminate inefficiencies produced by intermittent working on an activity, a certain quantity of resources must be available before it is feasible or economical to begin work. Figure 1.4 illustrates the dynamic functioning and interaction of the project resource flows necessary for each project work task and activity. The figure also indicates a number of management functions that must be performed in support of field activities and the need for a project organization.

1.5 CONSTRUCTION MANAGEMENT LEVELS

A number of significantly different areas of management concern can be identified in construction management. These areas can be differentiated because they occur at different levels in the construction management hierarchy and focus on the management of different types of resources. These management areas can be identified as (1) *project mission management*, (2) *project management*, and (3) *field management*. Although it is convenient to separate these

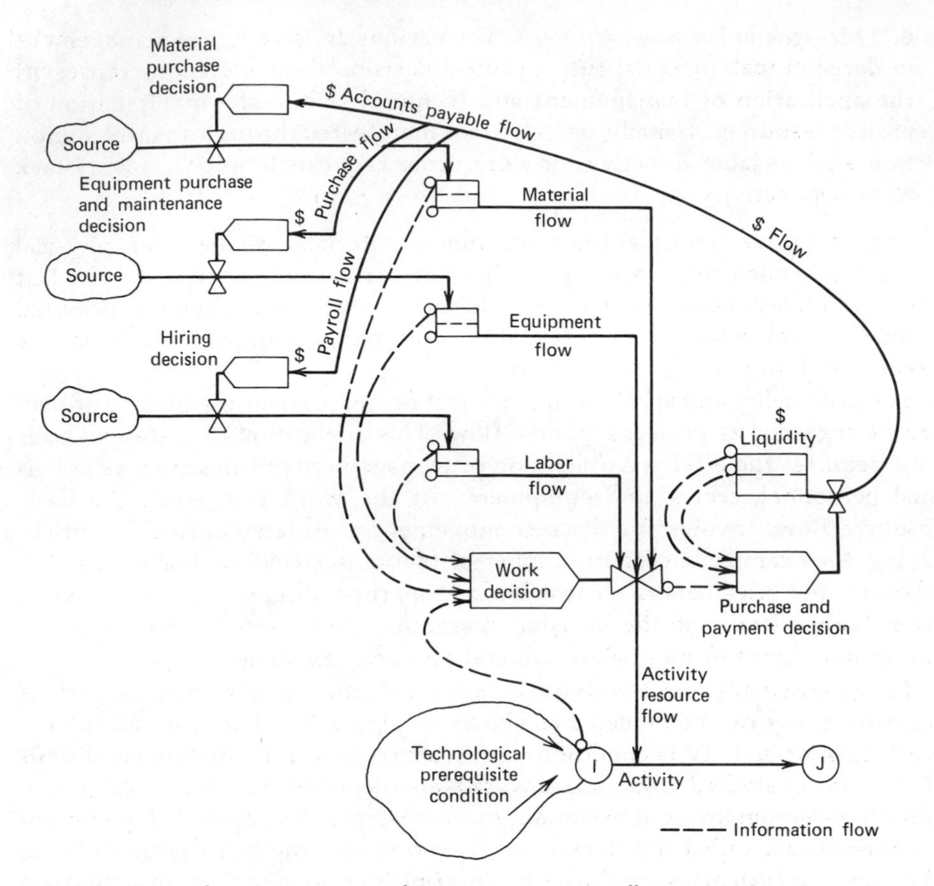

FIGURE 1.4 The interaction of project resource flows.

management areas because of the different problems, disciplines, and management techniques that are relevant to each, it must be understood that all levels of management concern interact and collectively comprise the management effort involved in project construction.

Project mission oriented construction management is concerned with the setting up and management control of the specific organizational form of the construction process thought necessary to produce the desired product. It relates to the interaction and coordination of the efforts and responsibilities of the owner and consultant, contractor, and contract administration agents. These agents effectively establish and control the construction process. They must reach an agreement (usually in terms of a contract) on how to work in relation to one another and on all details as to what is to be built. Each agent has a unique role to play in the construction management effort. Construction management at this level requires a knowledge of business, organizational, technical, legal, contractual, and financial matters and a high level of professional management skill.

Chapters bearing on areas relevant to *project mission* orineted construction management are: The Design and Construction Phases of the Project Cycle, Chapters Two and Three; Legal Structure, Chapter Five; Management Structure, Chapter Six; and the Construction Management Approach, Chapter Eighteen. Project Funding and Cash Flow, Chapters Seven and Eight; Construction Planning, Chapter Fifteen; and Project Scheduling, Chapter Sixteen, are contractor oriented. They contain sections relevant to considerations at the project mission level.

Project management comprises the planning of the construction effort and the organizing, monitoring, and management of the resource flows that support the construction effort. Project-oriented construction management is the professional role of the contractor. In the bidding stage, project management involves construction planning and estimating as well as financial and risk analyses of the project. On bid award, project management includes the scheduling, buying, procurement, and mobilizing of all the resources needed to initiate and maintain the construction effort. It is also concerned with monitoring of project status and progress, the preparation of progress payment claims, and the management of the cash flow needed to construct the project. The project management team requires a large variety of disciplines and specialists in its management of all the resource flows. Accordingly, a variety of planning-, analysis- and decision-oriented management techniques are required in project management.

Most of the following chapters are closely related to this area of construction management. The organization and functional breakout of the typical contractor company are covered in Legal Structure, Chapter Five, and Management Structure, Chapter Six. The management of the basic resources of manpower, machines, materials, and money are discussed in Labor Relations, Chapter Thirteen; Labor Cost and Productivity, Chapter Fourteen; Equipment Costs, Chapter Nine; Depreciation and Productivity, Chapters Ten and Eleven; Materials Management, Chapter Twelve. At the information flow level, specific chapters reflecting planning and control are Construction Planning, Chapter Fifteen, and Project Scheduling, Chapter Sixteen.

The general manner in which projects are developed and the sequential stages in project management are given in Chapters Two and Three, The Design and Construction Phases of the Project Life Cycle.

Field-oriented construction management is concerned with the manipulation of resources at the field level so that the orderly and effective implementation of the construction plan can be achieved. Field management focuses on the technical details of construction methods and operations and on equipment capabilities. It is concerned with the planning, scheduling, mobilization, and directing of the construction activity in terms of crews, available equipment, and materials. Construction management agents at this level must have considerable experience and be able to direct, command respect from, and motivate field labor.

Field-oriented construction management is considered in Chapter Thirteen, Labor Relations; and Chapter Seventeen, Safety. However, many of the chapters covering the management of the basic resources (men, machines, material) have sections directly related to the field management area.

REVIEW QUESTIONS AND EXERCISES

1.1 Give *three* important characteristics of the construction industry that make it unique in comparison with other industries. Use the heavy manufacturing industry as the basis for comparison.

1.2 Identify (by title and general areas of responsibility) the agents involved in a local construction project.

1.3 List the various trades commonly met in building construction and establish their hourly wage rates. In union areas name the relevant union locals and their business agents.

1.4 List the contractor and subcontractor organizations that are active in your region. Are these related to the general classifications of construction as indicated in the text and other publications such as *ENR*?

1.5 Identify (by business and professional names) the client, companies, and contractors involved in a local building or construction project and which are operating in the traditional organization of construction. Hence correlate their areas of responsibilities with Figures 1.1 and 1.2.

1.6 Identify a construction project that is being performed by a design-build contractor. Hence develop a schematic illustration (similar to Figure 1.2) of the design-build mode of construction project management. In what significant ways does the design-build contractor's mode of operation differ from that of the traditional contractor?

1.7 Visit a local architectural-engineering firm and establish the role they play in the project definition and contract administration phases of the building process.

1.8 Visit a local building or construction project and identify the prime contractor and the subcontractors. Hence prepare a work package dissection of the project that reflects the involvement of each contractor in the project. Why was the project construction effort divided in this manner?

1.9 Using *ENR* figures and a classification of construction work types prepare graphs of aggregated construction volumes by months over the past year. Can you explain the monthly and seasonal variations in these construction volumes? Are these graphs useful in predicting (or assessing) local construction work volumes?

1.10 Visit a local building or construction site and identify specific examples of each of the major project resource flows indicated in Section 1.4. Follow one specific resource through several stages in its use on the project.

1.11 Assume that you are a field investigator for a management consultant specializing in construction project management and that you have been asked to carry out the following investigations:

(a) An assessment of the quality of management efforts by the client's architect-engineer, the prime contractor, and the subcontractors.

(b) An assessment of the adequacy of field resources (labor work force, on-site equipment, the material procurement and handling process, and the ability and work load on the field management team) in relation to the work load and schedule requirements of the project.

(c) The current status of a construction project.

How would you go about carrying out the investigation and what checklists would you prepare as aids in your investigations?

CHAPTER TWO

Project Cycle—Design Phase

2.1 ORIENTATION

Construction activities are normally organized in project format. Most projects have a recognizable life cycle during which a facility is conceived, constructed, and then put into operation. Here and in the following chapter, the total Building Process and the traditional cycle as introduced in Chapter One will be amplified and examined in detail. The design and construction stages through which a project moves from the time project feasibility is considered to the point at which the owner accepts the project for occupancy will be presented in the context of the actual construction environment. In this chapter we consider the design phase of the project cycle.

2.2 PROJECT CONCEPT AND NEED

Any project, large or small, is generated by a need or concept, which initiates the requirement for a facility to carry out a given function. The motivation may be based on the desire to develop a facility for commercial profit. An example of the development-type project with profit as the motivating force is given in Chapter 7 (i.e., a 148-apartment-unit project). City, state, and federal government agencies are constantly developing projects for social welfare and to house the operations of government. In the United States, construction of buildings to house federal agencies is assigned to the Public Building Service (PBS), which is a subelement of the General Services Administration (GSA). Federal agencies such as the Army Corps of Engineers and the Bureau of

Reclamation are charged with the construction of water structures to enhance the navigability of streams, protect the populace from flooding, and provide water for irrigation. The Smallville Methodist Church may realize that the Sunday school building is too small and must be expanded. The church board, in this case, would consider and approve the expansion. On a different scale, a large international firm may identify a developing market for fertilizer in the Southeast Asia area. In order to meet this market, a plant construction in Singapore might be considered. The decision to commit $200 million to meet this commercial opportunity would be presented to the company board of directors for action much as the Sunday school addition is presented to the church board.

In the case of both Sunday school addition and the fertilizer plant construction, documentation to support the decision by the board must be developed. This documentation would normally consist of:

1. Cost/benefit analysis.
2. Conceptual drawings to include artist concept sketches and preliminary layout drawings.
3. Concept budget to identify the required level of funding.

Depending on the complexity and costliness of the project, the cost/benefit analysis can range from a very informal document to an expensive market analysis and feasibility study. In recent years, the requirement to reflect public safety and environmental impacts of new projects that are regulated or funded in part or totally by the government has led to ever-increasing preliminary studies. Typical of the imposing size of such studies is the Preliminary Safety Analysis Report (PSAR) required by the U.S. government before a nuclear power plant can begin construction. This documentation may be seven or eight large volumes in length and may cover only one apsect of the preliminary study.

Returning to the example of the Sunday school project, the church board will probably establish to cost/benefit relationship for the addition by retaining a local architect to prepare preliminary layout drawings and sketches and by having a concept budget developed on this basis. This budget would be accurate to the level of 1 ± 10% and is a preliminary estimate based on limited conceptual information. Once the cost and scope of the project have been reasonably defined, the benefit will be established by interviews with church officers and projections of Sunday school requirements.

Similarly, the economic basis for the fertilizer plant project must be established based on market studies projecting the demand for fertilizer and ancillary products across the planning horizon under consideration. In many cases, these studies recommend optimal time frames for the plant construction to meet the market in advance of competition. Plant site location, availability of labor and supporting resources such as energy, water, and shipping connections are also considered. This study is sometimes referred to as a *feasibility study*.

At this time also, engineers are called in to make a parametric analysis of the costs involved in construction of the plant. Based on gross parameters such as tons of fertilizer per day, square footage of storage space, and number of berths for shipping, very preliminary indications of the scope and cost of the project can be established. Preliminary layout and budgetary information are given to the corporate board of directors so that they can balance cost versus benefit.

Government projects at the federal level require similar supporting analysis and are submitted with budget requests each year for congressional action. Supporting documentation includes layout sketches and outline specifications such as those shown in Figure 2.1. A supporting budget for this project is shown in Figure 2.2. These projects are included as line items in the budget of the government agency requesting funding. In this case, the requestor would be the Post Engineer, Fort Campbell, Kentucky. This request would be consolidated with requests at the Army and Department of Defense level and forwarded to the Bureau of the Budget to be included in the budget submitted to congress.

It is of interest to note that since the Post Office project shown will not be constructed for a year (even if approved), a projection of cost is required. The projection is made using the *Engineering News Record* (*ENR*) indexes of basic construction cost. This is noted on the estimate. This index is printed each year in the First Quarterly Cost Roundup issue of the *ENR* magazine. The 1979 construction cost indexes are shown in Figure 2.3. The projected value used is 1100, which is taken from the Building Cost Index. A short explanation of the indexes is given in the text next to the figure. Other cost bases are also available for such projections.

The summary on the last page of the working estimate indicates the base cost (1) will be $460,600. The reserve of 10% for contingencies is $46,000. The amount allocated for construction supervision by the Corps of Engineers is $29,400. The design cost estimated as a percent of the total is $35,000. The total cost for construction excluding design is $536,000.

2.3 PRELIMINARY AND DETAIL DESIGN

Once the concept of the project has been approved, the *owner* desiring the construction retains an engineer, an architect, or a combination of the two called an architect/engineer. The end product of the design phase of project development is a set of plans and specifications that define the project to be constructed. The drawings are a graphical or schematic indication of the work to be accomplished. The specifications are a verbal or word description of what is to be constructed and to what levels of quality. When completed they

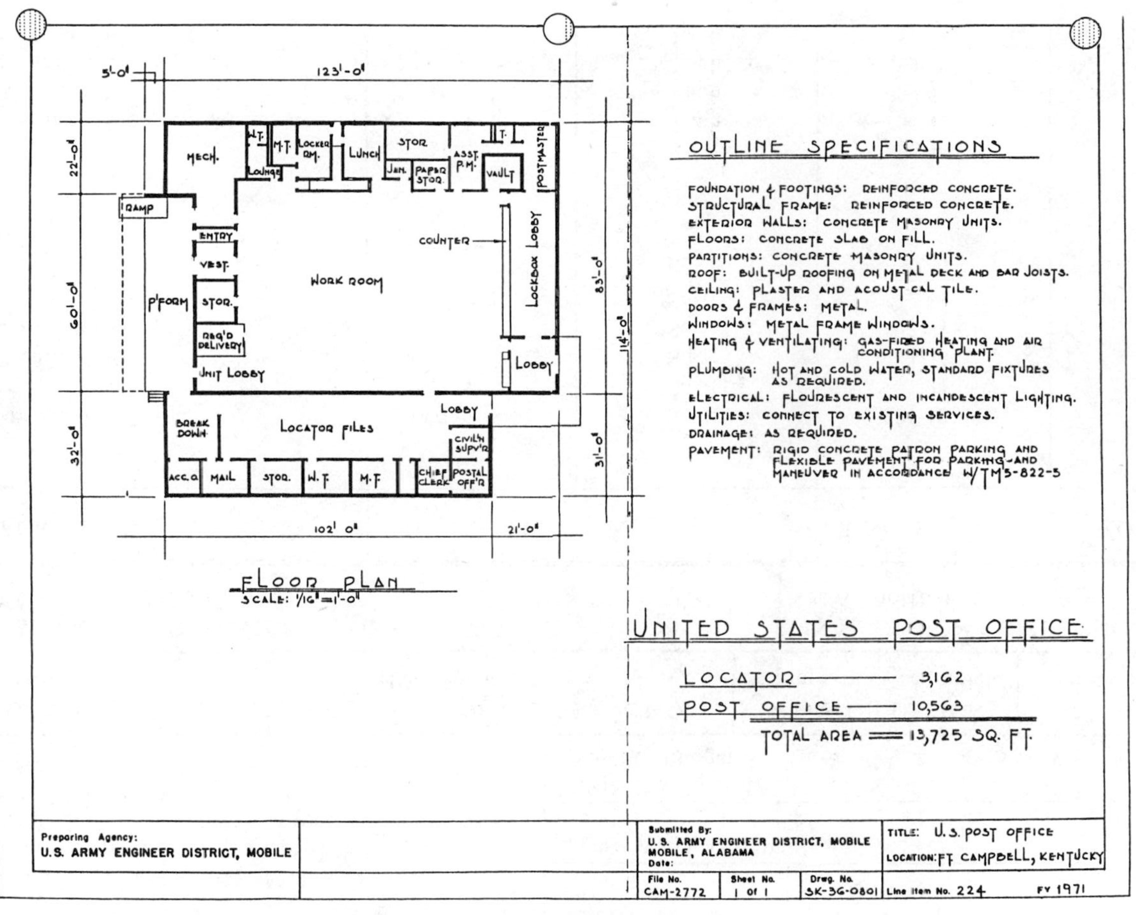

FIGURE 2.1 Project proposal: layout sketch and outline specifications.

TO: Chief of Engineers Department of the Army Washington 25, D.C.	FROM: Mobile District Corps of Engineers Mobile, Alabama 36601	FISCAL YEAR 1971	DATE PREPARED 14 Oct 69
		NAME AND ADDRESS OF A.E. N.A.	
		BASIS OF ESTIMATE Code "A" Budget Sketch & 1391	A.E. FEE N.A.

NAME AND LOCATION OF INSTALLATION	TYPE OF CONSTRUCTION	STATUS OF DESIGN	
Ft. Campbell, Kentucky	Permanent	Preliminary 0% complete	Final 0% complete

LINE ITEM NUMBER	DESCRIPTION OF FACILITY	FINAL DESIGN COMPLETION DATE
224	Post Office	Not Authorized

DESCRIPTION	QUANTITY	UNIT	UNIT PRICE	TOTALS ($000)
1. *Building*				
General Construction	13,725	Sq ft	$21.12	$289.9
Plumbing	13,725	Sq ft	1.21	16.6
Heating and Ventilating	13,725	Sq ft	1.34	18.4
Air Conditioning (50-ton)	13,725	Sq ft	3.81	52.3
Electrical	13,725	Sq ft	2.83	38.8
Subtotal	13,725	Sq. ft	30.31	416.0
2. *Utilities*				
a. *Electrical*				
Transformers	112.5	kVA	26.40	3.0
Poles with X-arms, Pins, Insulation, etc.	4	Each	356.40	1.4
Dead Ends	6	Each	39.56	0.2
Down Guys and Anchors	4	Each	89.10	0.4
Fused Cutouts and L.A.	6	Each	59.94	0.4

#6 Bare Cu. Conductor	2,400	lin ft	.30	0.7
#3/0 Neoprene Covered Service	160	lin ft	1.48	0.2
Parking Area Lights on Aluminum Pole	7	Each	933.41	6.5
3C #8 DB 600-V Cable	210	lin ft	1.70	0.4
2C #8 DB 600-V Cable	740	lin ft	1.27	0.9
3-in. Duct Conc. Encased U.G.	100	lin ft	4.75	0.5
Subtotal				15.0
b. *Water*				
3-in. Water Line	365	lin ft	4.30	1.6
3-in. Gate Valve and Box	1	Each	118.80	0.1
Fire Hydrants	2	Each	534.60	1.1
Connections to Existing Lines	3	Each	273.24	0.8
Subtotal				4.0
c. *Sewer*				
6-in. Sanitary Sewer	215	lin ft	5.94	1.3
8-in. Sanitary Sewer	375	lin ft	6.89	2.6
Manhole	2	Each	534.60	1.1
Connection to Exist. Manhold	1	Each	118.80	0.1
Subtotal				5.0
d. *Gas*				
1 1/4 in. Gas Line	1,000	lin ft	3.09	3.1
1 1/4 in. Plug Valve and Box	1	Each	118.80	0.1
Connect to Existing	1	Each	237.60	0.2
Street and Parking Area Crossing	280	lin ft	1.54	0.4
				4.0

FIGURE 2.2 Current working estimate for budget purposes. Costs are based on a projected 1972 level ENR of 1100.

DESCRIPTION	QUANTITY	UNIT	UNIT PRICE	TOTALS ($000)
3. *Site Work*				
Clearing and Grubbing	2.4	Acre	495.01	1.2
Borrow Excavation	10,000	cu yd	3.46	34.6
Remove B.T. Paving	1,070	sq yd	1.44	1.5
Subtotal				37.0
4. *Paving*				
Paving—1½ A.C. and 8-in. Stab.	3,950	sq yd	6.19	24.5
Aggr. Base	2,250	lin ft	4.70	10.6
6-in. P.C. Concrete Paving	380	sq yd	9.70	3.7
3-in. Painted Parking Lines	1,680	lin ft	0.27	0.5
Concrete Sidewalk	440	sq yd	8.10	3.6
Subtotal				43.0
5. *Storm Drainage*				
15-in. Concrete Cl. II Pipe	40	lin ft	8.91	0.4
15-in. Concrete Cl. III Pipe	20	lin ft	10.22	0.2
Reinf. Drainage Structure Concrete	8	cu yd	207.90	1.7
C.I. Grates and Frames	1,900	lb	0.37	0.7
Subtotal				3.0
6. *Landscaping*				
Sprigging and Seeding	1.6	Acre	945.00	1.5
Landscaping		Job		2.1
				4.0

7. *Communications*				
a. Telephone		LS	756.00	1.0
b. Support (Within Building)				
100 Pr. DB Pic Cable	600	LF	1.26	0.8
51 Pr DB Pic Cable	550	LF	0.72	0.4
Splicing Sleeves and Material		LS	$480.00	0.5
Labor		LS		2.4
				5.0
Total estimated cost (excluding design, but including reserve for contingencies and supervision and administration (S&A)				536.0
1. Estimated contract cost				460.6
2. Reserve for Contingencies	10	%		46.0
3. Supervision and administration (S&A)				29.4
Total estimated cost (excluding design, but including reserve for contingencies and supervision and administration				536.0
4. *Design*				
District Expenses (Preliminary and Final)				35.0
Subtotal				$35.0

FIGURE 2.2 (continued)

ENR indexes reflect material and labor trends

The ENGINEERING NEWS-RECORD construction cost index was created in 1921 to diagnose price changes that occurred during and immediately following World War I and to evaluate their effect on construction costs.

The index was designed as a general purpose construction cost index to chart basic costs. It is a weighted aggregate index of constant quanities of structural steel, portland cement, lumber and common labor. This hypothetical block of construction, repriced weekly, was valued at $100 in 1931 prices.

The original use of the common labor quantity in the construction cost index was based on the idea that it set the trend for all wage rates. In the '30s, however, wages plus fringe benefits climbed much faster for laborers than for skilled tradesmen, in percentage terms.

The ENGINEERING NEWS-RECORD building cost index was introduced in 1938 to weigh the impact of skilled labor on construction cost trends. For its labor component it uses skilled trades, an average of carpenter, bricklayer and structural ironworker wages. Its materials component is the same as that used in the construction cost index. The index also represents a hypothetical construction block valued at $100 in 1913 prices.

Because the indexes are designed to indicate basic underlying trends of construction costs in the U.S., they use construction materials least influenced by purely local conditions. Steel, lumber and cement were selected because they promised the most stable relation to the nation's economy and its price structure.

The relative importance of the elements in each index was based on U.S. annual production of each material and the number of nonfarm laborers and skilled workers. Although the quantities involved remain constant, the materials weights were revised on rare occasions to reflect the new price quoting bases dictated by changes in pricing systems of supplying industries. Nevertheless, those weight changes affected the materials component very little during the year of change.

Both the ENR construction and building cost indexes measure the effects of wage rate and materials price trends. They do not adjust for productivity, managerial efficiency, competitive conditions, automation, design changes, or other intangibles.

The two ENR cost indexes have developed a widening divergence over the years. This has resulted from the faster rate of increase in laborers' wage rates compared to skilled and from the use of nearly three times as many man-hours of labor in the construction cost index as in the building cost index. The common labor component of the construction cost index has increased much faster than the skilled labor component of the building cost index.

Actually, mechanization and more efficient construction methods have held labor bills for most construction to a smaller share of total expenditure for labor and material than the construction cost index indicates. Therefore, this index is appropriate only for construction involving an unusually large number of man-hours per $100 of labor and materials expenditure.

Over the years, both cost indexes have proved infallible as to direction. In normal times, they are also accurate as to degree of change. The building cost index is generally more applicable in measuring the degree of change because its skilled labor component is more representative of labor's share of the total cost of labor and materials in most types of construction.

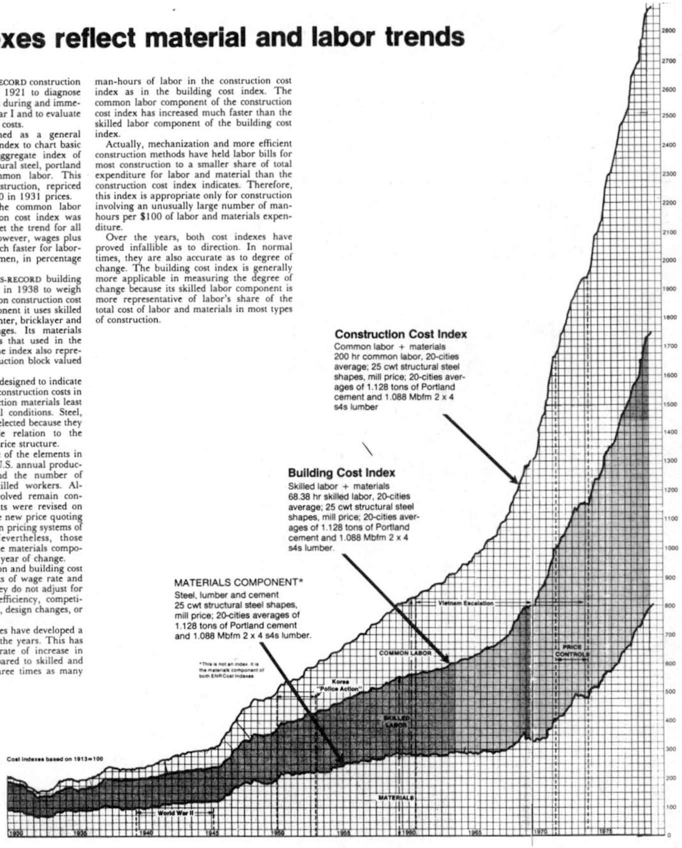

Construction Cost Index
Common labor + materials
200 hr common labor, 20-cities average; 25 cwt structural steel shapes, mill price; 20-cities averages of 1.128 tons of Portland cement and 1.088 Mbfm 2 x 4 s4s lumber

Building Cost Index
Skilled labor + materials
68.38 hr skilled labor, 20-cities average; 25 cwt structural steel shapes, mill price; 20-cities averages of 1.128 tons of Portland cement and 1.088 Mbfm 2 x 4 s4s lumber.

MATERIALS COMPONENT*
Steel, lumber and cement
25 cwt structural steel shapes, mill price; 20-cities averages of 1.128 tons of Portland cement and 1.088 Mbfm 2 x 4 s4s lumber.

*This is not an index. It is the materials component of both ENR Cost Indexes

Cost indexes based on 1913=100
Vietnam Escalation
COMMON LABOR
SKILLED LABOR
MATERIALS
Korea "Police Action"
World War II
PRICE CONTROLS

BUILDING COST INDEX HISTORY 1913-1978

How ENR builds the Index: 68.38 hours of skilled labor at the 20-cities average of bricklayers', carpenters' and structural ironworkers' rates, plus **25 cwt of standard structural steel shapes** at the mill price, plus **22.56 cwt (1.128 tons) of Portland cement** at the 20-cities average price, plus **1,088 feet of 2 x 4 lumber** at the 20-cities average price.

Year	Index	Year	Index	Year	Index	Year	Index	Year	Index	Year	Index	Year	Index
1913	100	**1920**	207	**1927**	186	**1934**	167	**1941**	211	**1948**	345	**1955**	469
1914	92	**1921**	166	**1928**	188	**1935**	166	**1942**	222	**1949**	352	**1956**	491
1915	95	**1922**	155	**1929**	191	**1936**	172	**1943**	229	**1950**	375	**1957**	509
1916	131	**1923**	186	**1930**	185	**1937**	196	**1944**	235	**1951**	401	**1958**	525
1917	167	**1924**	186	**1931**	168	**1938**	197	**1945**	239	**1952**	416	**1959**	548
1918	159	**1925**	183	**1932**	141	**1939**	197	**1946**	262	**1953**	431	**1960**	559
1919	159	**1926**	185	**1933**	148	**1940**	203	**1947**	313	**1954**	446	**1961**	568

1913 = 100	Monthly												
	Jan.	Feb.	Mar.	Apr.	May	June	July	Aug.	Sept.	Oct.	Nov.	Dec.	Annual Average
1962	571	573	575	576	579	580	583	586	586	586	584	584	**580**
1963	584	585	586	586	588	590	596	602	602	604	602	603	**594**
1964	604	604	606	607	609	612	615	616	617	617	617	617	**612**
1965	616	621	622	621	621	626	628	630	633	634	633	634	**627**
1966	635	641	643	649	652	656	653	655	656	655	655	655	**650**
1967	656	657	659	660	666	671	673	678	681	684	685	687	**672**
1968	692	695	698	701	710	718	721	729	741	747	747	755	**721**
1969	764	770	780	790	791	798	792	799	796	797	801	802	**790**
1970	802	801	802	813	827	834	848	851	857	862	866	866	**836**
1971	875	877	905	913	933	946	959	970	996	997	1001	1005	**948**
1972	1011	1016	1022	1027	1039	1047	1053	1057	1067	1070	1082	1090	**1048**
1973	1102	1114	1123	1135	1140	1138	1137	1144	1150	1156	1155	1158	**1138**
1974	1156	1154	1155	1177	1177	1199	1233	1240	1238	1246	1239	1240	**1204**
1975	1242	1265	1265	1269	1287	1307	1317	1330	1333	1351	1349	1354	**1306**
1976	1362	1370	1378	1391	1398	1416	1425	1455	1467	1476	1479	1484	**1425**
1977	1489	1499	1504	1506	1507	1521	1539	1554	1587	1618	1604	1607	**1545**
1978	1609	1617	1620	1621	1652	1663	1696	1705	1720	1721	1732	1734	**1674**
1979	1740	1740	1750										

CONSTRUCTION COST INDEX HISTORY 1905-1978

How ENR builds the Index: 200 hours of common labor at the 20-cities average rate, plus **25 cwt of standard structural steel shapes** at the mill price, plus **22.56 cwt (1.128 tons) of Portland cement** at the 20-cities average price, **plus 1,088 board feet of 2 x 4 lumber** at the 20-cities average price.

Year	Index	Year	Index	Year	Index	Year	Index	Year	Index	Year	Index	Year	Index
1906	95	**1914**	89	**1922**	174	**1930**	203	**1938**	236	**1946**	346	**1954**	628
1907	101	**1915**	93	**1923**	214	**1931**	181	**1939**	236	**1947**	413	**1955**	660
1908	97	**1916**	130	**1924**	215	**1932**	157	**1940**	242	**1948**	461	**1956**	692
1909	91	**1917**	181	**1925**	207	**1933**	170	**1941**	258	**1949**	477	**1957**	724
1910	96	**1918**	189	**1926**	208	**1934**	198	**1942**	276	**1950**	510	**1958**	759
1911	93	**1919**	198	**1927**	206	**1935**	196	**1943**	290	**1951**	543	**1959**	797
1912	91	**1920**	251	**1928**	207	**1936**	206	**1944**	299	**1952**	569	**1960**	824
1913	100	**1921**	202	**1929**	207	**1937**	235	**1945**	308	**1953**	600	**1961**	847

1913 = 100	Monthly												
	Jan.	Feb.	Mar.	Apr.	May	June	July	Aug.	Sept.	Oct.	Nov.	Dec.	Annual Average
1962	855	858	861	863	872	873	877	881	881	880	880	880	**872**
1963	883	883	884	885	894	899	909	914	914	916	914	915	**901**
1964	918	920	922	926	930	935	945	948	947	948	948	948	**936**
1965	948	957	958	957	958	969	977	984	986	986	986	988	**971**
1966	988	997	998	1006	1014	1029	1031	1033	1034	1032	1033	1034	**1019**
1967	1039	1041	1043	1044	1059	1068	1078	1089	1092	1096	1097	1098	**1070**
1968	1107	1114	1117	1124	1142	1154	1158	1171	1186	1190	1191	1201	**1155**
1969	1216	1229	1238	1249	1258	1270	1283	1292	1285	1299	1305	1305	**1269**
1970	1309	1311	1314	1329	1351	1375	1414	1418	1421	1434	1445	1445	**1385**
1971	1465	1467	1496	1513	1551	1589	1618	1629	1654	1657	1665	1672	**1581**
1972	1686	1691	1697	1707	1735	1761	1772	1777	1786	1794	1808	1816	**1753**
1973	1838	1850	1859	1874	1880	1896	1901	1902	1929	1933	1935	1939	**1895**
1974	1940	1940	1940	1961	1961	1993	2040	2076	2089	2100	2094	2101	**2020**
1975	2103	2128	2128	2135	2164	2205	2248	2274	2275	2293	2292	2297	**2212**
1976	2305	2314	2322	2327	2357	2410	2414	2445	2465	2478	2486	2490	**2401**
1977	2494	2505	2513	2514	2515	2541	2579	2611	2644	2675	2659	2669	**2577**
1978	2672	2681	2693	2698	2733	2753	2821	2829	2851	2851	2861	2869	**2776**
1979	2872	2877	2886										

FIGURE 2.3 Engineering News Record construction cost indexes (Courtesy of *Engineering News Record*).

are included as legally binding elements of the contract. The production of the plans and specifications usually proceeds in two steps. The first step is called preliminary design and offers the owner a pause in which to review construction before detail design commences. A common time for this review to take place is at 40% completion of the total design. The preliminary design extends the concept documentation. In most projects, a design team leader concept is utilized. The design team leader coordinates the efforts of architects and engineers from differing disciplines. The disciplines normally identified are architectural, civil and structural, mechanical, and electrical. The architect or architectural engineer, for instance, is responsible for the development of floor plans and general layout drawings as well as considerations such as building cladding and exterior effects. The mechanical engineer is concerned with the heating, ventilating, and air conditioning (HVAC), as well as service water systems. At preliminary design, decisions regarding size and location of air conditioning and heating units as well as primary water distribution components (e.g., pumps) are made. Similar decisions regarding the electrical system are made at this point by the electrical engineers. The structural and civil engineers develop the preliminary design of the structural frame and the subsurface foundation support. All of these designs are interlinked. The architectural layout impacts the weight support characteristics of the floor structure and, hence, the selection of structural system. The structural superstructure influences the way in which the foundation of the structure can be handled. The floor plan also determines the positioning of pipes and ducts and the space available for service mains.

Once the preliminary design has been approved by the owner, final or detail design is accomplished. For the architectural engineer this focuses on the interior finishes, which include walls, floors, ceilings, and glazing. Details required to install special finish items are designed. Precise locations and layout of electrical and mechanical systems as well as the detail design of structural members and fasteners are accomplished by the appropriate engineers. As noted above, the detail design phase culminates in the plans and specifications that are given to the constructor for bidding purposes. In addition to these detailed design documents, the architect/engineer produces a final "owner's" estimate indicating the total job cost minus markup. This estimate should achieve approximately $\pm$ 3% accuracy, since the total design is now available. The owner's estimate is used (1) to ensure the design produced is within the owner's financial resources to construct (i.e., the architect/engineer has not designed a gold-plated project), and (2) to establish a reference point in evaluating the bids submitted by the competing contractors. In some cases, when all contractor bids greatly exceed the owner's estimate, all bids are rejected and the project is withdrawn for redesign or reconsideration. Once detailed design is completed, the owner again approves the design prior to advertising the project to prospective bidders.

2.4 NOTICE TO BIDDERS

The document announcing to prospective bidders that design documents are available for consideration and that the owner is ready to receive bids is called the *notice to bidders*. Because of his commitment to the owner to design a facility that can be constructed within a given budget, the architect/engineer wants to be sure that the lowest bid price is achieved. To ensure this, the job is advertised to those contractors who are capable of completing the work at a reasonable price. A/E firms maintain mailing lists that contain qualified bidders. When design is complete, a notice to bidders, such as the one shown in Figure 2.4, is sent to all prospective bidders. The notice to bidders contains information regarding the general type and size of the project, the availability of plans and specifications for review, and the time, place, and date of the bid opening. Normally, sets of plans and specifications are available for perusal at the A/E office as well as at *plans rooms*, which are conveniently located and can be visited by contractors in large cities. These facilities have copies of plans and specifications for a large number of jobs being let for bid. They afford contractors the opportunity to go to a central location and look at several jobs without having to drive to the office locations of each architect/engineer. The expenditure on the part of the contractors in going to the A/E office or plans rooms to look at the contract documents amounts only to the price of gas and a small amount of time. If they should decide to bid on a particular job, their commitment increases sharply in terms of money and time invested.

In addition to the mailings made available by the A/E firm as the owner's representative, contractors have other methods of learning about jobs that are available for bid. In some large cities, a builder's exchange may operate to serve the contracting community and keep it apprised of the status of design and bid activity within a given area. In addition to operating plans rooms, these exchanges also regularly publish newsletters such as the one shown in Figure 2.5. These reports indicate what jobs are available for bidding, and architect/engineers make use of such facilities to gain maximum coverage in advertising their jobs. In addition to the basic information describing the job and the time and place of bid opening, these reports include, following each announcement, a statement that the plans are on file, along with the bin location.

Nationwide services such as the Dodge Reporting System also exist to provide information on projects being let for bid. For a subscription fee such services send information regarding jobs based on type of construction, geographical location, job size, and other parameters directly to the contractor. The information announcements indicate whether the job is under design, ready for bid, or awarded. In the cases of jobs that have been awarded, the low bid and other bid prices submitted are furnished so that the contractor can detect bidding trends in the market. A typical Dodge Reporting System announcement is shown in Figure 2.6.

2.5 THE BID PACKAGE

The documents that are available to the contractor and on which he must make his decision to bid or not to bid are those in the *bid package*. In addition to the plans and technical specifications, the bid package prepared by the architect/engineer consists of a *proposal* form, *general specifications* that cover procedures common to all construction contracts, and *special conditions*, which pertain to procedures to be used that are unique to this particular project. All supporting documents are included by reference in the proposal form. The bid package layout is shown schematically in Figure 2.7. The proposal form as designed and laid out by the architect/engineer is the document that, when completed and submitted by the contractor, indicates the contractor's desire to perform the work and the price at which he will construct the project. A typical example of a proposal is shown in Figure 2.8.

The proposal form establishes intent on the part of the contractor to enter into a contract to complete the work specified at the cost indicated in the proposal. It is an offer and by itself is not a formal contract. If, however, the owner responds by awarding the contract based on the proposal, an acceptance of the offer results and a contractual relationship is established. The prices at which the work will be constructed can be stated either as lump sum or as unit price figures. Only a portion of the price schedule (see Figure 2.8) is shown in this example proposal. As shown in the figure, both methods (lump sum and unit price) of quoting price are illustrated. Items 1 to 4 require the bidder to specify unit price (i.e., dollar per unit) for the guide quantities specified. Therefore, if the contractor will do the rock excavation for $20.00 per cubic yard, this price is entered along with the total price (550 × $20.00 = $11,000). Items 5, 6, and 7 require lump or stipulated sum quotations. Therefore, the contractor states a single price for the access road, finish grading, and so on.

In the proposal form shown, in Figure 2.8, the contract duration is also specified, although this is not always the case. In many instances, the project duration in working or calendar days is specified in the *special conditions* portion of the bid package. The proposal form indicates also that the contractor is to begin work within 10 calendar days after receipt of written notice of award of contract. Award of contract is usually communicated to the contractor in the form of a *notice to proceed*. Response by the owner to the contractor's *proposal* with a notice to proceed establishes a legally binding contractual relationship. Legal signatures (L.S.) by individuals empowered to represent (i.e., commit contractually) the firm making the proposal must be affixed to the proposal.

NOTICE TO BIDDERS
FOR
CONSTRUCTING SEWERAGE SYSTEM IMPROVEMENTS
CONTRACT "B"
CENTRAL STATE HOSPITAL
FOR THE
GEORGIA BUILDING AUTHORITY (HOSPITAL)
STATE CAPITOL—ATLANTA, GEORGIA

Sealed proposals will be received for Constructing Sewerage System Improvements, Contract "B," for the Georgia Building Authority (Hospital), State Capitol, Atlanta, Georgia, at Room 315, State Health Building, 47 Trinity Avenue, S.W., Atlanta, Georgia, until 2:00 P.M., E.S.T., February, 18, 19___, at which time and place they will be publicly opened and read. Bidding information on equipment in Section No. 10 shall be submitted on or before February 4, 19___.

Work to be Done: The work to be done consists of furnishing all materials, equipment, and labor and constructing:

Division One. Approximately 12,400 L.F. 36″ Sewer Pipe, 5,650 L.F. 30″ Sewer Pipe, 7,300 L.F. 24″ Sewer Pipe, 1,160 L.F. 15″ Sewer Pipe, 3,170 L.F. 12″ Sewer Pipe, 300 L.F. 8″ Sewer Pipe, 418 L.F. 36″ C.I. Pipe Sewer, 324 L.F. 30″ C.I. Pipe Sewer, 1,150 L.F. 30″ C.I. Force Main, 333 L.F. 24″ C.I. Force Main, 686 L.F. 24″ C.I. Pipe Sewer, and all other appurtenances for sewers.
Division Two. One Sewage Pumping Station—"Main Pump Station."
Division Three. One Sewage Pumping Station—"Fishing Creek Pump Station."
Division Four. One Sewage Pumping Station—"Camp Creek Pump Station."

Bids may be made on any or all Divisions, any of which may be awarded individually or in any combination.

Proposals. Proposals shall contain prices, in words and figures, for the work bid on. All Proposals must be accompanied by a certified check, or a bid bond of a reputable bonding company authorized to do business in the State of Georgia, in an amount equal to at least five (5%) percent of the total amount of the bid.

Upon the proper execution of the contract and required bonds, the checks or bid bonds of all bidders will be returned to them.

If Proposals are submitted via mail rather than delivery they should be addressed to Mr. Smith, Director, Department of Administration and Finance, Georgia Department of Public Health, Room 519, State Health Building, 47 Trinity Avenue, S.W., Atlanta, Georgia 30334.

Performance and Payment Bonds: A contract performance bond and payment bond, each in an amount equal to one hundred (100%) percent of the contract amount, will be required of the successful bidder.

Withdrawal of Bids: No submitted bid may be withdrawn for a period of sixty (60) days after the scheduled closing time for the receipt of bids.

Plans, Specifications, and Contract Documents: Plans, Specifications and Contract Documents are open to inspection at the Office of the Georgia Building Authority (Hospital), State Capitol, Atlanta, Georgia, or may be obtained from Wiedeman and Singleton, Engineers, P.O. Box 1878, Atlanta, Georgia 30301, upon deposit of the following amounts.

Division One: $45.00 for Plans and Specifications.
Divisions Two, Three and Four (Combined). $50.00 for Plans and Specifications.
Divisions One to Four, Inclusive. $75.00 for Plans and Specifications.
All Divisions. $20.00 for Specifications only.

Upon the return of all documents in undamaged condition within thirty (30) days after the date of opening of bids, one-half of the deposit will be refunded. No refunds will be made for plans and documents after thirty (30) days.

Wage Schedule: The schedule of minimum hourly rates of wages required to be paid to the various laborers and mechanics employed directly upon the site of the work embraced by the Plans and Specifications as determined by the Secretary of the U.S. Department of Labor, Decision No. AI-971, is included in the General Conditions of the Specifications. This decision, expiring prior to the receipt of bids, will be superceded by a new decision to be incorporated in the Contract before the award is made to the successful bidder.

Acceptance or Rejection of Bids: The right is reserved to accept or reject any or all bids and to waive informalities.

THIS PROJECT WILL BE FINANCED IN PART BY A GRANT FROM THE FEDERAL WATER POLLUTION CONTROL ADMINISTRATION AND WILL BE REFERRED TO AS PROJECT WPC-GA-157.

BIDDERS ON THIS WORK WILL BE REQUIRED TO COMPLY WITH THE PRESIDENT'S EXECUTIVE ORDERS NO. 11246 and NO. 11375. THE REQUIRE-QUIREMENTS FOR BIDDERS AND CONTRACTORS UNDER THESE ORDERS ARE EXPLAINED IN THE SPECIFICATIONS.

GEORGIA BUILDING AUTHORITY (HOSPITAL)

By: ______________________

Secretary - Treasurer

FIGURE 2.4 Notice to bidders (Courtesy of Georgia Building Authority).

DAILY BUILDING REPORT

THE ATLANTA BUILDERS' EXCHANGE

393 Techwood Drive, N. W., Atlanta, Georgia 30313 • Phone 522-5537

March 23, 1976

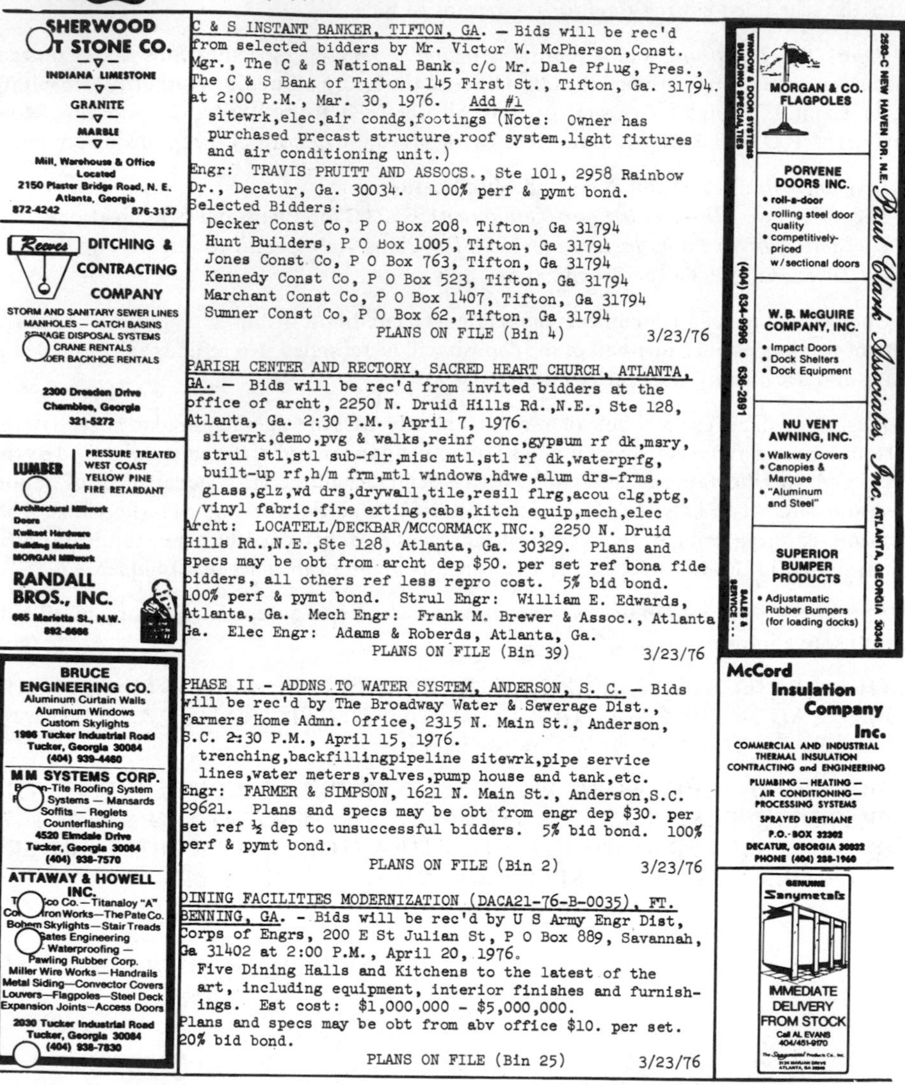

C & S INSTANT BANKER, TIFTON, GA. — Bids will be rec'd from selected bidders by Mr. Victor W. McPherson, Const. Mgr., The C & S National Bank, c/o Mr. Dale Pflug, Pres., The C & S Bank of Tifton, 145 First St., Tifton, Ga. 31794. at 2:00 P.M., Mar. 30, 1976. Add #1
sitewrk,elec,air condg,footings (Note: Owner has purchased precast structure,roof system,light fixtures and air conditioning unit.)
Engr: TRAVIS PRUITT AND ASSOCS., Ste 101, 2958 Rainbow Dr., Decatur, Ga. 30034. 100% perf & pymt bond.
Selected Bidders:
Decker Const Co, P O Box 208, Tifton, Ga 31794
Hunt Builders, P O Box 1005, Tifton, Ga 31794
Jones Const Co, P O Box 763, Tifton, Ga 31794
Kennedy Const Co, P O Box 523, Tifton, Ga 31794
Marchant Const Co, P O Box 1407, Tifton, Ga 31794
Sumner Const Co, P O Box 62, Tifton, Ga 31794
PLANS ON FILE (Bin 4) 3/23/76

PARISH CENTER AND RECTORY, SACRED HEART CHURCH, ATLANTA, GA. — Bids will be rec'd from invited bidders at the office of archt, 2250 N. Druid Hills Rd.,N.E., Ste 128, Atlanta, Ga. 2:30 P.M., April 7, 1976.
sitewrk,demo,pvg & walks,reinf conc,gypsum rf dk,msry, strul stl,stl sub-flr,misc mtl,stl rf dk,waterprfg, built-up rf,h/m frm,mtl windows,hdwe,alum drs-frms, glass,glz,wd drs,drywall,tile,resil flrg,acou clg,ptg, vinyl fabric,fire exting,cabs,kitch equip,mech,elec
Archt: LOCATELL/DECKBAR/MCCORMACK,INC., 2250 N. Druid Hills Rd.,N.E.,Ste 128, Atlanta, Ga. 30329. Plans and specs may be obt from archt dep $50. per set ref bona fide bidders, all others ref less repro cost. 5% bid bond. 100% perf & pymt bond. Strul Engr: William E. Edwards, Atlanta, Ga. Mech Engr: Frank M. Brewer & Assoc., Atlanta Ga. Elec Engr: Adams & Roberds, Atlanta, Ga.
PLANS ON FILE (Bin 39) 3/23/76

PHASE II - ADDNS TO WATER SYSTEM, ANDERSON, S. C. — Bids will be rec'd by The Broadway Water & Sewerage Dist., Farmers Home Admn. Office, 2315 N. Main St., Anderson, S.C. 2:30 P.M., April 15, 1976.
trenching,backfillingpipeline sitewrk,pipe service lines,water meters,valves,pump house and tank,etc.
Engr: FARMER & SIMPSON, 1621 N. Main St., Anderson,S.C. 29621. Plans and specs may be obt from engr dep $30. per set ref ½ dep to unsuccessful bidders. 5% bid bond. 100% perf & pymt bond.
PLANS ON FILE (Bin 2) 3/23/76

DINING FACILITIES MODERNIZATION (DACA21-76-B-0035), FT. BENNING, GA. - Bids will be rec'd by U S Army Engr Dist, Corps of Engrs, 200 E St Julian St, P O Box 889, Savannah, Ga 31402 at 2:00 P.M., April 20, 1976.
Five Dining Halls and Kitchens to the latest of the art, including equipment, interior finishes and furnishings. Est cost: $1,000,000 - $5,000,000.
Plans and specs may be obt from abv office $10. per set. 20% bid bond.
PLANS ON FILE (Bin 25) 3/23/76

FIGURE 2.5 Daily building report: The Atlanta Builders' Exchange.

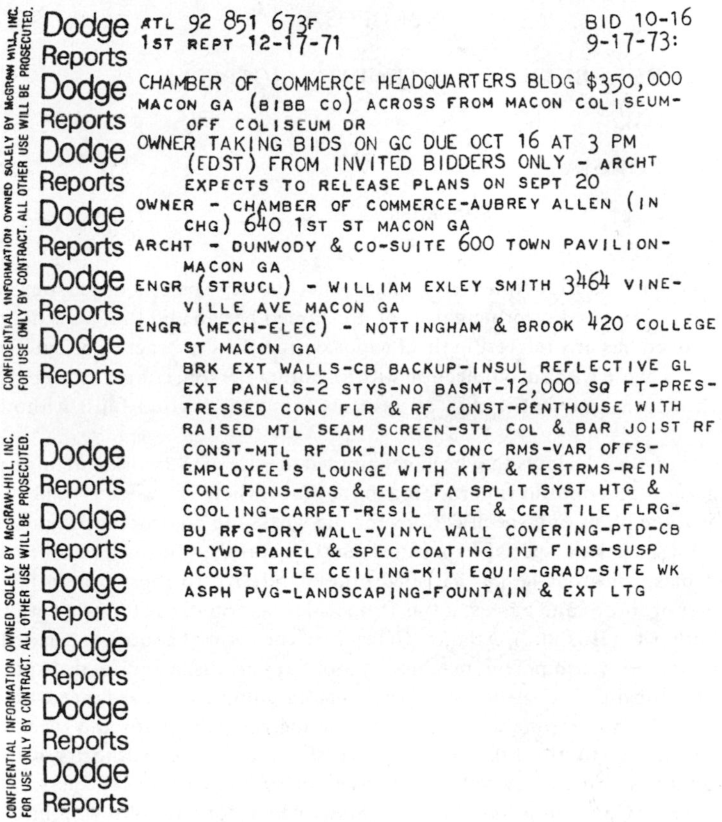

Dodge Reports

ATL 92 851 673F BID 10-16
1ST REPT 12-17-71 9-17-73:

CHAMBER OF COMMERCE HEADQUARTERS BLDG $350,000
MACON GA (BIBB CO) ACROSS FROM MACON COLISEUM-
OFF COLISEUM DR
OWNER TAKING BIDS ON GC DUE OCT 16 AT 3 PM
(EDST) FROM INVITED BIDDERS ONLY - ARCHT
EXPECTS TO RELEASE PLANS ON SEPT 20
OWNER - CHAMBER OF COMMERCE-AUBREY ALLEN (IN
CHG) 640 1ST ST MACON GA
ARCHT - DUNWODY & CO-SUITE 600 TOWN PAVILION-
MACON GA
ENGR (STRUCL) - WILLIAM EXLEY SMITH 3464 VINE-
VILLE AVE MACON GA
ENGR (MECH-ELEC) - NOTTINGHAM & BROOK 420 COLLEGE
ST MACON GA
BRK EXT WALLS-CB BACKUP-INSUL REFLECTIVE GL
EXT PANELS-2 STYS-NO BASMT-12,000 SQ FT-PRES-
TRESSED CONC FLR & RF CONST-PENTHOUSE WITH
RAISED MTL SEAM SCREEN-STL COL & BAR JOIST RF
CONST-MTL RF DK-INCLS CONC RMS-VAR OFFS-
EMPLOYEE'S LOUNGE WITH KIT & RESTRMS-REIN
CONC FDNS-GAS & ELEC FA SPLIT SYST HTG &
COOLING-CARPET-RESIL TILE & CER TILE FLRG-
BU RFG-DRY WALL-VINYL WALL COVERING-PTD-CB
PLYWD PANEL & SPEC COATING INT FINS-SUSP
ACOUST TILE CEILING-KIT EQUIP-GRAD-SITE WK
ASPH PVG-LANDSCAPING-FOUNTAIN & EXT LTG

FIGURE 2.6 Typical Dodge Reporting System announcement (COPYRIGHT © 1973 McGRAW-HILL, Inc.).

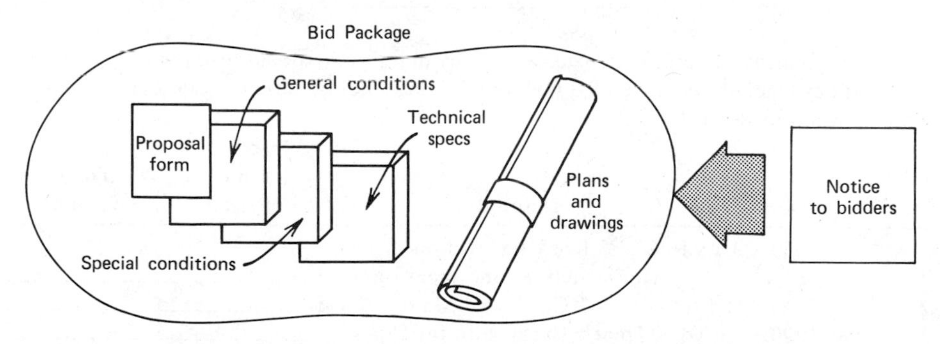

FIGURE 2.7 Bid package documents.

PROPOSAL

TO THE GEORGIA BUILDING AUTHORITY (HOSPITAL)
STATE CAPITOL
ATLANTA, GEORGIA

Submitted: ____________________, 1968

The undersigned, as Bidder, hereby declares that the only person or persons interested in the Proposal as principal or principals is or are named herein and that no other person than herein mentioned has any interest in this Proposal or in the Contract to be entered into; that this Proposal is made without connection with any other person, company, or parties making a bid or Proposal; and that it is in all respects fair and in good faith without collusion or fraud.

The Bidder further declares that he has examined the site of the work and informed himself fully in regard to all conditions pertaining to the place where the work is to be done; that he has examined the plans and specifications for the work and contractual documents relative thereto, and has read all Special Provisions and General Conditions furnished prior to the opening of bids; and that he has satisfied himself relative to the work to be performed.

The Bidder proposes and agrees, if this Proposal is accepted, to contract with the Georgia Building Authority (Hospital), Atlanta, Georgia, in the form of contract specified, to furnish all necessary material, equipment, machinery, tools, apparatus, means of transportation, and labor, and to finish the construction of the work in complete accordance with the shown, noted, described, and reasonable intended requirements of the plans and specifications and contract documents to the full and entire satisfaction of the Authority with a definite understanding that no money will be allowed for extra work except as set forth in the attached General Conditions and Contract Documents, for the following prices:

CAMP CREEK PUMP STATION

Section 1: Unit Price Work

(For part payment—except rock excavation—by unit prices, to establish price for variation in quantities. Include balance of quantities for these items—except rock excavation—in lump sum bid for Section 2.)

Item Number	Quantity	Unit	Description	Unit Price	Total Amount
1.	550	cu yd	Rock Excavation (for structures and pipes only)	$ ________	$ ________
2.	50	lin ft	8″ C.I. Force Main	$ ________	$ ________
3.	20	cu yd	Trench Excavation for Pipes	$ ________	$ ________
4.	200	sq yd	Paving	$ ________	$ ________

Subtotal, Section 1, Item Nos. 1 to 4, Inclusive ____________________

____

____________________ Dollars ($____________________)

FIGURE 2.8 A typical proposal.

Section 2: Lump Sum Work

Item No.	Description		Total Amount
5.	Excavation and Fill		
	(a) Access Roadway	$________	
	(b) Structure Excavation and Backfill	$________	
	(c) Finish Grading	$________	
	Total for Item No. 5		$________
6.	Paving		
	(a) Access Roadway	$________	
	(b) Station Area	$________	
	Total for Item No. 6		$________
7.	Concrete Work		$________

The Bidder further proposes and agrees hereby to commence work under his contract, with adequate force and equipment, on a date to be specified in a written order of the Engineer, and shall fully complete all work thereunder within the time stipulated, from and including said date, in 300 consecutive calendar days.

The Bidder further declares that he understands that the quantities shown in the Proposal are subject to adjustment by either increase or decrease, and that should the quantities of any of the items of work be increased, the Bidder proposes to do the additional work at the unit prices stated herein; and should the quantities be decreased, the Bidder also understands that payment will be made on the basis of actual quantities at the unit price bid and will make no claim for anticipated profits for any decrease in quantities, and that actual quantities will be determined upon completion of the work, at which time adjustment will be made to the Contract amount by direct increase or decrease.

The Bidder further agrees that, in case of failure on his part to execute the Construction Agreement and the Bonds within ten (10) consecutive calendar days after written notice being given of the award of the Contract, the check or bid bond accompanying this bid, and the monies payable thereon, shall be paid into the funds of the Georgia Building Authority (Hospital), Atlanta, Georgia, as liquidated damages for such failure, otherwise the check or bid bond accompanying his Proposal shall be returned to the undersigned.

Attached hereto is a bid bond by the ____________________

________________________________ in the amount of

____

____________________ Dollars ($________________)
made payable to the Georgia Building Authority (Hospital), Atlanta, Georgia, in accordance with the conditions of the Advertisement for Bids and the provisions herein.

Submitted: ____________________ L.S.

By: ____________________ L.S.

Title: ____________________

Note: If the Bidder is a corporation, the Proposal shall be signed by an officer of the corporation; if a partnership, it shall be signed by a partner. If signed by others, authority for signature shall be attached.

ADDRESS: ________________________________

____

2.6 GENERAL CONDITIONS

Certain stipulations regarding how a contract is to be administered and the relationships between the parties involved are often the same for all contracts. An organization that enters into a large number of contracts each year normally evolves a standard set of stipulations that establishes these procedures and applies them to all construction contracts. This set of provisions is normally referred to as the *general conditions*. Large government contracting organizations such as the U.S. Army Corps of Engineers, the Bureau of Reclamation, and the General Services Administration (Public Building Service) have a standard set of *general provisions*. For those organizations that enter into construction contracts on a less frequent basis, professional and trade organizations such as the American Institute of Architects (AIA) and the Associated General Contractors (AGC) publish standards that are commonly used in the industry. The topics that are considered in the general provisions are indicated in the cover sheet of the AIA standard shown in Figure 2.9. Among other topics the rights, privileges, and responsibilities that accrue to the primary contractual parties in any construction contract are defined. Therefore, sections pertaining to the (1) owner, (2) architect (or architect/engineer), (3) contractor, and (4) subcontractors are typically found in the general conditions. Most contractors become thoroughly familiar with the standard forms of general conditions (i.e., AIA, GSA, Corps of Engineers) and can immediately pick up any additions or changes. As noted on the AIA cover sheet, each of the provisions has legal implications, and the wording cannot be changed without careful consideration. The contract language embodied in the general conditions has been hammered out over the years from countless test cases and precedents in both claims and civil courts. The wording has evolved to establish a fair and equitable balance of protection for all parties concerned. In cases where a contractor finds considerable deviation from the standard language, he may decline the opportunity to bid fearing costs of litigation in clarifying contractual problems. In areas where small deviation is possible, the language of the AIA standard forms tends to be protective of the architect and to hold others responsible when gray areas arise. Predictably, the AGC standard subcontract protects the contractor in areas in which responsibility is unclear or subject to interpretation. The AIA General Conditions are presented in their entirety in Appendix A.

2.7 SUPPLEMENTARY CONDITIONS

Those aspects of the contractual relationship that are peculiar or unique to a given project are given in the *supplementary conditions*. Items such as the duration of the project, additional instructions regarding commencement of work, owner-procured materials, mandatory wage rates that are characteristic of

THE AMERICAN INSTITUTE OF ARCHITECTS

AIA Document A201

General Conditions of the Contract for Construction

THIS DOCUMENT HAS IMPORTANT LEGAL CONSEQUENCES; CONSULTATION WITH AN ATTORNEY IS ENCOURAGED WITH RESPECT TO ITS MODIFICATION

1976 EDITION

TABLE OF ARTICLES

1. CONTRACT DOCUMENTS
2. ARCHITECT
3. OWNER
4. CONTRACTOR
5. SUBCONTRACTORS
6. WORK BY OWNER OR BY SEPARATE CONTRACTORS
7. MISCELLANEOUS PROVISIONS
8. TIME
9. PAYMENTS AND COMPLETION
10. PROTECTION OF PERSONS AND PROPERTY
11. INSURANCE
12. CHANGES IN THE WORK
13. UNCOVERING AND CORRECTION OF WORK
14. TERMINATION OF THE CONTRACT

This document has been approved and endorsed by The Associated General Contractors of America.

FIGURE 2.9 Table of Articles: General conditions of the contract for construction (Reproduced with the permission of the American Institute of Architects under application number 80015). Further reproduction, in part or in whole, is prohibited. Because AIA Documents are revised from time to time, users should ascertain from the AIA the current edition(s) of the Document(s) reproduced herein.

the local area, the format required for project progress reporting (e.g., a network schedule, etc.), and the amount of liquidated damages are typical of the provisions included in the supplementary conditions. Regarding the supplementary conditions, the AIA Guide states:

Items contained in Supplementary Conditions are of two types:

1. Modifications to the basic fourteen Articles of the General Conditions in the form of additions, deletions, or substitutions.
2. Additional Articles of a contractual-legal nature which may be desirable or necessary for a particular project.

Since some of the provisions are extensions or interpretations of the general conditions, some of the major paragraph titles are similar to those used in the general conditions. The contents of a typical set of supplementary or special* conditions for a Corps of Engineers channel improvement project are shown in Figure 2.10.

2.8 TECHNICAL SPECIFICATIONS

The contract documents must convey the requirements of the project to potential bidders and establish a legally precise picture of the technical aspects of the work to be performed. This is accomplished visually through the use of drawings. A verbal description of the technical requirements is established in the *technical specifications*. These provisions pertain in large part to the establishment of quality levels. Standards of workmanship and material standards are defined in the specifications. For materials and equipment, this is often done by citing a specific brand name and model number as the desired item for installation. In government procurement, where competitive procurement must take place, a similar approach is utilized. Government specifications usually cite a specific brand or model and then establish the requirement that this or an equal item be used. The fact that equality exists must be established by the bidder.

Often the quality required will be established by reference to an accepted practice or quality specification. The American Concrete Institute (ACI), the American Welding Society (AWS), the American Association of State Highway Officials (AASHO), the American Society for Testing and Materials (ASTM), as well as federal procurement agencies publish recognized specifications and guides. A list of some typical references is given in Figure 2.11. The organization of the technical specifications section usually follows the sequence of construction. Therefore, specifications regarding concrete place-

*AIA in an attempt to standardize terminology supports the use of the term supplementary conditions as opposed to special conditions. However, both are commonly used.

Serial No. CIVENG-36-058-65-26

PART II
SPECIAL CONDITIONS
INDEX

FIGURE 2.10 Special conditions: Typical index of special conditions (Courtesy of the Army Corps of Engineers).

American Concrete Institute

ACI 211-65	Recommended Practice for Selecting Proportions for No Slump Concrete
ACI 211.1-70	Recommended Practice for Selecting Proportions for Normal Weight Concrete
ACI 211.2-69	Recommended Practice for Selecting Proportions for Structural Lightweight Concrete
ACI 214-65	Recommended Practice for Evaluation of Compression Test Results of Field Concrete
ACI 304-72	Recommended Practice for Measuring, Mixing, and Placing Concrete
ACI 305-72	Recommended Practice for Hot Weather Concreting
ACI 306-66	Recommended Practice for Cold Weather Concreting
ACI 311-64	Recommended Practice for Concrete Inspection
ACI 318-71	Building Code Requirements for Reinforced Concrete

National Ready Mix Concrete Association

Concrete Plant Standards of the Concrete Plant Manufacturers Bureau

Truck Mixer and Agitator Standards of the Truck Mixer Manufacturers Bureau

American Society for Testing and Materials

ASTM A36-70a	Structural Steel
ASTM A325-71	High Strength Bolts for Structural Steel Joints
ASTM A370-71	Mechanical Testing of Steel Products
ASTM A490-71	Quenched and Tempered Alloy Bolts for Structural Steel Joints
ASTM A615-70	Deformed Billet-Steel Bars for Concrete Reinforcement
ASTM C31-69	Making and Curing Concrete Compressive and Flexural Strength Test Specimens in the Field
ASTM C33-71a	Standard Specification for Concrete Aggregates
ASTM C39-71	Test for Compressive Strength of Molded Concrete Cylinders

American Welding Society

AWS D1.7-72	Structural Welding Code
AWS D12.1-61	Recommended Practice for Welding Reinforcing Steel, Metal Inserts and Connections in Reinforced Concrete Construction

American Association of State Highway Officials

T-26-70	Method of Test for Quality of Water to be Used in Concrete

American Institute of Steel Construction

AISC, 7th Edition	Manual of Steel Construction

American Society for Nondestructive Testing

SNTC-TC-1A, Third Edition	Recommended Practice

ASTM C40-66	Test for Organic Impurities in Sands for Concrete
ASTM C42-68	Obtaining and Testing Drilled Cores and Sawed Beams of Concrete
ASTM C88-71a	Test for Soundness of Aggregates by Use of Sodium Sulfate or Magnesium Sulfate
ASTM C94-72	Standard Specification for Ready Mixed Concrete
ASTM C109-70T	Standard Method of Test for Compressive Strength of Hydraulic Cement Mortar. (Using 2" Cube Specimen)
ASTM C117-69	Test for Materials Finer Than No. 200 Sieve in Mineral Aggregates by Washing
ASTM C123-69	Test for Lightweight Pieces in Aggregate
ASTM C131-69	Test for Resistance to Abrasion of Small Size Coarse Aggregate by Use

FIGURE 2.11 Typical references to structural, inspection, and testing standards.

ment precede those pertaining to mechanical installation. A typical index of specifications for a heavy construction project might appear as follows:

Section	
1	Clearing and grubbing
2	Removal of existing structures
3	Excavation and fill
4	Sheet steel piling
5	Stone protection
6	Concrete
7	Miscellaneous items of work
8	Metal work fabrication
9	Water supply facilities
10	Painting
11	Seeding

As with the general conditions, most contractors are familiar with the appearance and provisions of typical technical specifications. A contractor can quickly review the specifications to determine whether there appears to be any extraordinary or nonstandard aspects that will have an impact on cost. These clauses or nonstandard provisions are underlined or highlighted to be studied carefully.

2.9 ADDENDA

The bid package documents represent a description of the project to be constructed. They also spell out the responsibilities of the various parties to the contract and the manner in which the contract will be administered. These documents establish the basis for determining the bid price and influence the willingness of the prospective bidder to bid or enter into a contract. It is important, therefore, that the bid package documents accurately reflect the project to be constructed and the contract administration intentions of this owner or of the owner's representative. Any changes in detail, additions, corrections, and contract conditions that arise *before* bids are opened that are intended to become part of the bid package and the basis for bidding are incorporated into the bid package through *addenda*.

An *addendum* thus becomes part of the contract documents and provides the vehicle for the owner (or the owner's representative) to modify the scope and detail of a contract before it is finalized. It is important therefore that addenda details be rapidly communicated to all potential bidders prior to bid submission. Since addenda serve notice on the prospective bidder of changes in the scope or interpretation of the proposed contract, steps must be taken to ensure that all bidders have received all issued addenda. Consequently, addenda delivery is either documented through certified mail receipts or confirmed on bid submission through the bidder's submission of a signed document listing receipt of each duly identified addendum.

Once a contract has been signed future changes in the scope or details of a contract may form the basis for a new financial relationship between contracting parties. The original contract, in such cases, can no longer be accepted as forming the basis for a full description of the project. Such changes are referred to as *change orders* (see section 3.5).

2.10 DECISION TO BID

After investigating the plans and specifications at the architect's office or a plans room, the contractor must make a major decision — whether or not to bid the job. This is a major financial decision, since it implies incurring substantial cost which may not be recovered. Bidding the job requires a commitment of man-hours by the contractor for the development of the estimate.

Estimating is the process of looking into the future and trying to predict project costs and various resource requirements. The key to this entire process is the fact that these predictions are made based on past experiences and the ability of the estimator to sense potential trouble spots that will affect field costs. The accuracy of the result is a direct function of the skill of the estimator and the accuracy and suitability of the method by which these past experiences were recorded. Since the estimate is the basis for determining the bid price of a project, it is important that the estimate be carefully prepared. Recent studies reveal the fact that the most frequent causes of contractor failure are incorrect and unrealistic estimating and bidding practices.

The quantities of materials must be developed from the drawings by a quantity take-off man. The process of determining the required material quantities on a job is referred to as quantity surveying. Once quantities are established, estimators who have access to pricing information use these quantities and their knowledge of construction methods and productivities to establish estimates of the *direct costs* of performing each construction task. They then add to the totalled project direct costs those *indirect costs* that cannot be assigned directly to a particular estimating item. Finally, the bid price is established by adding the management and overhead costs, allowances for contingencies, and a suitable profit margin. Appendix B gives typical considerations affecting the decision to bid.

The cost of the time and effort expended to develop a total bid price and submit a proposal is only recovered in the event the contractor receives the contract. A common rule of thumb states that the contractor's estimating cost will be approximately 0.25% of the total bid price. This varies, of course, based on the complexity of the job. Based on this rule, an estimating cost of

$25,000 can be anticipated for a job in the vicinity of $10 million. Expending this amount of money to prepare and submit a bid with only a probability of being awarded the work is a major monetary decision. Therefore, most contractors consider it carefully. In order to recover bidding costs for jobs not awarded, contractors place a charge in all bids to cover bid preparation costs. This charge is based on their frequency of contract award. That is, if a contractor, on the average, is the selected bidder on one in four of the contracts he bids, he will adjust the bid cost included in each proposal to recover costs for the three in four jobs not awarded. In addition to the direct costs of bid preparation, the contractor is required by the architect to pay a deposit for each set of plans and specifications he uses. This is usually a nominal fee (e.g., $50.00 per set), which is refunded after the plans are returned. Other costs include telephone charges related to obtaining quotations from subcontractors and vendors and the administrative costs of getting these quotations in writing. A small fee must be paid for a *bid bond* (see Section 2.12) and the administrative aspects of submitting the proposal in conformance with the *instructions to bidders*.

2.11 PREQUALIFICATION

In some cases the complexity of the work dictates that the owner must be certain that the selected contractor is capable of performing the work described. Therefore, before considering a bid, the owner may decide to prequalify all bidders. This is announced in the instructions to bidders. Each contractor interested in preparing and submitting a bid is asked to submit documents that establish his firm's expertise and capability in accomplishing similar types of construction. In effect, the owner asks the firm to submit its "résumé" for consideration. If the owner has doubts regarding the contractor's ability to successfully complete the work, the owner can simply withhold qualification.

This is helpful to both parties. The contractor does not prepare a bid and incur the inherent cost unless he can qualify. On the other hand, the owner does not find himself in the position of being under pressure to accept a low bid from a firm he feels cannot perform the work. In the extreme case, a small firm with experience only on single-family residential housing may bid low on a complex radar-tracking station. If the owner feels the contractor will not be able to sucessfully pursue the work, he can fail to "prequalify" him.

2.12 SUBCONTRACTOR AND VENDOR QUOTATIONS/CONTRACTS

As noted above, estimating section personnel will establish costs directly for those items to be constructed by the prime contractor. For specialty areas such as electrical work, interior finish, and roofing, the prime contractor solicits quotations from subcontractors with whom he has successfully worked in the past. Material price quotations are also developed from vendors. These quotations are normally taken telephonically and included in the bid. It is good business practice to follow up this telephone quote with a written confirmation. This confirmation serves to bind the subcontractor or vendor to the quotation. The contractor integrates these quotations into the total bid price. Until the contractor has a firm subcontract or purchase order, he must carry the risk that the subcontractor might exercise the option to change his original quotation. If work is scarce, this seldom happens. If, however, subcontractors are busy, and no formal contract has been established, they may attempt to negotiate their initial quote to a higher level.

Following award of the contract, the prime contractor has his purchasing or procurement group move immediately to establish subcontracts with the appropriate specialty firms. Both the American Institute of Architects (AIA) and the Associated General Contractors (AGC) have standard forms for this purpose. The AGC form is shown in Appendix I.

2.13 BID BOND

A bond is a three-party instrument that protects one party from default on the part of a second party. In the event a default occurs, a third party is legally bound to offset any damages resulting from the default. In bonding terminology, the party in a position to be damaged by a default is called the obligee. The party who is in a position to default is the principal. The third party off-setting the damages is the *surety*. A default potential and three parties related as shown in Figure 2.12 are the ingredients of a bonding situation. In stores and shops, a sign near the cashier may note that the cashier is bonded. This means that if the cashier (as the principal) defaults by losing money, the store (as the obligee) can recover damages from a bonding

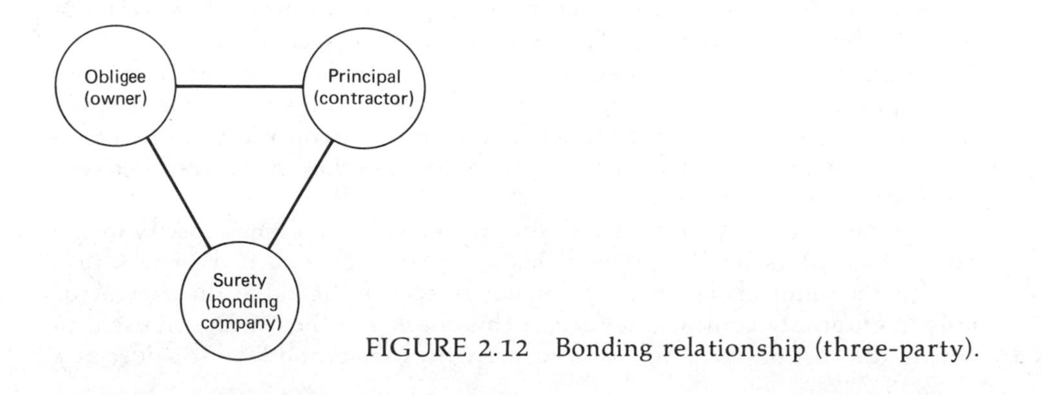

FIGURE 2.12 Bonding relationship (three-party).

company (the surety). This situation is similar to the case in which a friend asks you to cosign a note for her at the bank. If your friend fails to repay the note and interest as prescribed, the bank can recover its damage from you as the cosigner. In this case, your friend is the principal, the bank is the obligee, and you are the surety.

Owners require the submittal of a bid bond with the proposal. This bond protects the owner against failure by the contractor to enter into contract after having been awarded the contract. The contractor may get "cold feet" after being issued a notice to proceed and refuse to sign a formal contract. In such cases, the owner would incur a damage, since he would be forced to contract with the next lowest bidder. If, for instance, a contractor bidding $3 million refuses to enter into contract, and the next low bid is $3,080,000, the owner is damaged in the amount of $80,000. The AIA form for a bid bond is shown in Figure 2.13. Notice that the responsibility of the surety is indemnified (covered) by the principal. If the principal fails to enter into contract:

> . . . the Principal shall pay to the Obligee the difference not to exceed the penalty hereof between the amount specified in said bid and such larger amount for which the Obligee may in good faith contract with another party to perform the work . . . .

If the principal is unable to pay this amount, the surety must step in and cover the damage.

In most cases, the surety firm will not issue a bid bond unless it is very sure the assets of the principal will offset any default occurring due to failure to enter into a contract. Therefore, if issued at all, the bid bond is issued for a small administrative fee. From the bonding company's point of view, the importance of the bid bond is not the fee paid by the contractor for its issuance, but instead its implication that if the contract is awarded, the surety will issue performance and payment bonds. The bid bond is a "lead parachute," which pulls these two bonds out of the main pack. Typical performance and payment bonds are shown in Appendix C. The performance bond protects the owner against default on the contractor's part in completing the project in accordance with the contract documents. The payment bond protects the owner against failure on the part of the prime contractor to pay all subcontractors or vendors having outstanding charges against the project. If the surety fails to issue these bonds (required in the contract documents), the contractor is prevented from entering into contract and the surety could be forced to cover damages resulting from this default.

As an alternative to a bid bond, the owner will sometimes specify in the notice to bidders or the proposal his acceptance of a cashier's check in a specified amount made out to the owner to secure the bid. If the contractor fails to enter into contract, he forfeits this check and the owner can use it to defray the cost of entering into contract with the second lowest bidder at a

THE AMERICAN INSTITUTE OF ARCHITECTS

AIA Document A310

Bid Bond

KNOW ALL MEN BY THESE PRESENTS, that we
(Here insert full name and address or legal title of Contractor)

as Principal, hereinafter called the Principal, and
(Here insert full name and address or legal title of Surety)

a corporation duly organized under the laws of the State of
as Surety, hereinafter called the Surety, are held and firmly bound unto
(Here insert full name and address or legal title of Owner)

as Obligee, hereinafter called the Obligee, in the sum of

Dollars ($),

for the payment of which sum well and truly to be made, the said Principal and the said Surety, bind ourselves, our heirs, executors, administrators, successors and assigns, jointly and severally, firmly by these presents.

WHEREAS, the Principal has submitted a bid for
(Here insert full name, address and description of project)

NOW, THEREFORE, if the Obligee shall accept the bid of the Principal and the Principal shall enter into a Contract with the Obligee in accordance with the terms of such bid, and give such bond or bonds as may be specified in the bidding or Contract Documents with good and sufficient surety for the faithful performance of such Contract and for the prompt payment of labor and material furnished in the prosecution thereof, or in the event of the failure of the Principal to enter such Contract and give such bond or bonds, if the Principal shall pay to the Obligee the difference not to exceed the penalty hereof between the amount specified in said bid and such larger amount for which the Obligee may in good faith contract with another party to perform the Work covered by said bid, then this obligation shall be null and void, otherwise to remain in full force and effect.

Signed and sealed this day of 19

(Witness)	(Principal)	(Seal)
	(Title)	
(Witness)	(Surety)	(Seal)
	(Title)	

FIGURE 2.13 AIA form: Bid bond (Reproduced with the permission of the American Institute of Architects under application number 80015). Further reproduction, in part or in whole, is prohibited. Because AIA Documents are revised from time to time, users should ascertain from the AIA the current edition(s) of the Document(s) reproduced herein.

higher bid price. This method of bid security is indicated in the notice to bidders in Figure 2.4. The notice to bidders states:

All Proposals must be accompanied by a *certified check*, or a bid bond of a reputable bonding company authorized to do business in the State of Georgia, in an amount equal to at least five (5%) per cent of the total amount of the bid.

This procedure is further explained in the proposal form (Figure 2.8):

The bidder further agrees that, in case of failure on his part to execute the Construction Agreement* and the Bonds+ within ten (10) consecutive days after written notice being given of the award of the Contract, the check or bid bond accompanying this bid, and monies payable thereon, shall be paid into the funds of the [owner] . . . .

All government construction contracts require a bid bond that is normally for 20% of the bid price. Private construction agencies for which bid bonds are required generally designate that the bid bond be for 5 or 10% of the bid price. For this reason, residential and commercial construction contractors are different from public construction contractors to a surety. A contractor failing to enter into contract after acceptance of his low bid in public construction places a surety in greater risk because of the larger bid bond for government contracts.

2.14 PERFORMANCE AND PAYMENT BONDS

In the event the contractor is awarded the contract, performance and payment bonds are issued. A *performance bond* is issued to a contractor to guarantee an owner that the contract work will be completed and that it will comply with project specifications. If a contractor fails to perform the work, the surety must provide for completion of the work according to plans and specifications. A *payment bond* is issued to guarantee the owner freedom from any liens against his completed project. If the contractor does not pay subcontractors or suppliers, the surety must protect the owner from their claims.

The AIA standard bonding forms for performance and payment bonds are shown in Appendix C.

Because of the potential cost and trouble to take over the work of a contractor about to default, the surety may elect to negotiate short-term financing for a contractor who has current liquidity problems. The surety may grant loans directly or assist the contractor in getting additional loans for the construction. In the event of default, there is no surety payment until the contractor's funds are completely exhausted. Then the surety will normally relet the job, at a cost to itself and delay to the owner. For these

*See Section 3.2.

+Performance and payment bonds.

reasons, a surety will often seek to assist a contractor overcome temporary cash shortages.

Performance and payment bonds are issued for a service charge. The common rate is referred to as 1% and costs $10 per $1000 on the first $200,000 of contract cost. At higher contract costs, the rate is reduced incrementally. Actually, the surety is not in any great risk, since the bond includes an indemnity agreement on the part of the contractor. In other words, the contracting corporation, partnership, or proprietorship must pledge to pay back any monies expended by the surety on its behalf. Key personnel may be required to sign for their personal wealth in the case of closely held corporations or partnerships.*

The Miller Act (enacted in 1935) establishes the level of bonding required for federally funded projects. Performance bonds must cover 100% of the contract amount while payment bonds are required based on the following sliding scale: 50% if the contract is $1,000,000 or less; 40% if the contract is between $1,000,000 and $5,000,000; and a fixed amount of $2,500,000 if the contract is greater than $5,000,000.

A surety seeks to keep itself well informed of a contractor's progress with work and with his changing business and financial status. In order to help with this, the contractor makes periodic reports on the work in progress with particular attention to the costs, payments, and disputes associated with uncompleted work. Based on these reports the contractor's *bonding capacity* can be determined. This is normally a multiple of the *net quick* assets of the contractor as determined from his balance sheet. The net quick assets are the contractor's assets that can be quickly sold to cover default. The multiple is based on the contractor's performance over the years. New contractors with no "track record" may have a multiple of 3 or 4. Old and reliable contractors may have a multiple of 40 or greater. A reliable firm with net quick assets of $140,000 would have a bonding capacity of $5,600,000. This would allow the contractor's surety to permit him to undertake up to $5,600,000 in jobs at any particular time.

REVIEW QUESTIONS AND EXERCISES

2.1 What are the three major types of construction bonds? Why are they required? Name three items that affect bonding capacity?

2.2 In what major section of the contract is the time duration of the project normally specified?

2.3 Who are the three basic parties involved in any construction bonding arrangement?

*See Chapter 5 for a discussion of closely held corporations.

2.4 What type of bond guarantees that if a contractor goes broke on a project the surety will pay the necessary amount to complete the job?

2.5 What is the purpose of the following documents in a construction contract?
(a) General conditions.
(b) Special conditions.
(c) Addenda.
(d) Technical specifications.

2.6 Why is the contractor normally required to submit a bid bond when making a proposal to an owner on a competitively bid contract?

2.7 What is the Miller Act and what does it specify regarding government contracts?

2.8 What is the purpose of the notice to bidders?

2.9 List the various specialty groups that are normally involved in the design of a high-rise building project.

2.10 How much money is the contractor investing in an advertised project available for bid at the time of:
(a) Going to the architect/engineer's office to look at the plans and specifications?
(b) Deciding to take the drawings to his home office for further consideration?
(c) Deciding to make initial quantity takeoff?
(d) Full preparation of bid for submittal?

2.11 What are the major parameters to be considered in the prequalification assessment of a contractor? Investigate the local criteria used in the prequalification of both small housing and general contractors.

2.12 Obtain sample specification clauses relating to the quality of finish of an item such as face brick, exterior concrete, or paint surfaces. Who has the major responsibility for the definition, achievement in this field, and paid acceptance?

2.13 Read those clauses of the general conditions of the contract for construction that refer to the owner, architect, contractor, and subcontractor (see Appendix A). Hence list the major responsibilities of these agents with respect to the following:
(a) The definition, or attention to, the scope of the project.
(b) The financial transactions on the project.
(c) The finished quality of the work.

CHAPTER THREE

Project Cycle— Construction Phase

3.1 ACCEPTANCE PERIOD/WITHDRAWAL

In formal competitive bid situations, the timing of various activities has legal implications. The issuance of the notice to bidders opens the bidding period. The date and time at which the bid opening is to take place marks the formal end of the bidding period. Usually a bid box is established at some central location. Bids that have not been received at the bid box by the appointed hour and date are late and are normally disqualified. Prior to the close of the bidding (i.e., bid opening), contractors are free to withdraw their bids without penalty. If they have noted a mistake, they can also submit a correction to their original bid. Once bid opening has commenced, these prerogatives are no longer available. If bids have been opened and the low bidder declares a mistake in bid, procedures are available to reconcile this problem. If it can be clearly established that a mathematical error has occurred, the owner usually will reject the bid. However, if the mistake appears contrived to establish a basis for withdrawal of the bid, the owner will not reject. Then the contractor must enter into contract or forfeit his bid security. The chronology of the bid procedure is shown in Figure 3.1.

The bid security protects the owner from failure by the contractor to enter into a *formal construction agreement*. The contractor is protected by the *acceptance period*. The notice to bidders specifies a period following bid opening during which the proposed bids are to remain in force. The indication is that if the owner does not act in this period to accept one of the bids, then the contractors can withdraw or adjust their bids. This is indicated in the notice to bidders (Figure 2.4) as follows:

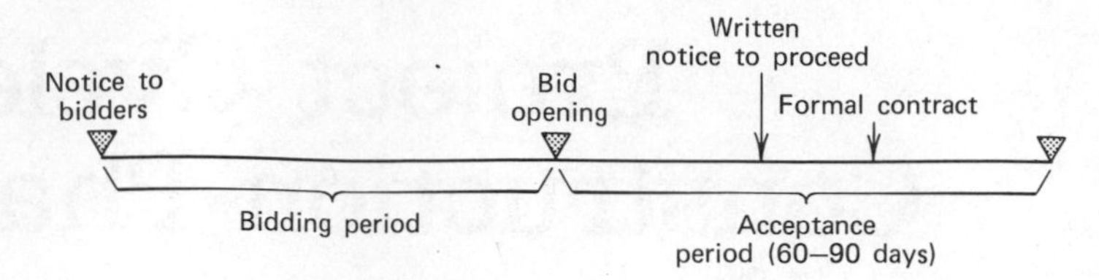

FIGURE 3.1 Chronology of bid procedure.

Withdrawal of Bids: No submitted bid may be withdrawn for a period of sixty (60) days after the scheduled closing time for the receipt of bid.

This is designed to protect the bidder, since otherwise the owner could hold the contractors to their bids for an unspecified period. If the expected financing or appropriation for the project does not materialize, the owner could, in theory, say "Wait until next year, and I will enter into contract with you at this price." This, of course, would be potentially disadvantageous to the bidder. Therefore, the owner must send written notice of award (e.g., notice to proceed) to the selected contractor during the acceptance period, or the bidders are released from their original proposals.

3.2 AWARD OF CONTRACT/NOTICE TO PROCEED

Notification of award of contract is normally accomplished by a letter indicating selection and directing the contractor to proceed with the work. This *notice to proceed* consummates the contractual relationship from a legal viewpoint despite the fact that a formal agreement has not been signed. The proposal (offer) - acceptance protocol of contractual law is satisfied by the issuance of this letter. The letter also implies that the site is free of encumbrances, and that the contractor can occupy the site for work purposes. Provisions of the contract usually direct that selected bidders commence work on the site within a specified period of time, such as 10 days.

The notice to proceed has an additional significance. The date of the notice to proceed establishes the reference date from which the beginning of the project is calculated. Therefore, based on the stipulated duration of the project as specified in the supplementary conditions, the projected end of the project can be established. As will be discussed later, time extensions may increase the duration of the project, but the end of project beyond which damages will be assessed for failure to complete the project on time is referenced to the date of the notice to proceed. This might be specified as follows:

Work shall be completed not later than one thousand fifty (1050) calendar days after the date of receipt by the Contractor of Notice to Proceed.

Calendar days are used since they simplify the calculation of the end-of-project date. In certain cases, the duration of the project is specified in working days. The general conditions normally specify working days as Monday through Friday. Therefore, each week contains five working days.

In some projects, all encumbrances to entry of the construction site have not been reconciled. Therefore, the owner cannot issue a notice to proceed, since he cannot authorize the contractor to enter the site. In such cases, in order to indicate selection and acceptance of a proposal, the owner may send the selected bidder a *letter of intent*. This letter will indicate the nature of encumbrance and establish the owner's intent to enter into contract as soon as barriers to the site availability have been removed.

3.3 CONTRACT AGREEMENT

Although the issuance of the notice to proceed establishes the elements of a contract, this is formalized by the signing of a *contract agreement*. In a legal sense, the formal contract agreement is the single document that binds the parties and by reference describes the work to be performed for a consideration. It pulls together under one cover all documents to include (1) the drawings, (2) the general conditions, (3) the supplementary conditions, (4) the technical specifications, and (5) any addenda describing changes published to these original contract documents. As with other bid package components, standard forms on the contract agreement for a variety of contractual formats are available from the American Institute of Architects. Forms for the *stipulated (lump) sum* and *negotiated (cost of work plus a fee)* type contract are given in Appendixes D and E.

3.4 TIME EXTENSIONS

Once the formal contract has been signed, certain aspects of the contractor's activity during construction must be considered. Often circumstances beyond the contractor's control, which could not have been reasonably anticipated at the time of bidding, lead to delays. These delays make it difficult or impossible to meet the projected completion date. In such cases the contractor will request an extension of time to offset the delay. These time extensions, if granted, act to increase the duration of the project. Time extensions are discussed in Article 8.3 of the AIA standard General Conditions. Claims for extension of time must be based on delays that are caused by the owner or the owner's agents or on delays due to acts of God. Delays that result from design errors or changes are typical of the owner-assignable delays and are not uncommon. A study of delay sources on government contracts indicated that a large percent of all delays can be traced to the reconciliation of design-related problems (see Table 3.1). Weather delays are typical of the so-called act of God type delay. Normal weather, however, is not justification for the granting of a time extension. Article 8.3.1 of the AIA General Conditions states specifically that only "adverse weather conditions not reasonably anticipatable" qualify as a basis for time extensions. This means that a contractor working in Minnesota in January who requests a 15-day time extension due to frozen ground that could not be excavated will possibly not be granted a time extension. Since frozen ground is typical of Minnesota in January, the contractor should have "reasonably anticipated" this condition and scheduled around it. Weather is a continuing question of debate, and many contractors will submit a request for time extension automatically each month with their progress pay request, if the weather is the least bit out of the ordinary.

Time extensions are added to the original duration so that if 62 days of time extension are granted to an original duration of 1050 days, the project must be completed by 1112 calendar days after notice to proceed. If the contractor exceeds this duration, liquidated damages (see Section 3.9) are assessed on a daily basis for each day of overrun. The question of what constitutes completion is examined in Article 8.1.3 of the AIA General Conditions. It states:

> The Date of Substantial Completion of the Work or designated portion thereof is the Date certified by the Architect when construction is sufficiently complete, in accordance with the Contract Documents, so the Owner can occupy or utilize the Work or designated portion thereof for the use for which it is intended.

This is often referred to as the beneficial occupancy date or BOD. Once the owner occupies the facility, he relinquishes a large portion of the legal leverage he has in making the contractor complete some outstanding deficiencies. Usually, a mutally acceptable date is established when substantial completion appears to have been reached. On this date an inspection of the facility is conducted. The owner's representative (normally the architect/engineer) and the contractor conduct this inspection recording deficiencies that exist and representing items for correction. Correction of these items will satisfy the owner's requirement for substantial completion. This deficiency list is referred to in the industry as the *punch* list. Theoretically, once the contractor satisfactorily corrects the deficiencies noted on the punch list, the owner will accept the facility as complete. If the rapport between owner and contractor is good, this phase of the work is accomplished smoothly. If not, this turnover phase can lead to claims for damages on both sides.

An indication of the amounts of time extension granted for various reasons on some typical government projects is given in Table 3.1. The types of delay sources categorized were due to (1) design problems, (2) owner modification, (3) weather, (4) strike, (5) late delivery, and (6) other. The percentages presented were calculated as % extension = no. of days of time extension granted ÷ originally specified project duration × 100.

Table 3.1 AVERAGE PERCENT EXTENSION BY EXTENSION TYPE

Facility	Design Problem	Owner Modification	Weather	Strike	Late Delivery	Other
Airfield paving/Lighting	7.2	1.3	2.2	0.0	10.5	4.9
Airfield buildings	12.1	2.3	3.7	3.2	0.8	29.9
Training facilities	6.2	20.8	2.9	0.0	0.6	4.6
Aircraft maintenance facilities	12.0	2.0	8.4	1.0	2.2	0.2
Automotive maintenance facilities	12.9	2.3	3.4	1.4	0.7	0.4
Hospital buildings	16.0	3.4	2.6	0.6	0.0	0.1
Officer housing	8.7	4.2	2.0	1.2	0.6	0.9
Community facilities	6.7	5.4	2.3	1.7	1.5	0.3

Source. From D. W. Halpin and R. D. Neathammer, "Construction Time Overruns," *Technical Report P-16*, Construction Engineering Research Laboratory, Champaign, Ill., August 1973.

3.5 CHANGE ORDERS

Since the contract documents are included by reference in the *formal agreement*, the lines on the drawings, the words in the technical specifications, and all other aspects of the contract documentation are legally binding. Any alteration of these documents constitutes an alteration of the contract. As will be discussed in Chapter 4 certain contractual formats such as the unit-price contract have a degree of flexibility. However, the stipulated or lump sum contract has virtually no leeway for change or interpretation. At the time it is presented to the bidders for consideration (i.e., is advertised), it represents a statement of the project scope and design as precise as the final drawings for an airplane or a violin. Changes that are dictated, for any reason, during construction represent an alteration of a legal arrangement and, therefore, must be formally handled as a modification to the contract.

These modifications to the original contract, which themselves are small augmenting contracts, are called change orders.

Procedures for implementing change orders are specified in Article 12 of the AIA General Conditions. Since change orders are minicontracts, their implementation has many of the elements of the original contract bid cycle. The major difference is that there is no competition, since the contractor has already been selected. Normally, a formal communication of the change to include scope and supporting technical documents is sent to the contractor. The contractor responds with a price quotation for performing the work, which constitutes his offer. The owner can accept the offer or attempt to negotiate (i.e., make a counteroffer). This is, of course, the classical contractual cycle. Usually, the contractor is justified in increasing the price to recover costs due to disruption of the work and possible loss of job rhythm. If the original contract documents were poorly scoped and prepared, the project can turn into a patchwork of change orders. This can lead to a sharpening of the adversary roles of the contractor and the owner, and can substantially disrupt job activities. An interesting example of this is presented in the article, "The Real Impact of Change Orders," by C. J. Collins.

3.6 CHANGED CONDITIONS

Engineering designs are based on the project site conditions as they are perceived by the architect/engineer or designer. For structural and finish items as well as mechanical and electrical systems above ground, the conditions are constant and easily determined. Variation in wind patterns leading to deviations from original design criteria may pose a problem. But normally, elements of the superstructure of a facility are constructed in a highly predictable environment.

This is not the case when designing the subsurface and site topographical portions of the project. Since the designer's ability to look below the surface of the site is limited, he relies on approximations that indicate the general nature of the soil and rock conditions below grade. His "eyes" in establishing the design environment are the reports from subsurface investigations. These reports indicate the strata of soil and rock below the site based on a series of bore holes. These holes are generally located on a grid and attempt to establish the profile of soil and rock. The ability of the below grade area to support weight may be established by a grid of test piles. The money available for this design activity (i.e., subsurface investigation) varies, and an inadequate set of bore logs or test piles may lead to an erroneous picture of subsurface characteristics. The engineer uses the information provided by the subsurface investigation to design the foundation of the facility. If the investigation is not representative, the design can be inadequate.

The information provided from the subsurface investigation is also the contractor's basis for making the estimate of the excavation and foundation work to be accomplished. Again, if the investigation does not adequately represent the site conditions, the contractor's estimate will be affected. The topographic survey of the site is also a basis for estimate and, if in error, will impact the estimate and price quoted by the contractor. If the contractor feels the work conditions as reflected in the original investigation made available to him for bidding purposes are not representative of the conditions "as found," he can claim a *changed condition*. For instance, based on the boring logs, a reasonable estimate may indicate 2000 cu yd of soil excavation and 500 cu yd of rock. After work commences, the site may be found to contain 1500 cu yd of rock and only 1000 cu yd of soil. This, obviously, substantially affects the price of excavation and would be the basis for claiming a changed condition.

In some cases, a condition may not be detected during design, and the assumption is that it does not exist. For instance, an underground river or flow of water may go undetected. This condition requires dewatering and a major temporary-construction structure to coffer the site and to construct the foundation. If this condition could not reasonably have been foreseen by the contractor, there would be no allowance for it in his bid. The failure of the bid documents to reflect this situation would cause the contractor to claim a changed condition.

If the owner accepts the changed condition, the extended scope of work represented will be included in the contract as a *change order*. If the owner does not accept the changed condition claim, the validity of the claim must be established by litigation or arbitration.

3.7 VALUE ENGINEERING

In the competitively bid contract format, which has been the basis for discussing the construction cycle in this chapter, two factors break down the transfer of information from contractor to designer. The first difficulty pertains to the adversary or "friendly enemy" attitude, which often develops in traditional competitively bid contract situations. The second is the sequencing of design and construction activity in competitive contracts.

Since the owner and the owner's representative, the architect/engineer, have interests that are, from time to time, at variance with those of the contractor, the competitively bid contract format often leads to an adversary relationship between engineer and contractor. The engineer can be likened to a composer and the constructor to an orchestra, which must play the composer's music. The composer wants the orchestra to play every note as written and complains about the performance. The orchestra, on the other hand, feels the music is poorly written and unnecessarily difficult to play.

Similarly, the engineer tends to control and criticize the construction while the contractor may return the criticism in kind. Just as some music is inherently less difficult to play, some designs are more easily constructed and less costly than others. The construction methods that are used to realize a given design in the field have a great impact on cost. Contractors are in a better position to know what materials are easiest to install and which designs are most constructable. This knowledge can greatly influence cost. Yet, because of the "friendly" opponent attitude between constructor and engineer, useful exchange of information regarding a design's cost effectiveness may be lost. It has been recognized that an incentive for better communication between architect/engineer and contractor regarding design is needed.

The fixed sequential order of design preceding construction also poses a problem. In the traditional bid sequence (Figure 3.1), the design and engineering are totally complete prior to selection of the contractor. Consequently, the contractor cannot comment on the design, and the designer has no access to the opinions of the contractor who will build the project. Certain negotiated contract formats consider this problem by ensuring selection of the contractor while design is in progress. In addition, the use of phased construction (or fast tracking as it is called) has engineering proceeding only slightly in advance of construction. Therefore, the constructor is able to feed back information regarding design constructability from the field. The classical versus the phased construction sequence is shown in Figure 3.2.

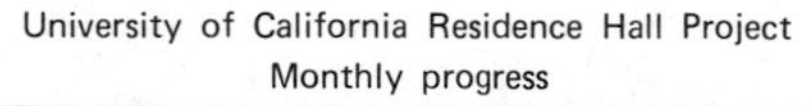

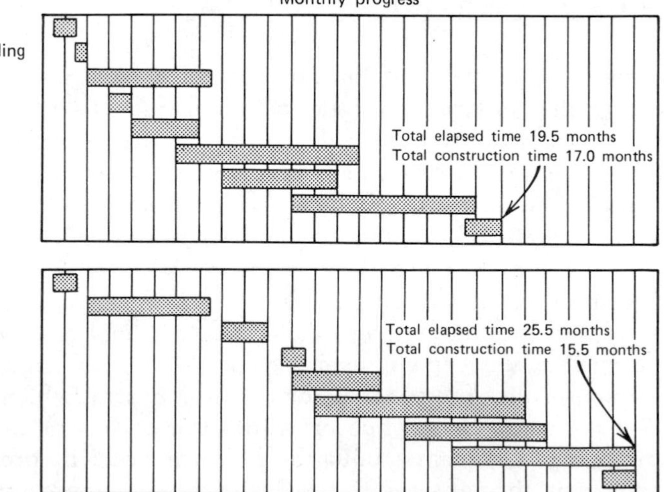

FIGURE 3.2 Traditional versus phased construction (May 4, 1972, *ENR*).

The idea behind *value engineering* is the improvement of design by encouraging the contractor to make suggestions. If at any point following his selection, a contractor feels he can make a suggestion to improve the cost effectiveness of the design, he is given a monetary incentive to do so. If his suggestion is accepted by the owner and leads to cost saving, a portion of the saving is paid as a reward. If, for instance, the contractor makes a suggestion that cuts the cost of the air conditioning system by $60,000 and an equal sharing value engineering clause is in the contract, he can receive an award of $30,000 for the suggestion. The guiding principal in making a suggestion is that the cost is reduced while the functionality remains the same.

Government agencies have led the way in the use of value engineering clauses in the general conditions, since under law they are required to use the competitively bid format. Therefore, they cannot, in general, enhance designer-constructor cooperation by using phased construction and other contractual formats. The distribution of savings between the owner and the contractor varies from contract to contract.

3.8 SUSPENSION, DELAY, INTERRUPTION

The standard General Conditions (Standard Form 23A) utilized for many government contracts provide that:

> The Contracting officer may order the Contractor in writing to suspend, delay, or interrupt all or any part of the work for such period of time as he may determine to be appropriate for the convenience of the government.

Interrupting or suspending work for an extended period of time may be costly to the contractor, since he must go through a demobilization-remobilization cycle and may confront inflated labor and materials at the time of restarting. In such cases, within the provisions of the contract, the owner (i.e., the government) is required to pay an adjustment for "unreasonable" suspensions as follows:

> An adjustment shall be made for any increase in cost of performance of this contract (excluding profit) necessarily caused by such unreasonable suspension, delay, or interruption and the contract modified in writing accordingly.

The amount of this adjustment is often contested by the contractor and can lead to lengthy litigation. Normally, the owner will attempt to avoid interruptions. Difficulties in obtaining continuing funding, however, are a common cause for these suspensions.

3.9 LIQUIDATED DAMAGES

Projects vary in their purpose and function. Some projects are built to exploit a developing commercial opportunity (e.g., a fertilizer plant in Singapore) while others are government funded for the good and safety of the public (e.g., roads, bridges, etc.). In any case, the purpose and function of a project are often based on the completion of the project by a certain point in time. To this end, a project duration is specified in the contract document. This duration is tied to the date the project is needed for occupancy and utilization. If the project is not completed on this date, the owner may incur certain damages due to the nonavailability of the facility.

For instance, assume an entrepreneur is building a shopping center. The project is to be complete for occupancy by 1 October. The projected monthly rental value of the project is $20,000. If all space is rented for occupancy on 1 October and the contractor fails to complete the project until 15 October, the space cannot be occupied and half a month's rental has been lost. The entrepreneur has been "damaged" in the amount of $10,000 and could sue the contractor for the amount of the damage. Contracts provide a more immediate means of recourse in liquidating or recovering the damage. The special conditions allow the owner under the contractual relationship to charge the contractor for damages for each day the contractor overruns the date of completion. The amount of the liquidated damage to be paid per day is given in the special or supplementary conditions of the contract. The clause SC-2 of the Special Conditions in Figure 2.10 is typical of the language used, and reads:

Liquidated Damages In case of failure on the part of the Contractor to complete the work within the time fixed in the contract or any extensions thereof, the Contractor shall pay the Government as liquidated damages the sum of $1,000 for each calendar day of delay until the work is completed or accepted.

The amount of the liquidated damage to be recovered per day is not arbitrary and must be a just reflection of the actual damage incurred. The owner who is damaged must be able if challenged to establish the basis of the figure used. In the example given above, the basis of the liquidated damage might be as follows:

Rental loss: $20,000 rent/month ÷ 30 days	= $667/day
Cost of administration and supervision of contract:	= $ 83/day
	$750/day

If a project overruns, the owner not only incurs costs due to lost revenues but also must maintain a staff to control and supervise the contract. This is the $83 cost for supervision above. The point is that an owner cannot specify an arbitrarily high figure such as $20,000 per day to scare the contractor into completion without a justification. The courts have ruled that such unsupported high charges are in fact not the liquidation of a damage but instead a penalty charge. The legal precedent established is that if the owner desires to specify a penalty for overrun (rather than liquidated damages), he must offer a bonus in the same amount for every day the contractor brings the project in early. That is, if the contractor completes the project three days late he must pay a penalty of $60,000 (based on the figure above). On the other hand, if the contractor completes the project three days early, he would be entitled to a bonus of $60,000. This has discouraged the use of such penalty-bonus clauses except in unusual situations.

Establishing the level of liquidated damages for government projects is difficult, and in most cases the amount of damage is limited to the cost of maintaining a resident staff and main office liaison personnel on the project beyond the original date of completion. In a claims court, it is impossible to establish the social loss in dollars of, for instance, the failure to complete a bridge or large dam by the specified completion date.

3.10 PROGRESS PAYMENTS AND RETAINAGE

During the construction period, the contractor is reimbursed on a periodic basis. Normally, at the end of each month, the owner's representative (usually the resident engineer at the site) and the contractor make an estimate of the work performed during the month and the owner agrees to pay a progress payment to cover the contractor's expenditures and fee or markup for the portion of the work performed. The method of making progress payments is implemented in language as follows:

> At least ten days before the date for each progress payment established in the Owner-Contractor Agreement, the Contractor shall submit to the Architect an itemized Application for Payment, notarized if required, supported by such data substantiating the Contractor's right to payment as the Owner or the Architect may require, and reflecting retainage, if any, as provided elsewhere in the Contract Documents.
>
> Section 9.3.1
AIA Document A201
General Conditions

This area is considered in greater detail in Chapter 8 in discussing cash flow. The owner typically retains or holds back a portion of the monies due the contractor as an incentive for the contractor to properly complete the project. The philosophy of retainage is that if the project is nearing completion and the contractor has received virtually all of the bid price, he will not be motivated to do the small closing-out tasks that inevitably are required to complete the project. By withholding or escrowing a certain portion of the monies due the contractor as *retainage*, the owner has a "carrot," which can be used at the end of a project. He can say essentially, "Until you have completed the project to my satisfaction, I will not release the retainage." Retainage amounts are fairly substantial and, therefore, the contractor has a strong incentive to complete the small finish items at the end of the project.

The amount of retainage is stated in the contract documents (e.g., general conditions) in the following fashion:

*In making progress payments, there shall be retained 10 percent of the estimated amount until final completion and acceptance of the work.**

Various retainage formulas can be used, based on the owner's experience and policy. If work is progressing satisfactorily at the 50% completion point, the owner may decide to drop the retainage requirement as follows:

*If the owner's representative (architect/engineer) at any time after 50 percent of the work has been completed, finds that satisfactory progress is being made, he may authorize any of the remaining progress payments to be made in full.**

If a project has been awarded at a price of $1,500,000 and 10% retainage is withheld throughout the first half of the job, the retained amount is $75,000. This is a formidable incentive and motivates the contractor to complete the details of the job in a timely fashion.

3.11 PROGRESS REPORTING

Contracts require the prime contractor to submit a schedule of activity and periodically update the schedule reflecting actual progress. This requirement is normally stated in the general conditions as follows:

Progress Charts *The contractor shall within 5 days or within such time as determined by the owner's representative, after date of commencement of work, prepare and submit to the owner's representative for approval a practicable schedule, showing the order in which the contractor proposes to carry on the work, the date on which he will start the several salient features (including procurement of materials, plant, and equipment) and the contemplated dates for completing the same. The schedule shall be in the form of a progress chart of suitable scale to indicate appropriately the percentage of work scheduled for completion at any time. The contractor shall revise the schedule as necessary to keep it current, shall enter on the chart the actual progress at the end of each week or at such intervals as directed by the owner's representative, and shall immediately deliver to the owner's representative three copies thereof. If the contractor fails to submit a progress schedule within the time herein prescribed, the owner's representative may withhold approval of progress payment estimates until such time as the contractor submits the required progress schedule.*†

This provision is fairly broad and could well be interpreted to require only grossly defined S-curves or bar charts. These bar charts may be based either on activities or percentage completion of the various work categories such as concrete, structural, electrical, and mechanical work. These reports are used at the time of developing the monthly progress payments and to ensure the

*This material is typical of the language used in implementing the retainage features of the contract.

†These clauses are typical of those used to implement the time control features of the contract and are based on contracts familiar to the authors.

contractor is making satisfactory progress. Figures 3.3 and 3.4 indicate sample reporting methods involving bar charts for work activities and S-curves of overall percentages complete.

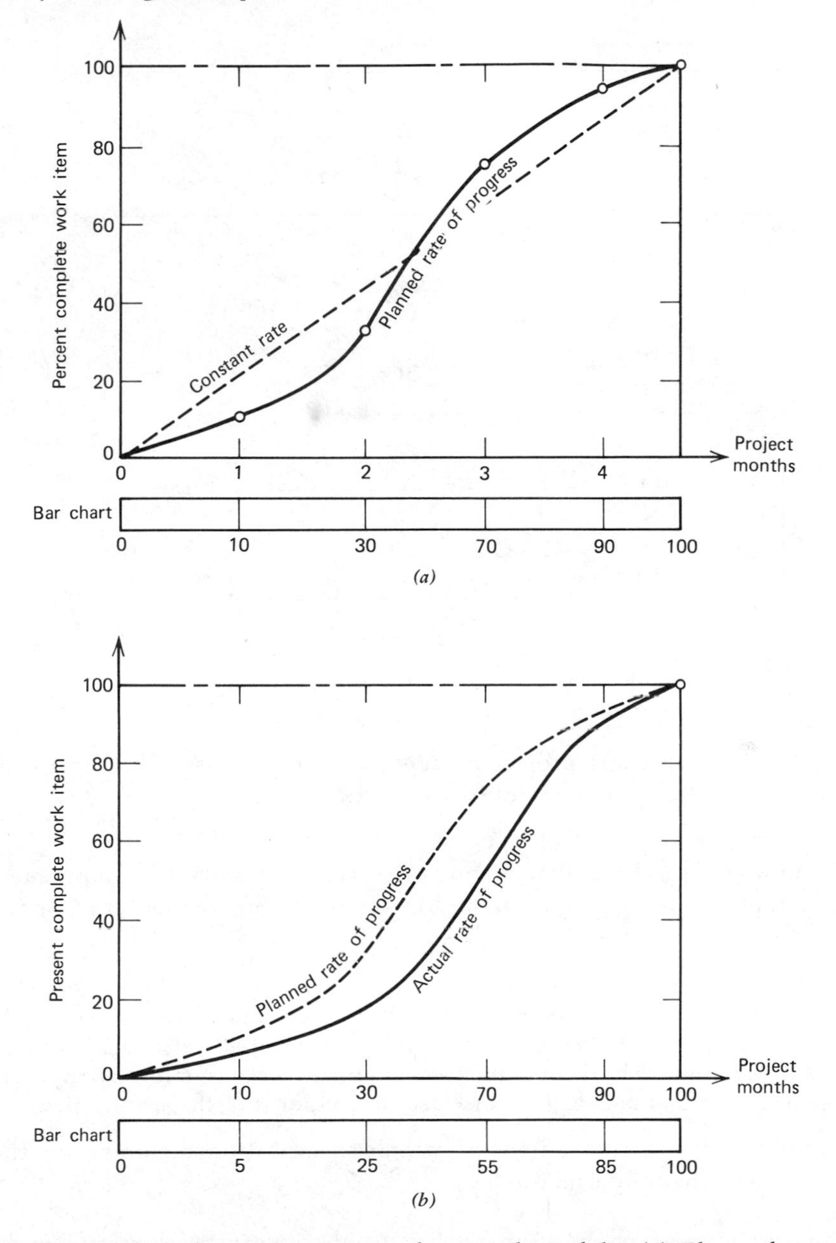

FIGURE 3.3 Bar chart planning and control models. (*a*) Planned rate of progress. (*b*) Actual rate of progress.

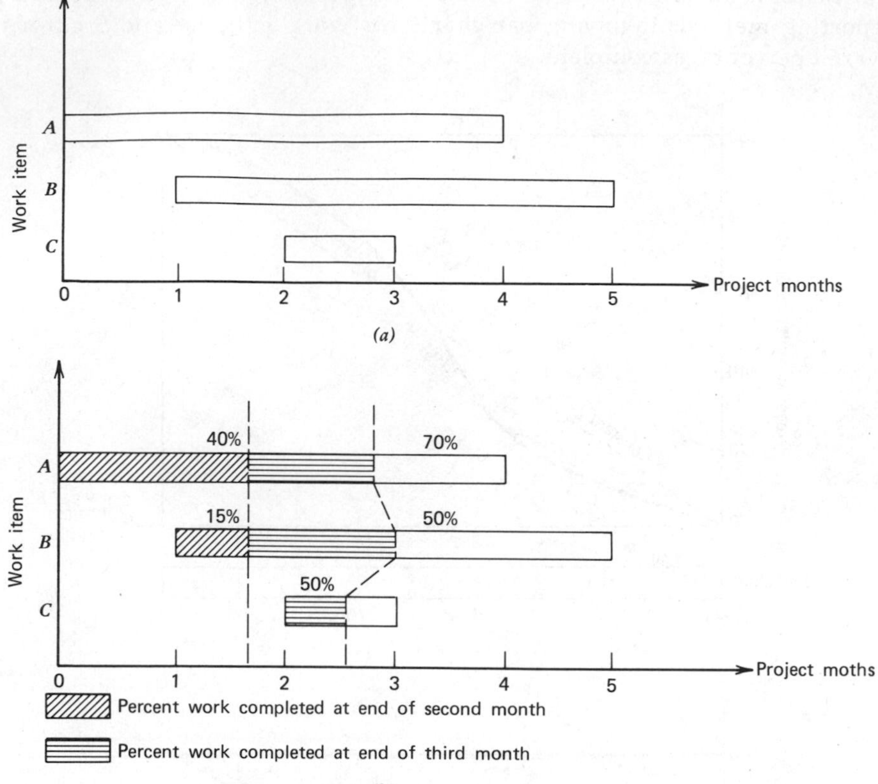

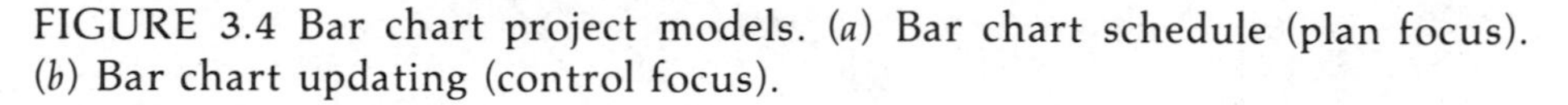

FIGURE 3.4 Bar chart project models. (*a*) Bar chart schedule (plan focus). (*b*) Bar chart updating (control focus).

Language for a clause that requires the use of network scheduling methods (see Chapter 16) is contained in the AIA Guide for Supplementary Conditions. The standard provision is as follows:

4.11.2* The progress schedule shall be prepared in the form of a network planning system for scheduling and controlling the Work. A consultant will be retained by the Owner to prepare this schedule and provide any required updating.† Details of the method for preparing the progress schedule as well as information regarding implementation and updating are included in Division 1 of the specifications.

Appendix F gives a more extensive sample specification implementing the use of network scheduling methods.

*Number of the section and subparagraph in the AIA Standard Supplementary Conditions.
†This sentence is optional.

Network methods provide greater detail and have the advantage during planning and scheduling of being oriented to individual activities and their logical sequence. From the owner's viewpoint they allow a more precise review of logic and progress during construction and acceptance periods. If the contractor is behind schedule on critical activities, a simple bar chart or S-curve will not highlight this. The network approach provides greater early warning of the impact of delays on total project completion.

3.12 ACCEPTANCE AND FINAL PAYMENT

Final acceptance of the project is important to all parties concerned. As noted above, it is particularly important to the contractor, since final acceptance means the release of retainage. Final acceptance of the project is implemented by a joint inspection on the part of the owner's representative and the contractor. The owner's representative notes deficiencies that should be corrected, and the contractor makes note of the deficiencies. These are generally detail items, and the list generated by the joint inspection is called the deficiency or "punch" list. It becomes the basis for accepting the work as final and releasing final payment (to include retainage) to the contractor. A similar procedure is utilized between the prime contractor and the subcontractors. When the subcontractor's work is complete, representatives of the prime and subcontractor "walk the job" and compile the deficiencies list for final acceptance of subcontract work. An example of a punch list between prime contractor and subcontractor is shown in Figure 3.5.

May 20, 1975

PUNCH LIST ITEMS ACME PLASTERING CO.
BARFIELD-400 PROJECT

LARGER BUILDING

1. Caulking required between stucco and brick on the lower level.
2. Streaks and cracking on stucco must be remedied.

SMALLER BUILDING

1. Very noticeable line of stucco in rear of building.
2. Streaks and cracking on stucco must be repaired.
3. Exterior bridge entrances: patch stucco must be made uniform.
4. Areas of excess spalling must be corrected.

FIGURE 3.5 Typical punch list.

3.13 CONCLUSION

This chapter has presented an overview of the cycle of activity that moves a project from the bid award stage through construction to acceptance by the owner or client. It is necessarily brief, but provides a general frame of reference indicating how the contractor receives the project and some of the contractual considerations he must be aware of during construction. The competitively bid type of contract and the bid sequence peculiar to this contractual format have been used as the basis for presentation. Other forms of contract will be discussed in Chapter 4. However, the basic chronology of events is the same. Having established this general mapping of the construction process, following chapters develop the details of the contractor's role in the construction team.

REVIEW QUESTIONS AND EXERCISES

3.1 What is the difference between liquidated damages and a penalty for late completion of the contract?

3.2 What is the purpose of retainage?

3.3 During what period can a contractor withdraw the bid without penalty?

3.4 As a contractor you have built a 100-unit apartment complex that rents for $300 per unit a month. For late completion you were assessed $1500 per day. Would you call the assessment liquidated damages or a penalty? If the contract had included a bonus of $500 per day for early completion would you expect to regain any assessment from court action? Why?

3.5 Describe the procedures to be followed for the receiving and opening of bids. If possible attend a bid opening and determine the number of bids that were submitted. For several unsuccessful bids determine the dollar amounts by which they exceeded the winning bid. Then calculate (relative to the winning bid), the percentages by which they exceed the winning bid. What do these figures tell about the strength of the current estimating and market environments? How much did the winning bidder "leave on the table?"

3.6 Scan a typical stipulated sum contract and identify those clauses that either prescribe, modify, or are related to time considerations. Then develop a time strip map (similar to Figure 3.1) for the contract that locates the times (or time zones) for which each of the clauses are relevant. Which clauses rigorously fix time constraints for the contract and which are dependent on acts of God or the owner for relevance?

3.7 Describe the procedure to be followed by the contractor who wishes to claim a time extension. What sort of documentation do you think is necessary to either refute or defend a time extension claim due to unusual weather? What sort of records do local contractors keep of weather conditions?

3.8 Must a contractor accept and perform all the work involved in each contract change order? Is there a limit to the number or magnitude of change orders that can be applied to a contract? When can a contractor refuse to accept a change order?

3.9 List the common causes of changed conditions in a building contract. What typical contract clauses bear on the problems caused by changed conditions? Suppose separate contracts are let for the building foundations and all remaining work. If you are the second contractor and you find that the foundations are incorrectly located, either in plan or elevation, would you be able to claim a changed condition?

3.10 Prepare a punch list of deficiencies or repairs that you consider necessary for your room, garage, or classroom. Can any of these items be related back to the original acceptance of the facility?

3.11 How would you go about either documenting a claim for a contractor's progress payment or its verification by the contract administrator for a typical building project in your locality?

CHAPTER FOUR

Construction Contracts

4.1 THE CONTRACT ENVIRONMENT

Construction contractors operate in a world structured by contracts. As described in Chapter 3, they contract for construction of a project with an owner by signing a construction agreement. They enter into subcontracts with specialty organizations such as electrical and mechanical subcontractors to have services performed that they do not handle with their in-house force. They buy materials for the project within the contractual format of purchase orders issued to vendors and suppliers (see Chapter 12). In many cases, contractors negotiate project or multiyear contracts with their labor force specifying wages, work rules, and fringe benefits (see Chapter 13). They establish contractual relationships with insurance carriers that provide a broad spectrum of required coverage such as workmen's compensation, public liability, and property damage insurance. They must provide the owner with various bonds, which contain the elements of a contract.

In general, an agreement between two or more parties to do something for a consideration establishes the basis for a contract. "A contract is a promise or a set of promises for the breach of which the law recognizes duty. This amounts to saying that a contract is a legally enforceable promise" (Jackson, 1973). This concept of a contract is fairly straightforward. Lawyers, however, spend a great deal of their time involved in the interpretation of this fairly simple idea. The courts often have to interpret who the parties are, what their promises are, as well as other aspects of contractual agreements. In general, a contractor renders services or goods to someone who is offering to exchange a rightful remuneration (consideration) for such goods or services. In order to eliminate as much interpretation as possible, construction contracting has evolved several fairly standard forms of contract for the building of projects.

The purpose of this chapter is to present and discuss the various contract forms presently being used for prime contracts between contractor and owner for the construction of a given facility or project.

4.2 MAJOR CONSTRUCTION CONTRACT TYPES

One of the areas in which the construction industry has been most dynamic in the past 15 years has been the development of new and innovative contractual formats for project construction. The first and most widely used form of contract is the competitively bid contract. The form of this contract has been perfected over the past 50 years. It is the required form of contract for all publicly funded construction, since it yields a low and competitive price which ensures taxpayers that their monies are being equitably and cost-effectively disbursed. The basic sequence of events associated with this type of contract is as described in Chapters 2 and 3. The two main categories of competitively bid contracts are (1) the lump or stipulated sum contract and (2) the unit price contract. The names of both of these contract formats refer to the method in which the price for the work is quoted.

The *negotiated* contract is the second major contract type in general use. This form of contract is also referred to as the *cost plus* contract. The contractor is reimbursed for the *cost* of doing the work *plus* a fee. The contractor's risk is greatly reduced, since the problem of being tied to a fixed price is not present.* The method of selecting the contractor is flexible and a function of the owner's policy. Usually a group of prequalified contractors† is selected. Each of the selected firms makes a presentation to further establish its credentials and the advantages it can offer. Here value judgment on the owner's part comes into play and individual preferences govern. This format is not well suited to public projects, since favoritism can play a major part in which contractor is selected. In the negotiated format, the owner is free to select the contractor on any criterion, and the firm that is the apparent low bidder is not necessarily the selected firm.

The *design-build* or *turnkey* type of contract is a natural evolutionary step from the negotiated contract. In this contractual format, the owner retains one firm to perform both design and construction functions. Its scope places responsibility for performance and coordinating both of these activities under the control of one contractor. Generally, only large firms with both design and construction capability can compete in this market. Similarly, since only a

*An exception to this is the guaranteed maximum price (GMP) provision in some types of negotiated contracts.

†See Section 2.11.

limited number of large firms compete in this area, the projects are large and complex. The owner wants to give the task of realizing the facility to one firm and to be called when the project is turnkey ready (i.e., complete and ready for the ribbon-cutting ceremony). Since only one firm is in charge of both project engineering and construction, the adversary relationship between designer and constructor is not present and the opportunity for phased construction (i.e., design and construction concurrently) is available.

The most recent innovation in contracting is the *construction management* form of contract. This format has become popular within the past 10 years and is described in great detail in several recent references (Heery, 1975; Goldhaber et al., 1977; and Barrie and Paulson, 1978). It is particularly attractive to organizations that periodically build complex structures (e.g., hospital authorities, municipalities, transit authorities, some power utilities) but do not desire to maintain a full-time construction staff to take the project through the cycle described in Chapters 2 and 3. In such cases, the owner retains a firm as *construction manager* to plan, develop, and coordinate the activities of an architect/engineer, trade contractors, vendors, and other interested parties such as public licensing and control bodies. The construction manager is the owner's direct representative and acts as a "traffic cop," ensuring that flows of drawings, equipment, and contract work are properly planned, scheduled, and controlled. Many variants of this concept are in use and will be described in Chapter 18.

4.3 COMPETITIVELY BID CONTRACTS

The mechanism by which competitively bid contracts are advertised and awarded has been described in Chapters 2 and 3. Essentially, the owner invites a quote for the work to be performed based on complete plans and specifications. The award of contract is generally made to the lowest *responsible* bidder. The word responsible is very important, since the contractor submitting the lowest bid may not, in fact, be competent to carry out the work. Once bids have been opened and read publicly (at the time and place announced in the instructions to bidders), an "apparent" low bidder is announced. The owner then immediately reviews the qualifications of the bidders in ascending order from lowest to highest. If the lowest bidder can be considered responsible based on his capability for carrying out the work, then further review is unnecessary.

The factors that affect whether a contractor can be considered responsible are the same as those used in considering a contractor for prequalification:

1. Technical competence and experience.
2. Current financial position based on the firm's balance sheet and income statement.

3. Bonding capacity.
4. Current amount of work under way.
5. Past history of claims litigation.
6. Defaults on previous contracts.

Because of shortcomings in any of the above areas, a contractor can be considered a risk and, therefore, not responsible. Owners normally verify the bidder's financial status by consulting the Dun and Bradstreet *Credit Reports* (Building Construction Division) to verify the financial picture presented in the bid documents.

Generally, the advantages that derive from the use of competitively bid contracts are two-fold. First, because of the competitive nature of the award, selection of the low bidder ensures that the lowest responsible price is obtained. This is only theoretically true, however, since change orders and modifications to the contract tend to offset or negate this advantage and increase the contract price. Some contractors, upon finding a set of poorly defined plans and specifications, will purposely bid low (i.e., zero or negative profit) knowing that many change orders will be necessary and will yield a handsome profit. That is, they will bid low to get the award and then negotiate high prices on the many change orders that are issued.

The major advantage, which is essential for public work, is that all bidders are treated equally and there are no favorites. This is very important since in the public sector political influence and other pressures could influence the selection of the contractor. Presently, public *design* contracts are not awarded by competitive bidding, and this practice has been greatly damaged due to apparent corrupt practices by politicians in the award of such contracts. The practice of negotiating *design* contracts is traditional and supported by engineering professional societies (e.g., the American Society of Civil Engineers and the National Society of Professional Engineers). Nevertheless it is being challenged by the U.S. Department of Justice. Recent rulings indicate that competitive procedures for award of design contracts are gaining support, and this may soon be as common in the *design* field as it is now in *construction*.

The competitive method of awarding construction contracts has several inherent disadvantages. First, the plans and specifications must be totally complete prior to bid advertisement. As discussed in Section 3.7, this leads to a sequentiality of design followed by construction and breaks down feedback from the field regarding the appropriateness of the design. Also, it tends to extend the total design-build time frame, since the shortening of time available by designing and constructing in parallel is not possible. In many cases, the owner wants to commence construction as quickly as possible to achieve an early completion and avoid the escalating prices of labor and materials. The requirement that all design must be complete before construction commences preempts any opportunity for commencing construction while design is still underway.

4.4 STIPULATED SUM CONTRACTS

A lump sum or stipulated sum contract is one in which the contractor quotes one price, which covers all work and services required by the contract plans and specifications. In this format, the owner goes to a set of firms with a complete set of plans and specifications and asks for a single quoted price for the entire job. This is like a musician going to a violin maker with the plans for an instrument and requesting a price. The price quoted by the violin maker is the total cost of building the instrument and is a lump sum price. Thus the lump sum must include not only the contractor's direct costs for labor, machines, etc. but also all indirect costs such as field and front office supervision, secretarial support, and equipment maintenance and support costs. It must also include profit.

In stipulated sum contracts the price quoted is a guaranteed price for the work specified in the plans and supporting documents. This is helpful for the owner since he knows the exact amount of money be must budget for the project barring any contingencies or change of contractual documents (i.e., change orders).

In addition, the contractor receives monthly progress payments based on the estimated percent of the total job that has been completed. In other contract forms precise field measurement of the quantity of work placed (e.g., cubic yards of concrete, etc.) must be made continuously, since the contractor is paid based on the units placed rather than on the percent of job completed. Since the percent of the total contract completed is an estimate, the accuracy of the field measurements of quantities placed need only be accurate enough to establish the estimated percent of the project completed. This means that the number and quality of field teams performing field quantity measurements for the owner can be reduced. The total payout on the part of the owner cannot exceed the fixed or stipulated price for the total job. Therefore, rough field measurements and observations, together with some "Kentucky windage," are sufficient support for establishing the amount of progress payment to be awarded.

In addition to the disadvantage noted above (requirement to have detailed plans and specifications complete before bidding and construction can begin), the difficulties involved in changing design or modifying the contract based on changed conditions are an important disadvantage. The flexibility of this contract form is very limited. Any deviation from the original plans and specifications to accomodate a change must be handled as a *change order* (see Section 3.5). This leads to the potential for litigation and considerable wrangling over the cost of contract changes and heightens the adversary relationship between owner and contractor.

The stipulated sum form of contract is used primarily in building construction where detailed plans and specifications requiring little or no modification can be developed. Contracts with large quantities of earthwork or subsurface work are not normally handled on a lump sum basis, since such contracts must be flexible enough to handle the imponderables of working below grade. Public contracts for buildings and housing are typical candidates for lump sum competitively bid contracts.

4.5 UNIT PRICE CONTRACTS

In contrast to the lump sum or fixed price type of contract, the unit price contract allows some flexibility in meeting variations in the amount and quantity of work encountered during construction. In this type of contract, the project is broken down into work items that can be characterized by units such as cubic yards, linear and square feet, and piece numbers (e.g., 16 window frames). The contractor quotes the price by units rather than as a single total contract price. For instance, he quotes a price per cubic yard for concrete, machine excavation, square foot of masonry wall, etc. The contract proposal contains a list of all work items to be defined for payment. Items 1 to 4 in Section 1 of Figure 2.8 provide a typical listing for unit price quotation. This section is reprinted for reference.

Item Number	Quantity	Unit	Description	Unit Price	Total Amount
1.	550	cu yd	Rock excavation (for structures and pipes only)	$ ____	$ ____
2.	50	lin ft	8-in. C.I. force main	$ ____	$ ____
3.	20	cu yd	Trench excavation for pipes	$ ____	$ ____
4.	200	sq yd	Paving	$ ____	$ ____

Four items of unit price work are listed. A guide quantity is given for each work item. The estimated amount of rock excavation, for example, is 550 cu yd. Based on this quantity of work, the contractor quotes a unit price. The total price is computed by multiplying the unit price by the guide quantity. The low bidder is determined by summing the total amount for each of the work items to obtain a grand total. The bidder with the lowest grand total is considered the low bidder. In true unit price contracts, the entire contract is divided into unit price work items.* Those items that are not easily expressed in units such as cubic yards are expressed in the unit column as "one job."

*In this respect, the listing of some items for unit price quotation (items 1-4) and some for lump sum quotation (items 5-7) in Figure 2.8 is not typical. All items are listed as unit price items.

Unit price quotations are based on the guide quantity specified. If a small quantity is specified, the price will normally be higher to offset mobilization and demobilization costs. Larger quantities allow "economies of scale," which reduce the price per unit. That is, if 100 sq ft of masonry brick wall is to be installed, the cost per square foot would normally be higher than the cost for 5000 sq ft. Mobilization and demobilization costs are spread over only 100 units in the first case, whereas in the second case these costs are distributed over 5000 units, reducing the individual unit cost.

Most unit price contracts provide for a price renegotiation in the event that the actual field quantity placed deviates significantly from the guide quantity specified. If the deviation exceeds 10% the unit price is normally renegotiated. If the field quantity is over 10% greater than the specified guide quantity, the owner or the owner's representative will request a price reduction based on economies possible due to the larger placement quantity. If the field quantity underruns the guide quantity by more than 10%, the contractor will usually ask to increase the unit price. He will argue that he must recover his mobilization, demobilization, and overhead costs, since the original quote was based on the guide quantity. He now has fewer units across which to recover these costs and therefore must adjust the unit price upward.

In developing the unit price quotation, the contractor must include not only directs costs for the unit but also indirect costs such as field and office overheads as well as the markup (see Figure 9.1).

In unit price contracts, the progress payments for the contractor are based on precise measurement of the field quantities placed. Therefore, the owner may have a good indication of the total cost of the project based on the grand total prices submitted. However, deviations between field measurement quantities and the guide quantities will lead to deviations in overall job price. Therefore, one disadvantage of the unit price contract form is that the owner does not have a precise final price for the work until the project is complete. In other words, allowances in the budget for deviations must be made. In addition, the precision of field measurement of quantities is much more critical than with the lump sum contract. The measured field quantities must be exact, since they are, in fact, the payment quantities. Therefore, the owner's quantity measurement teams must be more careful and precise in their assessments, since their quantity determinations establish the actual cost of the project.

Unit price contracts can also be manipulated using the technique called unbalancing the bid. The relationship between the contractor's expenditures and income across the life of a typical project is shown schematically in Figure 4.1. Because of delays in payment and retainage as described in Section 3.10, the income curve lags behind the expenditure curve and leads the contractor to borrow money to finance the difference. The nature and amount of this financing is discussed in detail in Chapter 8.

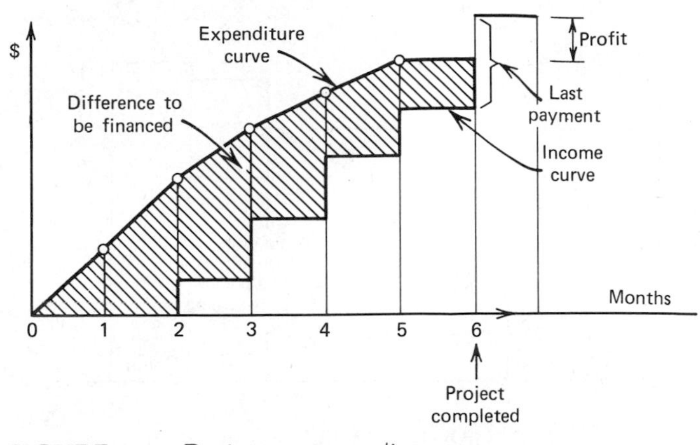

FIGURE 4.1 Project expense/income curves.

The cross hatched area in the figure gives an approximate indication of the amount of overdraft the contractor must support at the bank pending reimbursement from the client. In order to reduce this financing as much as possible, the contractor would like to move the income curve as far to the left as possible.

One way to achieve this is to unbalance the bid. Essentially, for those items that occur early in the construction, inflated unit prices are quoted. So, for instance, excavation that in fact costs $16 per cubic yard will be quoted at $24 per cubic yard. Foundation piles that cost $16 per linear foot will be quoted at $20 per linear foot. Since these items are overpriced, in order to remain competitive, the contractor must reduce the quoted prices for latter bid items. "Close-out" items such as landscaping and paving will be quoted at lower-than-cost prices. This has the effect of moving reimbursement for the work forward in the project construction period. It unbalances the cost of the bid items leading to front-end loading.

The amount of overdraft financing is reduced, as shown by the income and expense profiles in Figure 4.2. Owners using the unit price contract format are usually sensitive to this practice by bidders. If the level of unbalancing the quotations for early project bid items versus later ones is too blatant, the owner may ask the contractor to justify his price or even reject the bid.

Some contracts circumvent the practice of unbalancing the bid by allowing the contractor to quote a "mobilization" bid item. This essentially allows the bidder to request "front money" from the owner. The mobilization item moves the income curve to the left of the expense curve (see Figure 4.3). The contractor in the normal situation (Figure 4.1) will bid the cost of financing the income/expense difference into his prices. Therefore, the owner ulti-

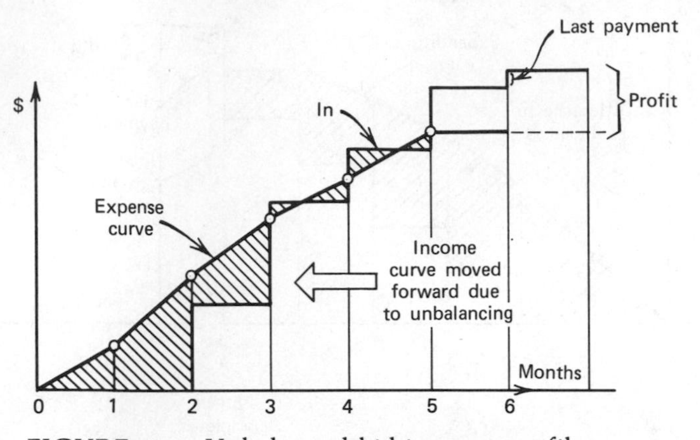

FIGURE 4.2 Unbalanced bid income profile.

mately pays the cost of financing the delay in payment of income. If the owner's borrowing rate at the bank is better than that of the contractor, he can save money by providing a mobilization item, thereby offsetting the contractor's charge for interim financing. Large owners, for instance, are often able to borrow at the prime rate (e.g., 8 or 9%),* while contractors must pay several percent above the prime rate (e.g., 11 or 12%). By providing a mobilization item, the owner essentially assumes the overdraft financing at his rate, rather than having the contractor charge him at the higher rate.

In addition to the flexibility in accomodating the variation in field quantities, the unit price contract has the added advantage to the contractor that his quantity takeoff need only verify the guide quantities given in the bid item list. Therefore, the precision of the quantity takeoff need not be as exact as that developed for a fixed price (lump sum) contract. The leeway for quantity deviation about the specified guide quantities also normally reduces the number of change orders due to the automatic allowance for deviation.

Because of its flexibility, the unit price contract is almost always used on heavy and highway construction contracts where earthwork and foundation work predominate. Industrial complexes also typically are contracted using the unit-price contract form with bid item list for price quotation.

4.6 NEGOTIATED CONTRACTS

An owner can enter into contract with a constructor by negotiating the price and method of reimbursement. A number of forms of contract can be concluded based on negotiation between owner and contractor. It is possible, for

*The prime rate is the interest rate charged to preferred customers by the bank for borrowed money.

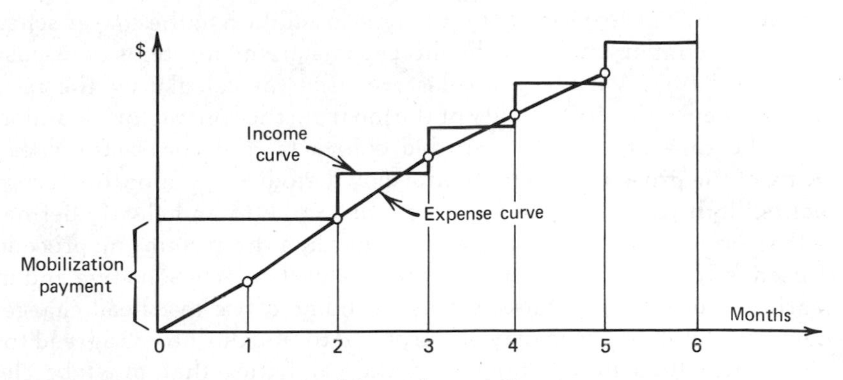

FIGURE 4.3 Income profile with mobilization payment.

example, to enter into a fixed price or unit price contract after a period of negotiation. In some cases, public owners will negotiate with the low three bidders on prices, materials, and schedule.

The concept of negotiation pertains primarily to the method by which the contractor is selected. It implies flexibility on the part of the owner to select the contractor on a basis other than low bid. Therefore, a contractor competing for award of contract in the negotiated format cannot expect to be selected solely on the basis of low bid. This affects the bid cycle and the completeness of plans and specifications that must be available at the time of contractor selection. The owner invites selected contractors to review the project documentation available at the time of negotiation. This documentation may be total and complete design documentation as in the case of competitively bid contracts or only concept level documentation. Based on the documentation provided, the contractor is invited to present his qualifications to perform the work and to indicate his projected costs and fee for completing the work. Since the level of the design documentation can vary from total detail to preliminary concept drawings, the accuracy of the cost projections will also vary. Within this presentation format, the owner evaluates the experience, reputation, facilities, staff available, charge rates, and fee structures of the various bidders participating. Based on this evaluation, the field is reduced to two or three contractors, and negotiations are opened regarding actual contract form and methods of reimbursement.

Since in most cases, the design documentation is not complete at the time of negotiation, the most common form of contract concluded is the COST + FEE. In this type of contract, the contractor is reimbursed for expenses incurred in the construction of the contracted facility. The contract describes in detail the nature of the expenses that are reimbursable. Normally, all direct expenses for labor, equipment, and materials as well as overhead charges required to properly manage the job are reimbursable. In addition, the contractor receives a fee for his expertise and the use of his plant in support of the job. The fee is essentially a profit or markup in addition to the cost reimbursement. The level and amount of fee in addition to the charge schedule to be used in reimbursement of the direct costs are major items of discussion during negotiation. Various formulas are used for calculating the fee and strongly influence the profitability of the job from the contractor's standpoint.

As in the case of competitively bid contracts, the contractor does the financing of the project and is reimbursed by periodic (e.g., monthly) progress payments. Both parties to the contract must agree to and clearly define the items that are reimbursable. Agreement regarding the accounting procedures to be used is essential. Areas of cost that are particularly sensitive and must be clearly established are those relating to home office overhead charges. If the owner is not careful, he may be surprised to find out he has agreed to pay for the contractor's new computer. Other activities that must be clearly defined for purposes of reimbursement are those pertaining to award and control of subcontracts as well as the charges for equipment used on the project.

Four types of fee structure are common. They lead to the following cost plus types of contract:

1. Cost + percent of cost.
2. Cost + fixed fee.
3. Cost + fixed fee + profit sharing clause.
4. Cost + sliding fee.

The oldest form of fee structure is the *percent of cost* form. This form is very lucrative for the contractor but is subject to abuse. There is little incentive to be efficient and economical in the construction of the project. Just to the contrary, the larger the cost of the job, the higher the amount of fee that is paid by the owner. If the cost of the job is $4 million and the fee is 2%, then the contractor's fee is $80,000. If the costs increase to $4.2 million the contractor's fee increases by $4000. Abuse of this form of contract has been referred to as the "killing of the goose which laid the golden egg."

In order to offset this flaw in the percent cost approach, the *fixed fee* formula was developed. In this case, a fixed amount of fee is paid regardless of the fluctuation of the reimbursable cost component. This is usually established as a percent of an originally estimated total cost figure. This form is commonly used on large multiyear power plant projects. If the projected cost of the plant is $500 million, a fixed fee of 1% of that figure is specified and does not change due to variation from that original estimated cost. Therefore, the contractor's fee is fixed at $5 million. This form gives the contractor an incentive to get the job done as quickly as possible in order to recover his fee over the shortest time frame. Because of the desire to move the job as quickly as possible, however, the contractor may tend to use expensive reimbursable materials and methods to expedite completion of the project.

The fixed fee plus profit sharing formula provides a reward to the contractor who controls costs, keeping them at a minimum. In this formula it is common to specify a target price for the total contract. If the contractor brings the job in under the target, the savings are divided or shared between owner and contractor. A common sharing formula provides that the contractor shares by getting 25% of this underrun of the target. If, for instance, the target is $15 million and the contractor completes the job for $14.5 million, he receives a bonus of $125,000. The projection of this underrun of the target and the percent bonus to be awarded the contractor are used by some construction firms as a measure of the job's profitability. If the contractor exceeds the target, there is no profit to be shared.

In some cases, the target value is used to define a Guaranteed Maximum Price (GMP). This is a price that the contractor guarantees will not be exceeded. In this situation, any overrun of the GMP must be absorbed by the contractor. The GMP may be defined as the target plus some fraction of the target value. In the example above, if the target is $15 million, a GMP of $16 million might be specified.

In this form, a good estimate of the target is necessary. Therefore, the plans, and concept drawings and specifications must be sufficiently detailed to allow determination of a reasonable target. The incentive to save money below the target provides an additional positive factor to the contractor. The owner tends to be more ready to compromise regarding acceptance of the project as complete if the job is under target. Additional work on punch list costs the contractor 25 cents, but it costs the owner 75 cents. The quibbling that is often present at the time the punch list is developed is greatly reduced to the contractor's advantage.

A variation of the profit sharing approach is the *sliding fee*, which not only provides a bonus for underrun but also penalizes the contractor for overrunning the target value. The amount of the fee increases as the contractor falls below the target and decreases as he overruns the target value. A typical formula for calculating the contractor's fee based on a sliding scale is

$$\text{Fee} = R(2T - A)$$

where: T - target price
R - base percent value
A - actual cost of the construction

Negotiated contracts are most commonly used in the private sector, where the owner wants to exercise a selection criterion other than low price alone. The negotiated contract is used only in special situations in the public sector, since it is open to abuse in cases where favoritism is a factor. Private owners are also partial to the negotiated format of contracting because it allows the use of phased construction in which design and construction proceed simultaneously. This allows compression of the classical "design first-then construct" sequence. Since time is literally money, every day saved in occupying the facility or putting it into operation represents a potentially large dollar saving. The cost of interest alone on the financing of a large hotel complex has run as high as $50,000 a day. Delays on nuclear power facilities are estimated at between $250,000 and $500,000 a day. Quite obviously, any compression of the design-build sequence is extremely important.

Large and complex projects have durations of anywhere from 2 to 3 up to 10 years. For such cases, *cost plus* contracts are the only feasible way to proceed. Contractors will not bid fixed prices for projects that continue over many years. It is impossible to forecast the price fluctuations in labor, material, equipment, and fuel costs. Therefore, negotiated cost plus fee contracts are used almost exclusively for such complex long-duration projects.

Often private entrepreneurs undertake "short fuse" projects with very sketchy design documentation. This can result from the need to quickly exploit an emerging market opportunity. In such cases, the cost plus fee approach is the most appropriate solution to contracting. Even governmental agencies will enter into cost reimbursement type contracts in emergency situations such as a flood or a natural disaster. During the Vietnam period, the government (U.S. Navy) operated on a cost plus fee basis with a joint venture of U.S. contractors for the construction of facilities in the theater of operations. These projects ranged from construction of warehousing and troop barracks to the construction of airfields.

REVIEW QUESTIONS AND EXERCISES

4.1 Name and briefly describe each of the two basic types of competitively bid construction contracts. Which type would be most likely used for building the piers to support a large suspension bridge? Why?

4.2 If you were asked to perform an excavation contract competitively with limited boring data, what type contract would you want and why?

4.3 Name three ways the construction contract can be terminated.

4.4 Name two types of negotiated contracts and describe the method of payment and incentive concept.

4.5 What is meant by unbalancing a bid? What type of contract is implied? Give an example of how a bid is unbalanced.

4.6 Why is a cost plus a percentage of cost type contract not used to a great extent?

4.7 Under what circumstances is a cost plus contract favorable to both owner and contractor?

4.8 Valid contracts require an offer, an acceptance, and a consideration. Identify these elements in the following cases:
(a) The purchase of an item at the store.
(b) The hiring of labor.
(c) A paid bus ride.
(d) A construction contract.
(e) The position of staff member in a firm.

4.9 Suppose you are a small local building contractor responsible for the construction of the small gas station in Appendix L. List the specialty items that you would subcontract.

4.10 Visit a local building site and ascertain the number and type of subcontracts that are involved. How many subcontracts do you think may be needed for a downtown high-rise building? Why would there be more subcontracts in a building job as opposed to a heavy construction job?

4.11 From the point of view of the owner's contract administrator, each different type of contract places different demands on supervision. List the significant differences that would impact the complement (number) of field personnel required to monitor the contract.

4.12 Visit a local contractor and determine the proportion of contracts that are negotiated against those that are competitively bid awards. Is this percentage likely to change significantly with small building contractors? Is there a difference between building contractors and heavy construction contractors?

CHAPTER FIVE

Legal Structure

5.1 TYPES OF ORGANIZATION

One of the first problems confronting an entrepreneur who has decided to become a construction contractor is that of deciding how best to organize the firm to achieve the goals of profitability and control of business as well as technical functions. When organizing a company, two organizational questions are of interest. One relates to the *legal organization* of the company and the second focuses on the *management organization*. The legal structure of a firm in any commercial undertaking, be it construction or dairy farming, is extremely important, since it influences or even dictates how the firm will be taxed, the distribution of liability in the event the firm fails, the state, city, and federal laws that govern the firm's operation, and the firm's ability to raise capital. Management structure establishes areas and levels of responsibility in accomplishing the goals of the company and is the road map that determines how members of the firm communicate with one another on questions of common interest. The types of company legal organization will be considered here. The management organizational structure will be discussed in Chapter 6.

5.2 LEGAL STRUCTURE

At the time an entrepreneur decides to establish a company, one of the first questions to be resolved is which type of legal structure will be used. The nature of the business activity may point to a logical or obvious legal structure. For instance, if the entrepreneur owns a truck and decides to act as a free agent in hauling materials by contracting with various customers, the entrepreneur is acting alone and is the proprietor of his own business. In situations where a single person owns and operates a business activity and makes all of the major decisions regarding the company's activity, the company is referred to as a *proprietorship*. If the business prospers, the entrepreneur may buy additional trucks and hire drivers to expand his fleet, thereby increasing business. The firm, however, remains a proprietorship even if he has 1000 employees so long as the individual retains ownership and sole control of the firm.

If a young engineer with management experience and a job superintendent with field experience decide to start a company together, this firm is referred to as a *partnership*. The size of a partnership is not limited to two persons and may consist of any number of partners. Law firms as well as other professional companies (e.g., engineering firms) are often organized as partnerships consisting of as many as 10, 12, or more partners. If two or three individuals decide to form a partnership, the division of ownership is decided by the initial contribution to the formation of the company on the part of each partner. The division of ownership may be based solely on the monetary or capital assets contributed by each partner. Therefore, if three individuals form a partnership with two contributing $20,000 and a third contributing $10,000, the division of ownership among the partners is 40%, 40%, and 20%. In other cases, one of the partners may bring a level of expertise that is recognized in the division of ownership. For instance, in the example just cited, if the partner contributing the $10,000 was the expert in the area of business activity to be pursued, his expertise could be valued at the nominal level of $10,000, making his overall contribution to the firm $20,000. Therefore, ownership would be equally divided among the three partners. The actual division of ownership is usually specified in the charter of the partnership. If no written charter exists, and the partnership was concluded by verbal mutual agreement only, the assumption is that the division of ownership among the partners is equal.

In some business activity the risk of failure or exposure to damage claims may be such that a corporate structure is deemed appropriate. This form of ownership recognizes the company itself as a legal entity and makes only those assets that belong to the firm attachable for settlement of claims in the event of bankruptcy or damage claims. This allows principals or stockholders in a corporation to protect their personal and private assets from being called in to settle debts or claims arising out of the firm's operation or insolvency. Therefore, if a stockholder in a corporation has private assets of $100,000 and the corporation declares bankruptcy, the $100,000 cannot be attached to settle debts of the corporation.* Other desirable features of corporate structure that cause firms to select this legal structure will be discussed later in this chapter.

*In certain situations, stockholders may by ancillary agreement, such as a bond, waive some of the protection offered by the corporate structure and find that their personal assets are subject to attachment.

Two types of corporations are commonly encountered. Corporations in which a small number of persons hold all of the stock in the firm are referred to as *close* or *closely held* corporations. This form of ownership is very common in the construction industry, since it offers risk protection and also allows a small group of principals to control company policies and functions. A *public* corporation in contrast to a *closely held* corporation allows its stock to be bought and sold freely. The actual ownership of the stock varies daily as the stock is traded by brokers, in the case of large corporations, on the stock market. Figure 5.1 gives a graphical indication of the forms of legal ownership utilized by a set of building construction companies located throughout the southeastern United States. In this example, the companies have been grouped according to the volume of work done using fixed price contracts versus that done using negotiated contracts. The groups were defined as follows:

Group A. Contractors doing 25% or less of their volume in negotiated contract format.
Group B. Contractors doing between 25 and 50% negotiated work.
Group C. Contractors doing between 50 and 75% negotiated work.
Group D. Contractors doing more than 75% of their work in negotiated contract format.

The figure indicates that the close corporation format is very popular.

The last form of organization that has legal implications is the *joint venture.* This is not a form of ownership, but instead a temporary grouping of existing firms defined for a given period to accomplish a given task or project. A *joint*

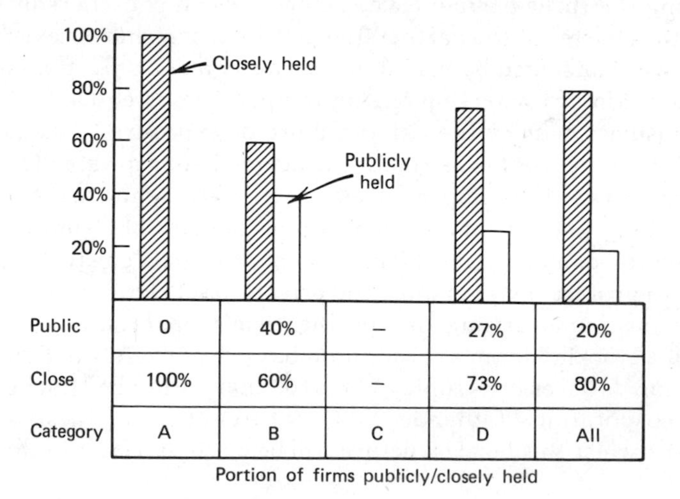

Category	A	B	C	D	All
Public	0	40%	–	27%	20%
Close	100%	60%	–	73%	80%

FIGURE 5.1 Forms of legal ownership in the construction industry (Study by T. Gibb, Georgia Institute of Technology, 1975).

venture organizational structure is used when a very large project is to be constructed and requires the pooling of resources or expertise from several companies. Typically the companies establish a basis for division of responsibility on the job and cooperate toward the end of successfully completing the project. They are bound together for a period of cooperation by a legal agreement that defines the nature of the relationship. Joint venturing first became popular during the construction of large dams such as the Grand Coulee and Boulder Dam in the western United States and has since been used for a wide variety of large construction tasks. As mentioned at the end of Chapter 4, a large, joint-venture RMK-BRJ* was established during the time U.S. forces were operating in Vietnam to construct a diverse variety of structures supplementing the efforts of Army troop units.

5.3 PROPRIETORSHIP

The simplest form of legal structure is the proprietorship. In this form of business ownership, an individual owns and operates the firm, retaining personal control. The proprietor makes all decisions regarding the affairs of the firm. The assets of the firm are held totally by one individual and augment the individual's personal worth. All cash income to the firm is personal cash income to the proprietor, and all losses or expenses incurred by the firm are personal expenses to the proprietor. The proprietor is, therefore, taxed as an individual and there is no separate taxation of the firm. Consider Uncle Fudd who has a small contracting business. The firm generated $147,000 in total volume during the calendar year. The firm has $100,000 in expenses, so that the before-tax income of the firm is $47,000. Uncle Fudd declares this income on his personal income tax return. Assuming this is his total income (i.e., he received no income from other sources) and that he has $4000 in deductions and exemptions, his taxable income is $43,000.

Since the owner's capital and that of the firm are one and the same, the credit that the firm can obtain and its ability to generate new capital are limited by the personal assets of the proprietor. Furthermore, any losses incurred by the firm must be covered from the personal assets of the proprietor. Any liabilities incurred by the firm are the owner's liability and he must cover them from his personal fortune. Therefore, bankruptcy of the firm is personal bankruptcy. Since there is no limitation of liability, high-risk businesses do not normally use the proprietorship form of structure.

The life of the *proprietorship* corresponds to that of the owner. Upon the death of the owner, the proprietorship ceases to exist. Assets of the proprietorship are normally divided among the heirs to the proprietor's estate.

*This joint venture consisted of four large construction firms: (1) R = Raymond International, (2) MK = Morrison-Knudson, (3) BR = Brown and Root, and (4) J = J. A. Jones Company.

5.4 PARTNERSHIP

The partnership is similar to the proprietorship in the sense that liabilities of the firm are directly transmitted to the partners. That is, there is no limitation of liability. However, in this case, since there are two or more partners, the liability is spread among several principals. The reason for forming a partnership is based on the principles of division of risk and pooling of management and financial resources. The ownership of the firm is shared among the partners to a degree defined in the initial charter of the partnership. Since several persons come together to form a partnership, the capital base of the firm is broadened to include the personal assets of the partners involved. This increase in assets increases the line of credit available to a partnership as opposed to a proprietorship. Control of the firm, however, is divided among the principals who are called *general partners*. Partners share the profits and losses of the firm according to their degree of wonership as defined in the partnership agreement, but since the liability of each of the partners is not limited, one partner may carry more liability in the case of a major loss. Assume that Carol, Joan, and Bob are partners in a small contracting business. The personal fortunes and percent ownership of the three principals are as follows:

Carol	$1,400,000	40% ownership
Joan	800,000	30% ownership
Bob	100,000	30% ownership

The firm loses $1,000,000 and must pay this amount to creditors. The proportionate shares of this loss are:

Carol	$400,000
Joan	$300,000
Bob	$300,000

However, since Bob can only cover $100,000, the remaining $900,000 must be carried by Carol and Joan in proportion to their ownership share.

A *limited partnership* as the term implies provides a limit to the liability that is carried by some partners. This concept allows the general partners to attract capital resources to the firm. The limited partner is liable only to the extent of his or her investment. Assume that Tom comes into the partnership described above as a *limited partner*. He makes $200,000 available to the capitalization of the firm. The percentages of ownership are redefined to provide Tom with 15% ownership. He, therefore, shares in the profit and loss of the firm in this proportion. Nevertheless, his level of loss is limited to the $200,000 he has invested. No amount beyond this investment can be attached from his personal fortune to defray claims against the firm. This provides the general partners with a mechanism to attract wealthy investors who desire liability limitation but profit participation. Limited partners have the position of a stockholder in a corporation in that this loss is limited to the amount of their investment.

Limited partners have no voice in the management of the firm. Therefore, the *general partners* retain the same level of control, but increase the capital and credit bases of the firm by bringing in limited partners. There must be at least one general partner in any partnership. The limited form of partnership (i.e., a partnership that includes limited partners) is more difficult to establish and subject to more regulation by state chartering bodies (usually the Office of the Secretary of State of the state in which the partnership is chartered). This is because limited partnerships realize some of the advantages available in the corporate legal structure. Corporations are subject to close control by state chartering bodies.

The contribution made by the limited partner must be tangible. That is, the limited partner cannot contribute a patent, copyright, or similar instrument. The contribution must have a tangible asset value (i.e., equipment, cash, notes, shares of stock in a corporation, etc.).

Any partnership is terminated in the event of the death of one of the partners. However, arrangements can be made to provide for the continuity of the partnership should one of the partners die. An agreement can be made among the partners that in the event of the death of a partner the remaining principals will purchase the ownership share of the deceased partner. Usually a formula that recognizes the fluctuating worth of the partnership is adopted in this agreement. The remaining partners pay this amount to the estate of the deceased partner.

General partners who are actively involved in the day-to-day management of the firm may decide to pay themselves a salary. In this way, the time and level of expertise contributed to the operation of the partnership are recognized. This level of day-to-day participation may be different from the level of initial contribution made in capitalizing the firm. In the case of Carol, Joan, and Bob, the levels of ownership were 40, 30, and 30%, respectively. If Bob is most active in the management of the partnership, he may be paid a full-time salary to recognize his commitment. Carol and Joan being active only on a part-time basis will be paid proportionately smaller or part-time salaries. Taxation, in any case, will be on both salary and earnings deriving from the operation of the partnership.

The action of one partner is binding on all partners. For instance, in the partnership described, if Joan enters into a contract to construct a building for a client, this agreement binds Bob and Carol as well. In this sense, a partnership is a "marriage," and any partner must be able to live with any commitment made on behalf of the partnership by another partner. On the other hand, it is not proper for a partner to sell or mortgage an asset of the partnership without the consent of the other partners. If the partner sells the asset, the income accrues to the partnership. If the partner utilizes a partnership asset to secure a personal note or loan, the other partners could advise the noteholder that they contest the use of this asset as security.

5.5 CORPORATION

A corporation is a separate legal entity and is created as such under the law of a state in which it is chartered. In most states corporations are established by applying to the office of the Secretary of State or similar official. The office of the Secretary issues a chartering document and approves the initial issuance of shares of stock in the corporation to establish the level of ownership of initial stockholders. As in the case of a partnership, the initial stockholders contribute financial capital and expertise as well as other intangible assets such as patents and royalty rights. The level of contribution is recognized by the number of shares of stock issued to each of the founding stockholders. If, in the partnership just described, Carol, Joan, and Bob decided to incorporate and the level of ownership was to remain the same, shares in the proper proportion would be issued to each principal. The number of shares and the share value defined at the initialization of a corporation are arbitrary and are selected to facilitate the recognition of ownership rather than actual value of the corporate assets. If the Carol-Joan-Bob (CJB) corporation is established by the issuance of 1000 shares of stock, Carol would receive 400 shares (40%), and Joan and Bob would receive 300 shares each (30%). For simplicity, each share could have a par value of one dollar. This assignment of one dollar per share simplifies the unit (i.e., share value) used to recognize ownership. On the other hand, the initial capital contributed to the formation of the corporation might have been $50,000. Therefore, the book value of each share of stock would be $50 per share. The book value of each share of a corporation is the asset value or net worth of the corporation divided by the number of shares issued. In this case, 1000 shares are issued and the asset value is $50,000. Therefore, each share has a book value of $50.

In addition to the par and book values associated with a share of stock in a corporation, each share has a traded or market value. This is the value that is listed on stock exchanges for those publicly traded corporation shares and that is printed in the newspaper. It indicates what the general public or stock traders are willing to pay for a share of ownership in the corporation. If the future looks good, traders will anticipate an increase in the value of the corporation's stock and will pay to own a stock that is increasing in value. If the corporation is about to experience a loss, the market price of the stock may indicate this by declining in value. To illustrate, if CJB, Incorporated wins a contract that promises to net the corporation an after-tax profit of $10,000, the market price of the stock will tend to move up. In fact, as noted above, most construction firms hold their stock closely and do not trade it publicly. Therefore, the market value of the stock is of interest primarily to the giant construction firms that are publicly traded.

Because of the legal procedures required, the corporation is the most complicated form of ownership to establish. A lawyer is normally retained to prepare the proper documents, fees must be paid to cover actions by the chartering body (i.e., office of the Secretary of State), printed stock is prepared, and formal meetings by the principals are required. Since the corporation can sell further stock to raise capital, it has an advantage in this respect over the proprietorship and the partnership. This power to sell stock can be and has been abused. Once a corporation is established, it may sell stock to unsuspecting buyers based on an idea or concept that is not properly presented or explained. For this reason and others, the corporation is closely controlled by the chartering agency in regard to its issuance and sale of additional stock. Federal law also dictates certain aspects of the presentation of corporate stock for sale.

The most desirable aspect of the corporate structure to businesses that are exposed to high risk such as the construction industry is its limitation of liability. Since the corporation is a legal entity of itself, only the assets of the corporation are subject to attachment in the settling of claims against and losses incurred by the corporation. This means that stockholders in a corporation can lose the value of their investment in stock, but that is the limit of their potential loss. Other assets that they own outside of the corporation cannot be impounded to offset debts against the corporation.

One disadvantage associated with the corporation is the double-taxation feature. Since the corporation is a legal entity, it is subject to taxation. The same profit that is taxed within the corporation is taxed again when it is distributed to stockholders as a dividend. This distributed profit becomes taxable as personal income to the individual stockholders. Assume that CJB Corporation has a before-tax profit (e.g., income - expense) of $100,000 during the corporation's first year of operation. Corporations are taxed by the Internal Revenue Service (IRS) at the rate of 17% on the first $25,000 of profit. For all profit above $100,000 the corporate rate is 46%. The corporation would be taxed $26,750 for $100,000 of before tax profit.* The after-tax income would be $73,250. Assume the CJB decides to distribute $30,000 to the three stockholders. That is, Carol, Joan and Bob as directors of their closely held corporation distribute $30,000 to themselves and retain $43,250 of these earnings within the corporation as working capital. In this case Carol, the major stockholder, receives a dividend of $12,000. Joan and Bob would receive $9,000 each. If we assume that each stockholder pays approximately 25% on personal taxable income, Carol will pay $3000 in tax on this dividend, and Joan and Bob will pay $2250. In other words, the federal tax at the corporation and stockholder levels combined will be $26,750 plus $7,500 or $34,250.

*Based on Revenue Act of 1978.

The double-taxation feature does not always prove to be a disadvantage. Returning to the situation of Uncle Fudd who is organized as a proprietorship, assume his before-tax income with the proprietorship is \$47,000. Assume that Uncle Fudd decides to incorporate his proprietorship and become Fudd Associates, Inc. As president of this corporation, Uncle Fudd pays himself a salary of \$17,000. At this salary level, Uncle Fudd is taxed at 20% of his taxable income (i.e., his gross income minus deductions and exemptions). In the proprietorship format, his tax would be 32% of \$47,000 minus \$4,000 in deductions and exemptions (see Section 5.3). He will pay \$13,760 in tax. In the corporate format, Uncle Fudd's tax will be:.

\$ 47,000	
- 17,000	Fudd's salary = expense
\$ 30,000	gross income of corporation

Corporate
tax = (\$25,000 × 0.17) + (5,000 × 0.20) = \$4250 + \$1000 = \$5250

Personal tax = \$17,000
- 4,000 (deductions and exemptions)
0.20 (\$13,000) = \$2600

Therefore, Uncle Fudd's tax in the corporate format will be \$7,850. In this case, the corporate form of ownership yields a lower tax payment despite the double taxation. For this reason, a good tax consultant is a very valuable advisor when deciding which form of ownership is most appropriate.

Certain states provide for a special corporate structure that avoids the double-taxation feature of a normal corporation but retains the protection of limited liability. This is referred to as a subchapter "S" corporation. In a subchapter S corporation, the principals are taxed as if they were members of a partnership. That is, corporate income is taxed only once as personal income. The corporate shareholders are, however, still protected and their loss is limited to the value of the stock they possess.

As noted above, the corporation is very advantageous when attraction of additional capital is of interest. Figure 5.2 shows a typical stock certificate as issued at time in incorporation. The certificate indicates that 250 shares of stock are represented. In addition, the corporation has authority to issue a total of 50,000 shares. Therefore, the directors of a corporation can decide to raise money for capital expansion by selling stock rather than borrowing money. This provides for the generation of additional capital by distributing ownership. It has the advantage that the money generated is not subject to repayment and therefore is not a short-term or long-term liability on the company balance sheet.

The corporation also has a continuity that is independent of the stockholders. Unlike the proprietorship or partnership in which the firm is terminated on the death of one of the principals, the corporation is perpetual.

SPECIMEN

NUMBER 2

SHARES 250

HALCO CONSTRUCTORS, INC.

The Corporation is authorized to issue 50,000 Shares — Par Value $1 Each

This Certifies that Daniel W. Halpin is the owner of Two Hundred Fifty fully paid and non-assessable Shares of the above Corporation transferable only on the books of the Corporation by the holder hereof in person or by duly authorized Attorney upon surrender of this Certificate properly endorsed.

In Witness Whereof, the said Corporation has caused this Certificate to be signed by its duly authorized officers and to be sealed with the Seal of the Corporation.

Dated August 17, 1976

SPECIMEN

Daniel W. Halpin SECRETARY

SPECIMEN

Daniel W. Halpin PRESIDENT

FIGURE 5.2 Typical stock certificate.

Unless the corporation is bankrupt or the corporate charter lapses, the corporation continues in existence until all stockholders agree to dissolve it. In most states also, clauses can be included in the corporate charter that in effect allow the control of sale of stock outside of the circle of present stockholders. That is, any stockholder who wishes to sell a block of stock must first offer the stock for sale to the other stockholders. They have an option to purchase it before it is sold to others. This allows the closed nature of a closely held corporation to be maintained. If a stockholder should die, the stockholder's heirs are committed to offer it to the present stockholders before selling it to others. The heirs can, of course, decide simply to retain the stock.

Two disadvantages that are inherent in the corporate form of ownership are the reduced level of control exercised in management decision making and certain restrictions that can be placed on the corporation when operating outside of its state of incorporation. The larger a corporation becomes the more decentralized the ownership becomes. On questions of dividend levels, the issuance of stock to generate capital, and other critical operational decisions, agreement of all stockholders must be obtained. In large corporations, this leads to involved balloting to establish the consensus of the ownership. This process is cumbersome and greatly reduces the speed with which corporations can respond to developing situations. In small closely held corporations, however, this presents no more of a problem than it does in a partnership.

When a corporation operates in a state other than the one in which it is incorporated, it is referred to as a *foreign* corporation. For instance, a corporation incorporated in Delaware is considered a foreign corporation in Georgia. Corporations in certain industries may encounter restrictive laws when operating as a foreign corporation. They must establish legal representation in states in which they operate as foreign corporations. Restrictive legislation of this type cannot be applied to proprietorships and partnerships, since these entities consist of individuals who are legally recognized. The individual is protected by equal treatment under the Constitution, and what is a legal restriction when placed on a corporation is illegal when applied to a proprietorship or a partnership.

5.6 COMPARISON OF LEGAL STRUCTURES

The decision to choose a particular legal structure for the firm hinges on seven major considerations. The pluses and minuses of each type of structure are summarized in Table 5.1. These considerations have already been introduced in general form. Specifically, an owner contemplating a legal structure for the firm must consider:

1. Taxation.
2. Costs associated with establishing the firm.

3. Risk and liability.
4. Continuity of the firm.
5. Administrative flexibility and impact of structure on decision making.
6. Laws constraining operations.
7. Attraction of capital.

The question of taxes to be paid in each organizational format is mixed, and the anticipated balance sheet and cash flow of each firm must be studied to arrive at a "best" solution. The corporation has the disadvantage that the firm is taxed twice, once on corporate profit and a second time when the stockholders must pay tax on the dividends received as distributed income. The subchapter S type of firm mentioned above circumvents this to a degree in that the shareholders are taxed as individuals as if the firm were a partnership. The normal proprietorship and partnerships have the disadvantage that all income is taxed whether or not it is withdrawn from the firm. Thus, as in the example of Uncle Fudd, incorporating yields a benefit despite the double-taxation feature.

Costs and procedures associated with establishing the firm are generally minimal for a proprietorship, slightly more involved for a simple partnership, and a major financial consideration for limited partnerships and corporations. Normally whatever costs and procedures are associated with local, state, and federal tax registration and the purchasing of a license are all that must be considered in establishing a proprietorship or simple partnership. These as well as significant legal costs ($500 to $2000) must be considered in establishing limited partnerships and corporations. These costs may be justified, however, based on the limitation of liability achieved and the benefits of medical, health, and insurance plans that can be implemented in a corporate format.

Corporations and limited partnerships limit the level of loss in the event of a default or bankruptcy to the level of investment. That is, stockholders cannot lose more than the value of their shares. The loss of a limited partner cannot exceed the amount of his initial investment. If he initially invested $20,000, he can lose this amount, but his other assets cannot be attached in the event of bankruptcy or default. The assets of stockholders in a subchapter S corporation are similarly protected. Personal assets can be used to pay creditors in the event a proprietorship or simple partnership defaults and is forced into bankruptcy.

Proprietorships have the disadvantage that they terminate when the proprietor dies. This may present a problem, particularly if the firm as an asset must be divided among several heirs. It can be circumvented in part by willing the firm and its market and "good will" to one heir (who will carry on the business) and providing that that heir will compensate the other heirs for their share. If a partner dies, the partnership is dissolved. Again, however, provisions in the partnership agreement can provide the means for surviving

Table 5.1. CONSIDERATIONS IN CHOOSING LEGAL STRUCTURE

	Proprietorship	Partnership	Corporation
Tax	Tax on personal income—tax on earnings whether or not they are withdrawn.[a]	Tax on personal salary and earnings	Lower taxes in some cases[a] Dividends are not deductible—double taxing. Taxes on dividends—that is, money actually received
Costs and procedures in starting	No special legal procedure—apply for licenses; register with IRS	General: Easy—oral agreement Limited: More difficult—closely adhere to state law	More complex and expensive. Meeting must be held.
Size of risk	Personal liability	Personal liabilities. Extent of personal fortune. Limited: each partner is protected; loss of limited partner cannot exceed initial investment	Limited to assets of corporation
Continuity of the concern	No continuity on death of proprietor	Dissolution No continuity on death of partner. Surviving partners can buy share if in agreement.	Perpetual (charter can expire)

[a] See Fudd Associates, Inc., example.

Table 5.1 (continued)

	Proprietorship	Partnership	Corporation
Adaptability of administration	Simplicity of organization—direct control	Decisions and policies implemented by oral agreement.	Directors—good if involved. Policy decision predefined by bylaws—stockholders can't bind company.
Influences of applicable laws	Laws are well defined—no limit on doing business in various states.	Laws are also well defined—a license may be required.	Foreign corporation status—requires legal counsel on permanent basis.
Attraction of additional capital	Limited potential for capital expansion. —Borrowing —Line of credit —Personal fortune investment	Better—more capital—limited partner concept	Issue Securities.—Collateral corporate assets —Issue stock

partners to purchase the deceased partner's share from his estate. Corporations are perpetual and the stock certificates are transferred directly as assets to heirs of the estate.

Policy decisions are relatively simple in the proprietorship and partnership formats. Principals make all decisions. In the corporate format, certain decisions must be approved by the stockholders, which may impact the corporation's ability to react to a developing need or situation. In closely held corporations, however, this is no problem since the partners are able to call ad hoc board meetings to react quickly. Corporations with large numbers of shareholders are not as flexible in this regard. The chief operating officer or president handles the day-to-day decision making. A board of directors is charged with intermediate-range and strategic planning and decision making. Major decisions, however, such as stock expansion and acquisition of other firms or major assets, must often be approved by all stockholders in a formal vote.

Local laws may encourage the formation of small and local businesses by placing restrictive constraints and burdensome additional cost on out-of-state or foreign corporations. These discriminating practices must be investigated when bidding construction work in a state other than the one in which the construction corporation is chartered. Special licenses and fees are sometimes required of foreign corporations. Proprietorships and partnerships that consist of individuals are protected against these discriminatory practices by the Constitution and enabling "Equal Rights" legislation.

In raising capital, proprietorships and simple partnerships must rely on the personal borrowing of the principals to generate capital for expansion. The unique feature of the corporation that permits it to sell stock allows corporate entities to attract new capital by further distributing ownership. The corporate assets as well as future projections of business provide a collateral basis to attract new stockholders. This mechanism is not always viable, however. During 1973-1974, large corporations were unable to sell large issues of stock for capital expansion and were forced to go to the commercial banks to borrow. In periods of economic uncertainty, this method of attracting capital may be limited.

Good information regarding the advantages and disadvantages of various legal forms of organization are contained in the Small Business Administration *Management Aids* series.

REVIEW QUESTIONS AND EXERCISES

5.1. Name the *three* principle forms of business ownership in construction and state the liability limits of the owners in each case.

5.2. Which legal structure is most difficult to establish and why?

5.3. Name three types of *partnerships*.

5.4. Describe briefly two advantages and two disadvantages of a corporate form of business organization as compared to a partnership.

5.5. Jack Flubber, who owns Son of Flubber Construction Co. and runs it as a proprietorship, had gross profits last year of $46,000. His personal and family expenses are $20,000 and he has $7,000 in exemptions and deductions. He paid $17,000 in taxes. If he paid himself a salary of $24,000 taxed at 20%, would it be advantageous for him to incorporate as a close corporation? Explain.

5.6. What is meant by the term foreign corporation?

5.7. What would be the advantages of organizing as a subchapter S corporation?

5 8. Is it possible to characterize the legal structures of local contractors using the yellow pages as a guide?

5.9. What steps must be taken to set up a partnership? Now can a partnership be dissolved?

5.10. In problem 5, what taxes would Carol pay if she organized as a closely held corporation (as described) and, after paying her salary, also issued herself a dividend of $10,000?

5.11. What is the difference between par and book values of corporate stock? If an incorporated construction company wins a large cost plus-fixed fee contract, what impact might this have on the market value of the company's stock?

5.12. Uncle Fudd has decided to sell his ownership in the Cougar Construction, Inc. to Cousin Elmer. How would the legal firm handling this transaction determine a fair price per share?

CHAPTER SIX

Management Structure

6.1 THE MANAGEMENT ENVIRONMENT

Construction companies must continuously generate new project management organizations that support field effort during project start-up and field construction and reabsorb key personnel back into the home office (for later reassignment) while final punch list, wrap-up, and turnover aspects of the project are being finalized. These management and organizational problems are magnified for remote or inconveniently located projects. Contractors operate with as few as one or two and as many as twenty projects in a continuously changing portfolio of projects. At any one time, projects can be at different stages in their life cycles. The contractor must continuously cope with complex and dynamic organizational, resource availability, and project management problems in the course of running his business. Hence the contractor must devise an effective but flexible management approach and organization for each project, yet maintain a general business, management, and professional capability for the company as a single entity. A common and practical solution is to locate all general business functional groups in a head office, or conveniently located regional office, together with the relevant professional project-oriented staff support groups, and to locate all field construction and field management functional groups at the project site. The individual projects are then integrated with the construction company through the performance of project management functions. These project management functions can be performed by head office or field-based agents, or by a coordinator who is both head office and field based. Each alternative exists in practice and creates its own unique project management environment. The selection of a particular form is a critical management decision significantly influencing the management environment, the project team structure, and the professional and decision roles of the project team agents.

This chapter introduces concepts relating to management functions and their roles in defining a management organization. The development of the project team structure and its relation to common organizational forms are discussed. On this basis, project management decision processes are identified and related to the professional roles of project management agents.

6.2 GENERAL MANAGEMENT FUNCTIONS

Once management has been empowered to act, its initial efforts must be directed to organizing itself, in deciding its scope of action and in formulating the policies under which it will act. When this has been accomplished, the project management effort can be directed toward the completion of the project. In general, management must *establish an organization*, *staff it* with suitable people, *plan* what is to be done and how it is to be accomplished, *direct* others to do whatever has to be done, and *monitor* work progress in such a way that effective project control can be realized. These management functions are necessary for the management of any project and consequently are called general management functions.

The general management functions can be divided into two groups: (1) the *organizing* and *staffing* group, which is associated with establishing the project team, and (2) the *planning, directing, monitoring,* and *controlling* group, which is associated with the activity of managing the project, that is, project management. Figure 6.1 indicates schematically the relationship between these management functions.

The determination of the organizational structure and size of the project team must be based on a full understanding of the project, the scope of management effort required, and a technical knowledge of the work required to perform each management function involved in the project. Once this assessment has been formulated, it is possible to define general areas of responsibility for the project team, to develop job descriptions, and to prescribe lines of communication. In this way, meaningful description of the project organizational structure can be developed.

The general management function that involves organizing can be described as the process of determining what project management team positions must be created, defining the responsibilities of each position, and establishing the relationship of one position with another. Once an organization has been formulated, it requires staffing with suitably qualified decision agents before it can begin to function. Management's role is to build up the organizational structure in such a way that an efficient and effective team results with highly motivated staff.

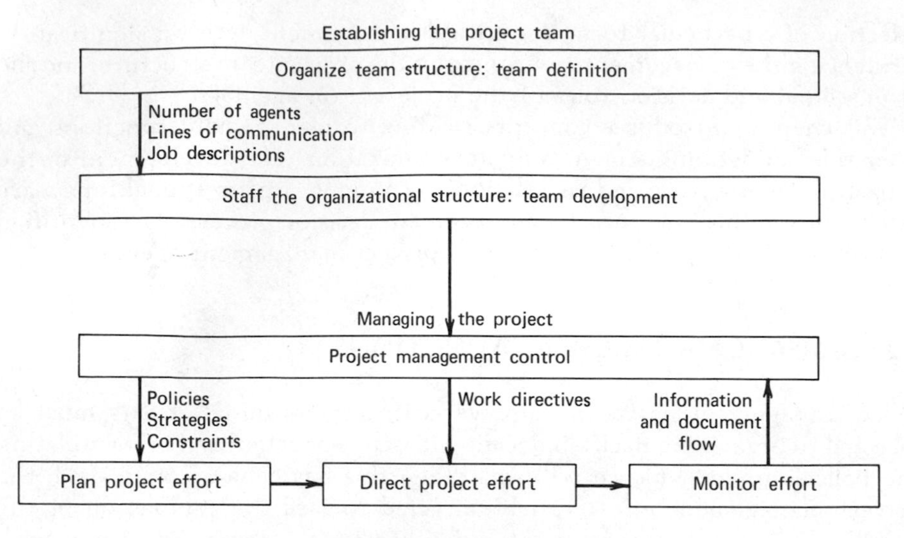

FIGURE 6.1 Relationship between general management functions.

The general management functions associated with project management are performed by project team agents within the organizational structure adopted for the project. The particular breakout of these functions depends on the nature of the project. The diagram shown in Figure 6.2 breaks planning into two components, one of which is associated with *planning what to do*; the other with *scheduling when to do it*. Similarly, directing is dissected into the active allocation of staff and labōr to a job and the directing of them in their work tasks. Finally, the control function has been associated with decision making, based on information gathered in the monitoring and recording of current project status and work progress, and with the consequential impact of decision making on the planning, scheduling, allocation, and work functions of the project effort. These general management functions show a progressive movement of management concern from the planning and enumeration of required resources to the commitment, use, and evaluation of effective use of these project resources.

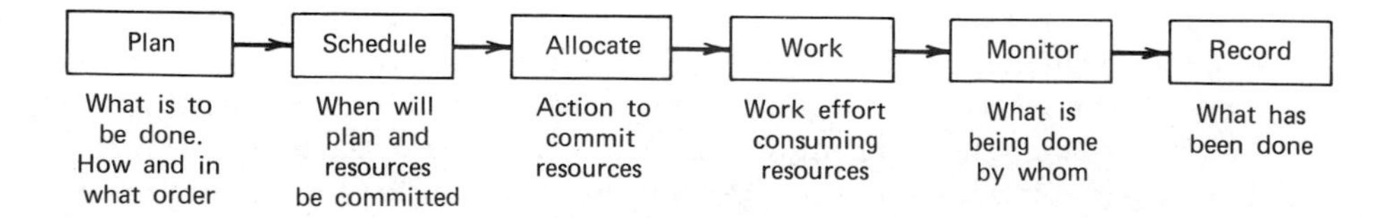

FIGURE 6.2 General project management functions.

6.3 ORGANIZATIONAL CONCEPTS

As mentioned previously the general management function, organizing, involves: (1) the determination of the number of management and work positions to be established for a particular enterprise or project; (2) the definition of the duties and responsibilities for each individual agent position and of groups within the organization; and (3) the identification of the lines through which authority flows in the organization, which establishes the relationship of each position and agent with others. It is natural therefore to describe and portray an organization in terms of its organizational structure (with its implied lines of communication and authority) and its breakdown into functional groups (with its implied grouping of professional expertise and common work effort focus) and position titles that reflect both professional area of interest and level of authority. In this section we introduce concepts relating to the development of the organizational structure of an organization. In later sections, material relating to the functional treatment of construction management and the development and portrayal of job descriptions will be presented.

A number of basic management principles must be considered when planning, designing, and establishing an organizational structure for a project management effort. These principles rest on two basic concepts: (1) a belief in or acceptance of leadership; and (2) the delegation of authority. The first basic principle indicates the permissive environment without which an organization cannot develop. The second ensures that agents at various levels within the organization can be readily responsive to the needs of the project. If a single authority exists for the different links of the organization, decisions and actions are delayed pending communication of problems and issues to the sole management authority. Thus delegation of authority provides the organization with the ability to respond quickly to problems at all levels.

Any meaningful organizational structure must portray clearly the lines of authority running from the top to the bottom of the organization. The line concept of management embodies delegation of authority from the highest executive in the organization to the employee who has least responsibility in the organization and no authority over others. In delegating authority one must ensure that line agents have responsibilities commensurate with their authority.

A necessary consequence of delegated authority is that top management must be held responsible for the acts of subordinates. This accountability of top management for agents with delegated authority highlights the need for management to carefully define the authority of team members and to ensure that these agents are properly briefed on management policies and objectives.

Fundamental concepts associated with the development and structure of

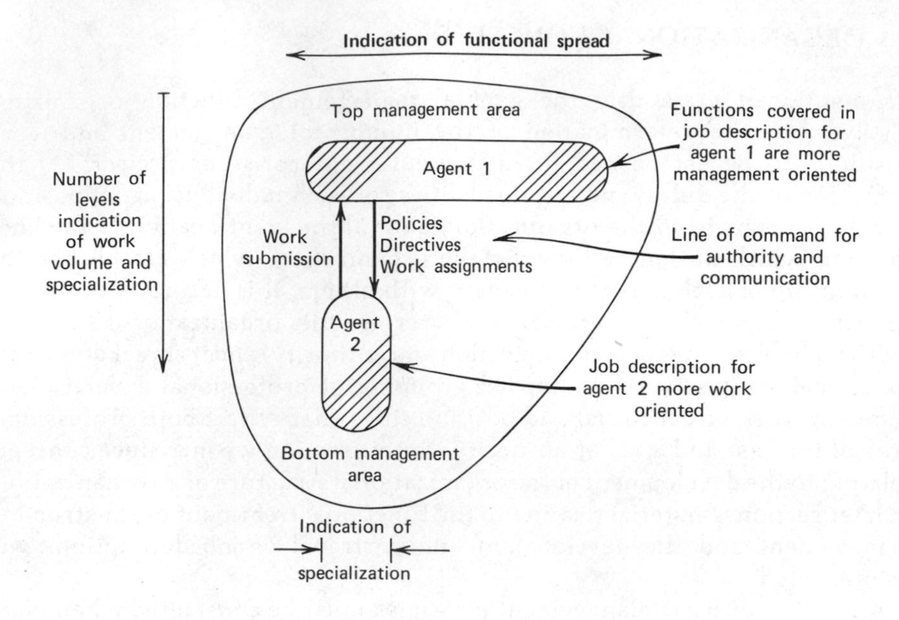

FIGURE 6.3 Organizational concepts: Management and work-oriented agents. (Taken from an unpublished manuscript by Anderson and Woodhead, "Project Manpower Management.")

organizations are shown in Figures 6.3 to 6.6. The general relationship between work- and management-oriented agents in an organization is shown in Figure 6.3, wherein the management-oriented agent performs more management-type functions (i.e., this job description covers a wide functional spread) while the technical or work-oriented agent performs a deeper set of technical functions (i.e., this job description covers a narrower functional spread). This basic hierarchical structure between management and technical

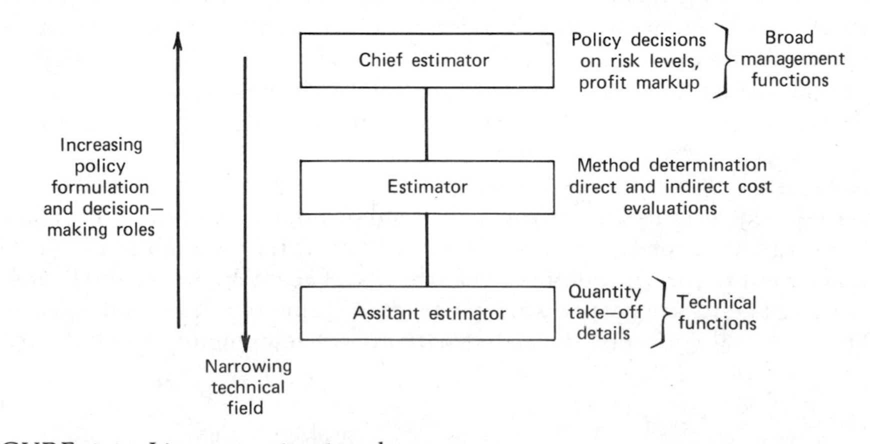

FIGURE 6.4 Line organizational structure.

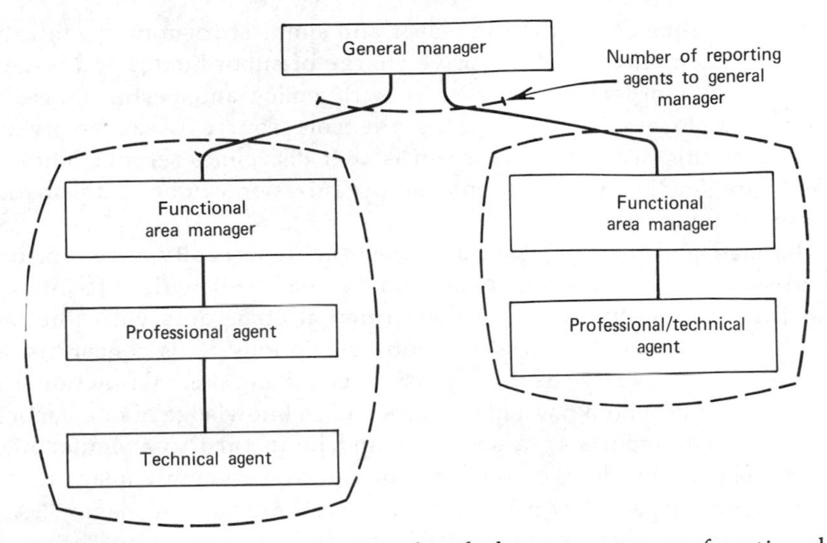

FIGURE 6.5 Organizational growth of departments on functional and work-volume lines.

or work-oriented agents is further developed in Figure 6.4 in terms of a line structure for agents involved in the estimating and bidding process. The line structure shown indicates increasing involvement in policy formulation and decision making with agents located up the management line and increasing work orientation in a narrowing technical field with agents located down the management line.

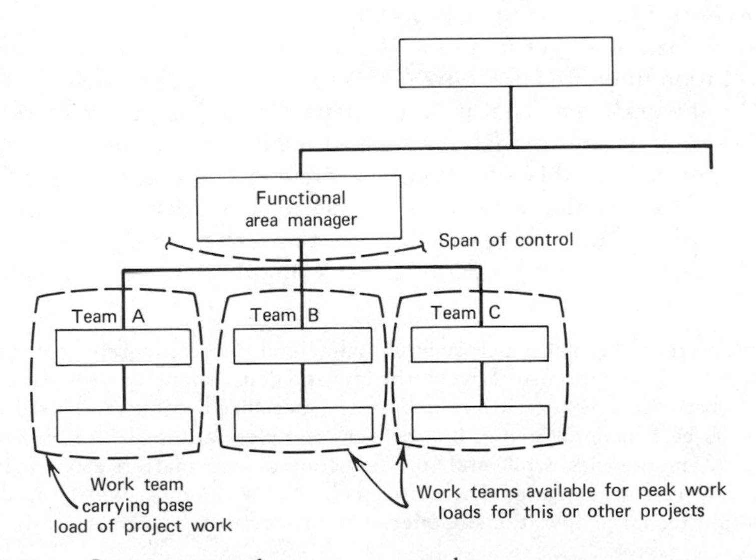

FIGURE 6.6 Organizational structure: work team concepts.

Individual or line control is the oldest and simplest form of organization. The superior is in direct and exclusive charge of subordinates and issues all instructions and orders having to do with policy and performance. The superior's employees regard him as the sole source of authority. The advantage of this organizational form is that discipline, responsibility, and authority are simple and direct, and the organization can be established in a minimum of time.

As the management work load increases the opportunity arises to funnel work to specialized functional areas. Thus, it may be beneficial to introduce special functionally oriented line departmental structures with line levels graded by technical and professional job descriptions. This concept is illustrated in Figure 6.5. "Here authority is divided among several functional area managers, each of whom has a highly specialized knowledge of one particular field. Each agent reports to a specialist and he in turn to a similar higher specialist. The various lines of authority do not converge until near the top of the organization where they meet under a general executive manager."* As the number of functional areas or line structures and, by implication, the number of area managers, increases in an organization, the necessity that they report to a general manager consumes more and more of their time. Eventually, with increasing work load, a functional (i.e., work) overload develops for the general manager and consequently his management grasp and control weakens.

In this organizational framework, there is a limit to the number of staff agents that can be coordinated by a single executive. This limitation is referred to as the span of control and the concept is also illustrated in Figure 6.5. Some managers have a larger span of control than others, which is related to the individual's capacity to manage. Many organizations, however, establish strict span of control limits for their key staff.

As work load increases in a *particular* functional area a common organizational solution (indicated in Figure 6.6) is to set up multiple copies of the basic hierarchical work team. That is, in the departmental structure of Figure 6.6, multiple sets of organizational components similar to that shown in Figure 6.5 are incorporated. In this situation, the area manager can assign different segments of work to the various teams. As project work load increases for one and as work volume decreases for another, the manager can reassign individual teams to similar functional work on other projects, as is illustrated by the following.

> Finally an organizational structure built on line and staff control concepts [as shown in Figure 6.7] is a compromise between the line and departmental extremes and divides authority between line (executive) and staff (specialized) officers. This division of authority is best made by giving line officers complete authority for execution and handling of emergencies, while staff officers control such matters as standards, procedures, etc. This organizational structure is probably the most widely used form of organization for large operations. Generally considered to be a conservative form of

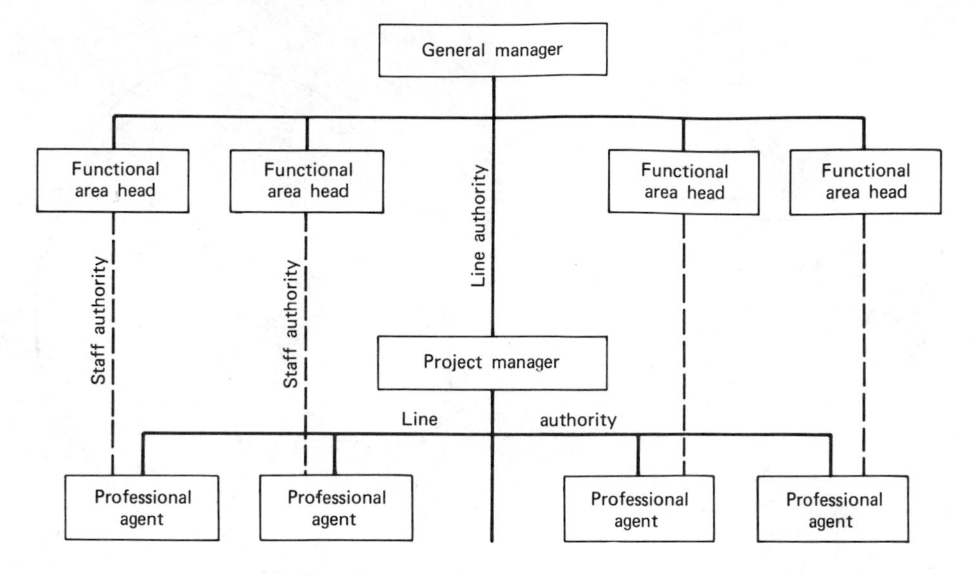

FIGURE 6.7 Line and staff organizational structure.

organization, it is highly effective for either concentrated or widely scattered situations.*

6.4 PROJECT TEAM CONCEPTS†

The planning, field start-up, construction, monitoring, and management of a construction project represent different stages in the life of a project and in its management. They require the combined talents and efforts of a variety of professional agents at all levels in the construction management hierarchy. Some agents perform specialized functions concentrated wholly within one stage of a project or management process. Thus, for example, estimators perform functions during the quantity takeoff and planning phases of a project, and consequently their work effort is quickly transferred from project to project. Other agents, however, are involved for long periods of time on the project and become identified with its project management. Thus the nature of the construction and management processes establishes the basic framework within which the various construction management agents act and interact

*With acknowledgement to Rubey, Logan and Miller, The Engineer and Professional Management, The Iowa State University Press, 1970.

†The material of this section is based on an unpublished manuscript by Anderson and Woodhead, "Project Manpower Management."

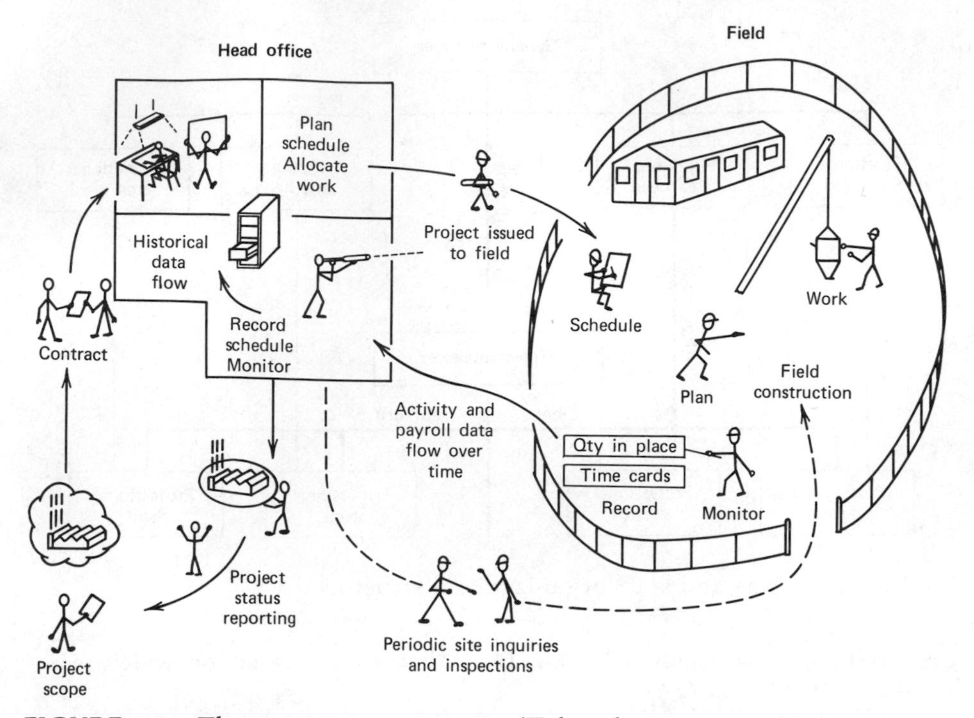

FIGURE 6.8 The project team concept. (Taken from an unpublished manuscript by Anderson and Woodhead, "Project Manpower Management.")

management implemented for the project and characterize its *project team structure*.

In its most elemental form, the project team is composed of separate head office and field management *components* that are almost functionally autonomous, and consequently their interaction can be described as sequential over time. The *head office component* is functionally responsible for pre-job *project definition* (e.g., planning, estimating and scheduling) and broad-based *project management* after construction commences. The *field component*, however, is only functionally responsible for on-site *construction management* activities. A schematic illustration of the relationship between these three components and of the project team concept is given in Figure 6.8.

When referenced against time, the initiation of project effort begins when the head office allocates and releases personnel to the project, defines its scope, and plans its execution. The project is carried to the field when head office management issues it to the field component for on-site construction with each other over the life of a project. Within this basic framework, the number of agents involved, the type and extent of the functions they perform, and the organizational structure adopted define the nature of construction start-up. Finally, on project completion, project effort returns to the head office for contractual finalization, and the facility is released to the client. Because functional autonomy exists, the interaction between these project management areas is generally weak at the *component level* of the *project team structure.*

The business, management, and construction environment that both produces and ensures the success of this management organizational structure is based on the heavy commitment of specialized resources (especially of management capacity) to projects involving familiar work processes and small intimate crews. In this way, the head office can safely entrust projects to field management at the superintendent and foreman levels, providing projects are assigned to agents within their area of expertise and past experience. Thus, within this environment and the guidance, decisions, and directives of the head office project *manager, exercised through management-by-exception techniques, the* three management phases, project definition, field construction, and project management, are autonomous.

The advantage of this form of project team structure is that it clearly separates the construction management functional areas so that different professional agents can adequately cover the project management sphere of action by complementing each others' talents. Thus the dual partnership firm in which one agent predominantly handles business and office matters and the other acts primarily as a construction agent is a typical example of this project team component structure. A representation of this project team structure is given in Figure 6.9. Notice the clear separation of functions between the office and field agents. In this elemental project team structure, the field partner spends most of his time in field supervision situations and interacts with the office partner in an ad hoc manner through frequent discussions as problems

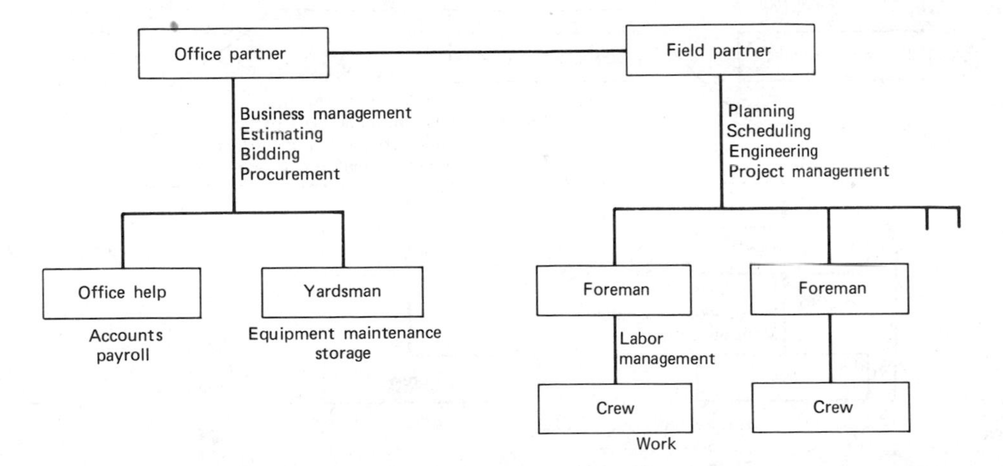

FIGURE 6.9 Dual partnership project team.

arise. As indicated, the organizational structure and division of functions could be representative of the small contractor.

Another common situation illustrating the component type of project team structure is the foreman- or superintendent-run project. In this case head office agents perform general project-related business functions and, because of heavy office commitments and function overloads, can only monitor field activity and progress through field reports. As shown in Figure 6.10 the job superintendent is the dominant field agent who receives periodic advice and directives from a head-office-based project manager. The job superintendent and foremen constitute the field component of the project team.

A basic problem of construction management, however, arises out of the physical separation of the field work site from the head office and the need to establish effective interaction between head office and field operations. The problem is aggravated with increase in project size and complexity, since these factors place greater demands on management effort, causing management to become increasingly aware of the need to strengthen links with the field.

A common management solution to this problem is to introduce a decision agent who can operate at both head office and field levels and, through this

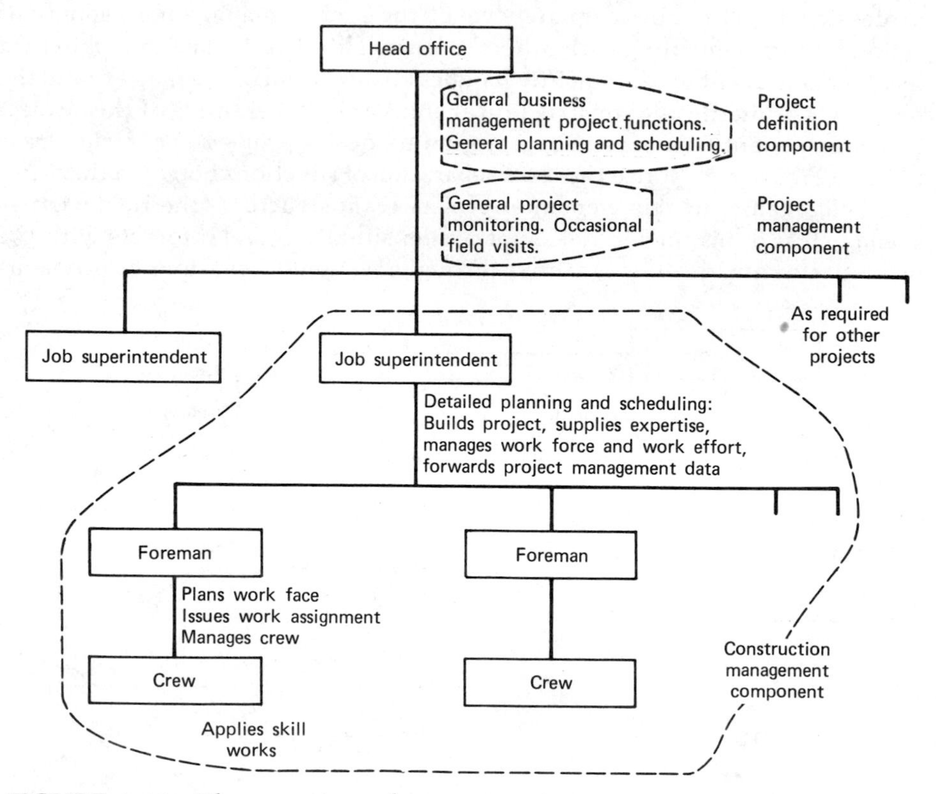

FIGURE 6.10 The superintendent-run project.

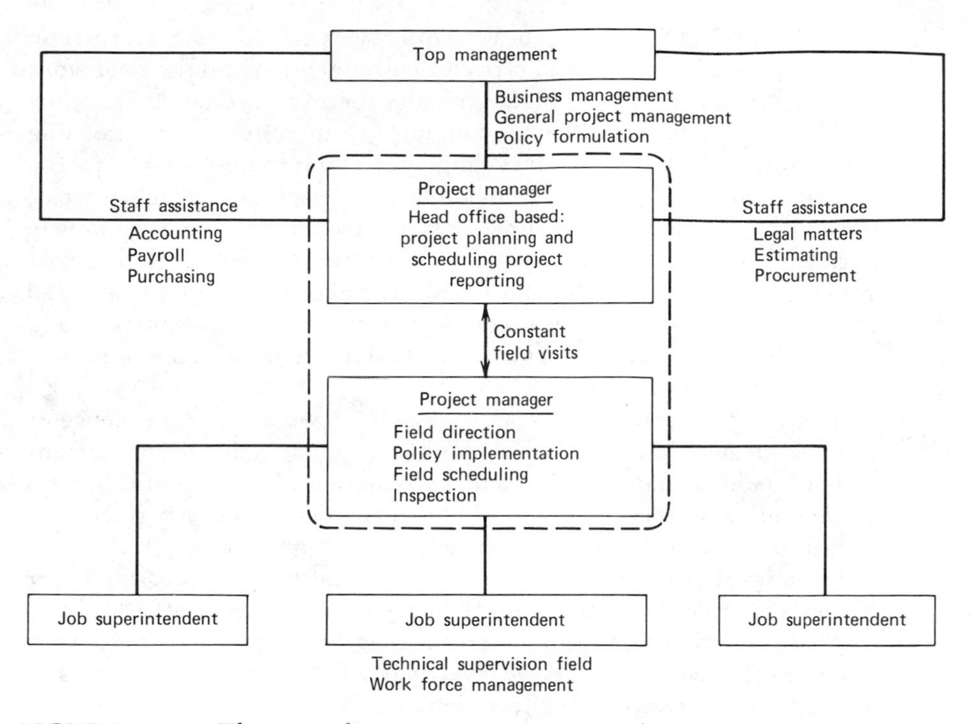

FIGURE 6.11 The traveling project manager. (A term first coined by Anderson and Woodhead.)

management role, knit together head office and field components, thereby compromising each component. This agent is responsible to, but yet acts independently of, top level management, while maintaining full authority over field operations. In order to accomplish this, the decision agent must continuously interface with both top-level management and the field. This interaction is accomplished by frequent travel between the head office and field site and suggests that the agent be described as a *traveling project manager*.* An organizational structure illustrating the traveling project manager form of the project team is shown in Figure 6.11. This project team structure emerges at the medium firm level of activity in the construction industry. Notice that in Figure 6.11 the project manager's management functions have been divided into those performed in the head office and those performed in the field. At this level of construction activity, it is common for the *project manager* to be responsible for three or more projects (depending on dollar volume, size and complexity). The Traveling Project Manager receives head office staff assistance from two different and emerging professional groups, namely the estimating/procurement group and the accounting/purchasing/payroll group.

*A term first coined by Anderson and Woodhead.

The traveling project manager acts as an interface point for data, status, and requirements passed between head office top management and the field work environment. While the field staff is responsible for project construction, they carry out the scheduling, allocation, and monitoring directives of the traveling project manager. Similarly, the traveling project manager reports to top management on project status and relevant labor management problem areas. In this way, head office and field functions are linked together continuously over time. The advantage of this type of project team structure is that project problems can be handled almost instantaneously by both top management and field agents. In this way, project status reports, policy statements, and decisions are readily available, and a continuous and integrated project information flow becomes possible.

The management approach that can be characterized by the emergence of the traveling project manager integrates the construction management process. The business, management, and construction environment that both produces and ensures the success of this management approach requires a heavy commitment of management capacity to each project.

At a certain level of project size and complexity, it becomes desirable (for remote sites essential) both to strengthen field management effort and to broaden project team responsibilities. This management need stems from the requirement to cope with an increased level of detailed project information and the desirability of increasing the effectiveness of field management by relieving key field personnel of management function overloads.

A characteristic management solution is to assign relevant project-oriented head office functions for field-based execution by incorporating into an enlarged field project team structure, agents capable of performing the transferred head office functions (see Figure 6.12). In this way a complete field project team structure emerges, with dual lines of communication, that is, internally to the field-based project management team headed by a full-time on-site project manager, and externally direct to each relevant head office functional area. Since more management functions now appear and are staffed at the field level, a more balanced and capable field organization emerges than that achieved by the project team component or the traveling project manager management approaches. In larger or more complex projects the field project team develops distinct business-, technical-, and construction-oriented components that correspond to a similar division of functions in the head office companies (see Figure 6.13). The field managers of these organizational components together with the full-time resident project manager comprise the field project decision team.

The project management approach indicated above represents a significant step in project team development toward the fully integrated and autonomous project team. The fully integrated and autonomous project team emerges as a viable management approach for contractors involved in large remote projects or for projects performed in the construction management team or phased construction management environment.

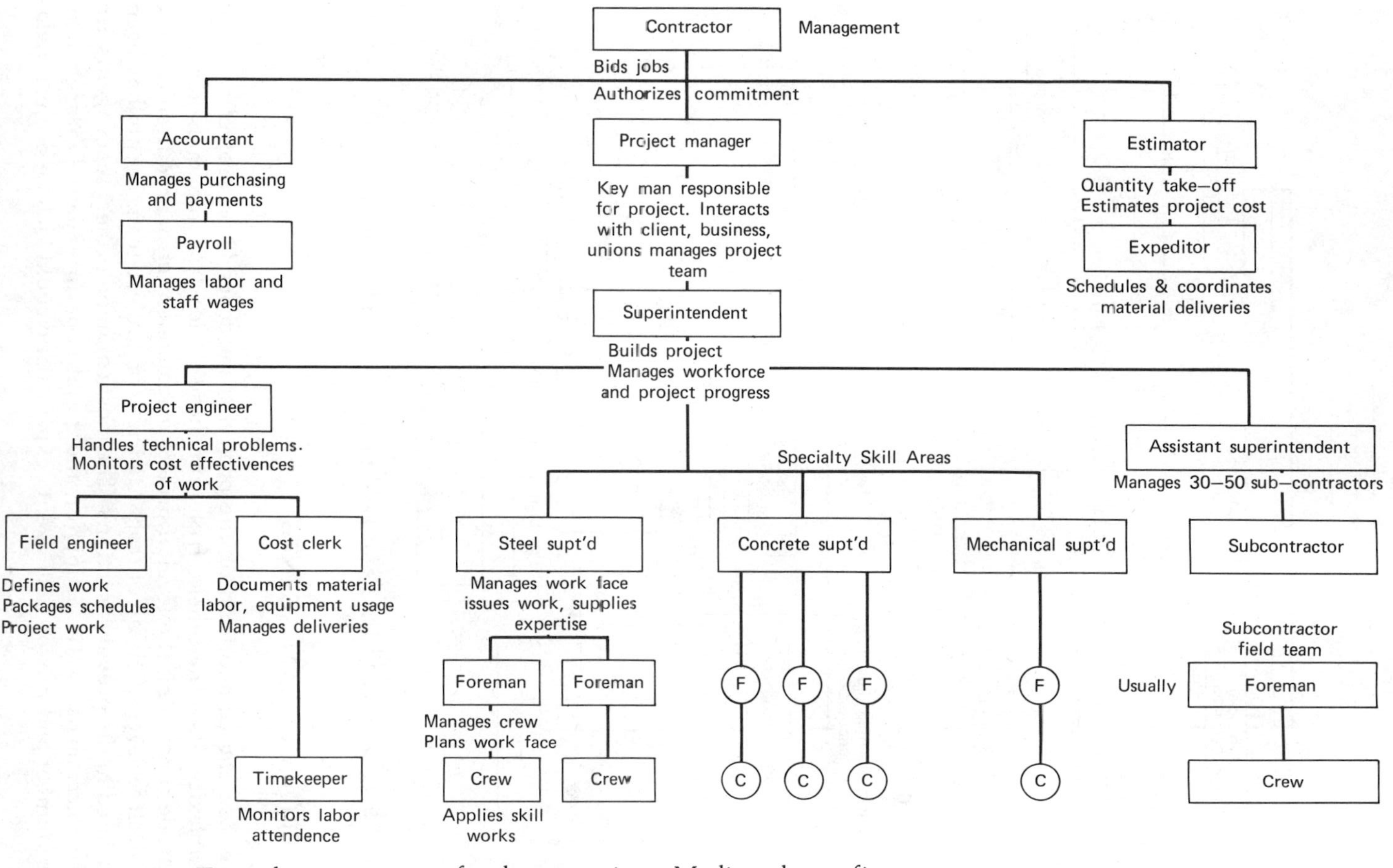

FIGURE 6.12 Typical project team for large project: Medium-large firm.

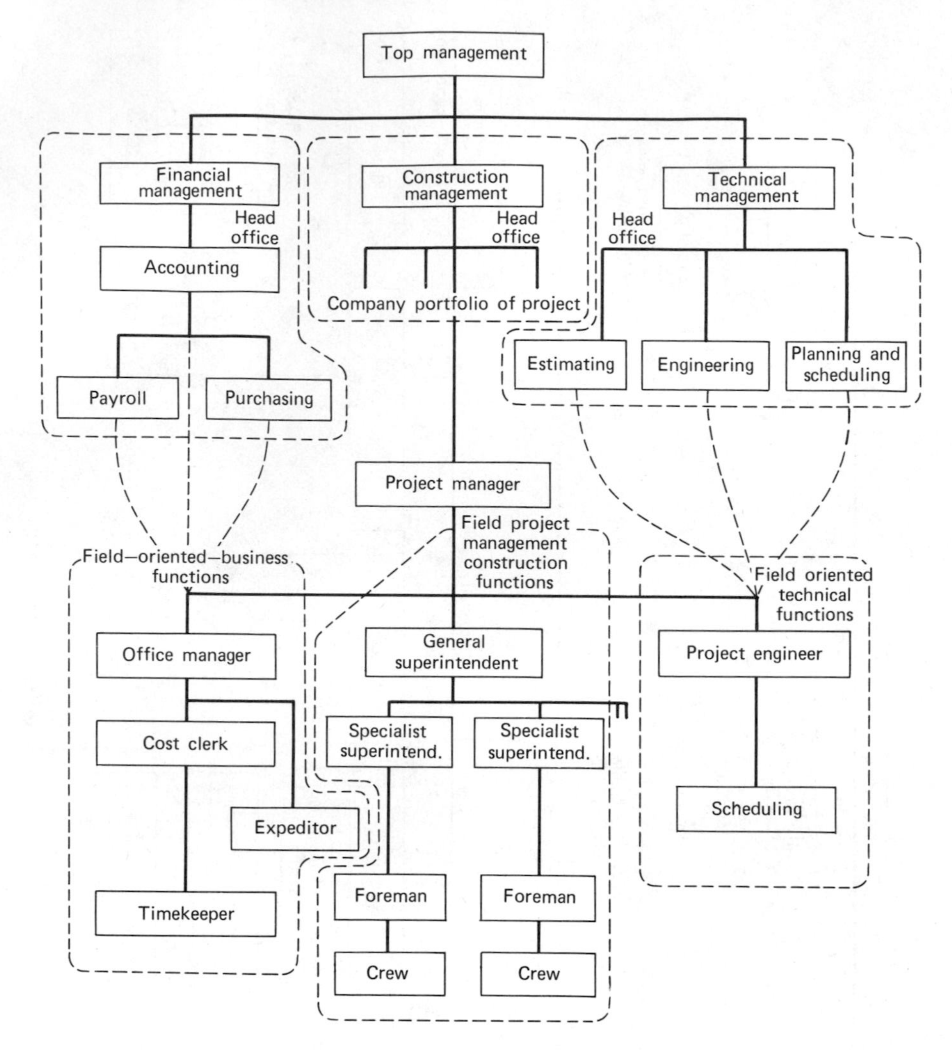

FIGURE 6.13 Head office and field project team structure.

6.5 PROJECT TEAM AGENTS

The size and composition of the project team depends on project size and complexity, its remoteness from the head office, and top management's attitude concerning the level of project management effort required to meet its desired level of field control. A variety of project team agents has been indicated in the previous sections. The most common project team agents are the project manager, the estimator, and the job superintendent. On larger jobs the number and mix of field agents expands and could include project, field, and scheduling engineers, assistant and specialty superintendents, time-keepers, expeditors, and cost accountants. As more agents are included in the project team, there is a definite trend toward the identification and performance of specialized functions. This generally means that on large projects, management focuses formally on a greater number of project status, progress, and performance indicators, whereas on smaller projects these are either ignored or handled in an ad hoc manner. This follows because in most cases contractors view field management as a cost center with field staff very visible as overhead items. In these situations, top management often reduces the field management team size as a cost-paring device to the extent that field management agents become functionally overloaded and effective performance, or consideration, of many management functions is jeopardized. Thus agent job descriptions that prescribe duties (i.e., management functional areas requiring performance) must always be interpreted or related to the management environment that exists for each specific project.

A widespread difficulty relates to the usage of commonly accepted titles for field agents which, however, cover a great variety of job descriptions. Thus the "project manager" may be referred to as the superintendent, general superintendent, or project manager, or may in fact be a company's vice-president or estimator. Clearly the functions performed will depend on the full job description of the agent involved and will vary from company to company and even, within a company, from project to project.

Typical general-purpose job descriptions have been compiled by a number of agencies and contractor groups. Examples of these job descriptions for project manager, estimator, expeditors, equipment superintendent, field superintendent, mechanical superintendent, scheduling engineer, and timekeeper, as seen by the Builders Association of Chicago, Inc., are given in Appendix H. Clearly these job descriptions have been designed to have broad rather than specific coverage and consider only building construction as seen by both small and medium-sized contractors. They do, however, give a good overall view of the role of typical agents.

REVIEW QUESTIONS AND EXERCISES

6.1 Visit the head offices of several local contractors and ascertain their organizational structures. Are there clear distinctions between field-oriented and head office business agents?

6.2 On the basis of the information gained in answering the above question is it possible to correlate organizational size with annual construction volumes? If so, then develop approximate construction volume/agent ratings for typical head offices and field agents.

6.3 Visit a field construction site and ascertain the size of the work force and its classifications. Then develop a field organizational chart for the site. What sort of contact does the field have with its head office?

6.4 For a local construction job identify agent responsibilities in terms of the general management functions (as shown in Figure 6.2).

6.5 Using the job descriptions of Appendix H as a frame of reference develop job descriptions for at least two field agents on a local construction site. Explain any differences you may encounter.

6.6 Identify a number of line or staff relationships that commonly exist between construction management agents in both head office and field situations. Then develop a functional description similar to that shown in Figure 6.4.

6.7 Attempt to quantify the span of control limits of the following agents:

The construction foreman (in relation to crew size),
Job superintendent (in relation to field staff),
Project manager (in relation to the number of field projects that he supervises).

What factors most influence the span of control limits for these agents?

6.8 What project team would you propose for the small gas station project of Appendix L? If you are the head-office-based project manager for this project what field inspection routines would you propose if site construction is being handled by:
(a) A competent job superintendent.
(b) A new foreman.

6.9 Assume you are an architect/engineer supervising a construction project from its inception to completion. Develop a schematic portrayal of your functions and responsibilities in a form similar to that of Figure 6.8.

CHAPTER SEVEN

Project Funding

7.1 MONEY: A BASIC RESOURCE

The essential resource ingredients that must be considered in the construction of a project are usually referred to as the four M's. These basic construction resources are (1) money, (2) machines, (3) material, and (4) manpower. The are presented in this order since this is the sequence in which they will be examined in the next few chapters. Here, the first of these resources to be encountered in the construction process, money, is considered. Money (i.e., actual cash or its equivalent in monetary or financial transactions) is a cascading resource that is encountered at various levels within the project structure. The owner or developer must have money available to initiate construction. The contractor must have cash reserves available to maintain continuity of operations during the time he is awaiting payment from the owner. The major agents involved in the flow of cash in the construction process are shown in simple schematic format in Figure 7.1.

Rising construction costs have increased the pressure on the construction industry to carefully monitor and control the flow of money at all levels. As a result, more emphasis is being placed on cash-flow and cost-control functions in construction management than ever before. In the planning phases, more thorough investigations and more accurate cost estimates are being required for those seeking financial backing. To remain competitive, contractors are being forced to monitor their cost accounts more closely and to know where losses are occurring. In this chapter, the methods by which the owner/entrepreneur acquires project funding will be considered. The relationship between the flow of money from owner to contractor

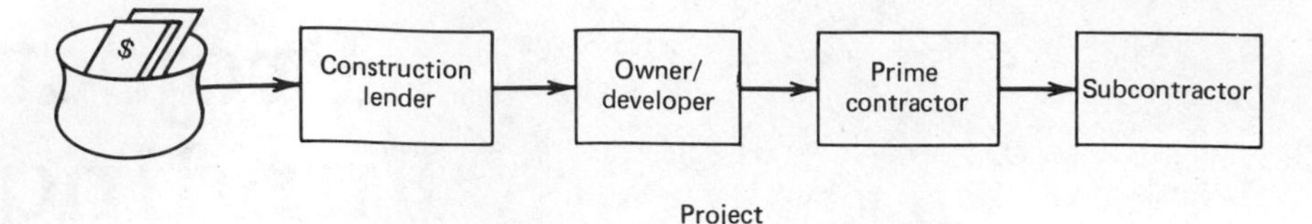

FIGURE 7.1 Project money flow.

and its impact on the contractor's project financing will be developed in Chapter Eight.

7.2. THE CONSTRUCTION FINANCING PROCESS

The owner's financing of any significant undertaking typically requires two types of funding, *short-term* (construction) funding and *long-term* (mortgage) funding. The short-term funding is usually in the form of a construction loan, whereas the long-term financing involves a mortgage loan over from 10 to 30 years.

The short-term loans may provide funds for items such as facility construction, land purchases, or land development. Typically these short-term loans extend over the construction period of the project. For large and complex projects, this can extend up to six to eight years as in the case of utility power plants. A short-term loan is provided by a lending institution, based on the assurance that it will be repaid with interest, by some other loan. This following *mortgage loan* constitutes the long-term financing. Therefore, the first objective of any entrepreneur is to seek a commitment for long-term or permanent financing from a mortgage lender. Regardless of the type of project, this commitment will permit the construction loan, and any other funding required, to be obtained with relative ease or, at least, more easily.

Unless he is in a position to raise the funds required directly by the issue of his own securities, the entrepreneur will seek to obtain a commitment from one of several alternate sources, including real estate investment trusts (REITs), investment or merchant banks, commercial banks, savings and loan associations, insurance companies, governmental agencies (VA, FHA) or, in special cases, from one of the international development banks. Public institutions often raise project construction funds by the sale of bonds. The choice of lender often depends on the type and size of project. The choice of the form of security employed depends on a number of factors such as relative cost, the time period for which the funds will be available, the degree of flexibility involved (the freedom to pay out or refinance) as to whether there are any restrictions involved and whether there is any sacrifice of control to the lender. The funding of some larger projects may be consortium handled by a consortium of international bankers.

Lending institutions are cautious; they are not interested in financing failures, or in owning partially completed buildings. Therefore, they will undertake a great deal of research and evaluation prior to a commitment for funding. As a minimum, an entrepreneur will be expected to provide the following as part of the loan application:

1. A set of financial statements for the firm.
2. Personal financial statements from the principals of the firm.
3. Proof of clear title to the land and that it has an appropriate zoning.
4. Preliminary floor plans and elevations for the project.
5. Preliminary cost estimates.
6. A market research study to verify expected income.
7. A detailed pro forma indicating projected income and expenses throughout the life of the mortgage loan.

An example of a long-term finance pro forma for a venture involving the construction and leasing of a 148-unit apartment complex is shown in Figure 7.2*a*. This document indicates that the annual income from the proposed apartment complex project will be $306,830. The requested loan is $2,422,000 and the annual debt service (i.e., interest) on this amount is $236,145, realizing an income after debt service of approximately $70,000. The ratio between income and debt service is 1.3. Lenders normally wish this ratio to be *below* 1.3. The basis for the loan amount is given in Figure 7.2*b*. Items 1 to 34 are construction-related items and are developed from standard references (e.g., R. S. Means, *Building Construction Cost Data*, Dodge Construction Pricing and Scheduling Manual, etc.) based on unit measures such as square footage. The lender normally has a unit price guide for use in verifying these figures. Items 35 to 46 cover nonconstruction costs that are incurred by the entrepreneur. It should be noticed that the interest for the construction loan is included in the costs carried forward to the long-term financing.

The method used to calculate the actual dollar amount of the loan is of great interest to the entrepreneur. The interest the developer pays for the use of the borrowed money is an expense, and it is generally considered prudent business policy to minimize expenses. One way to minimize the interest expense would be to borrow as little as possible. This is not, however, the way the developer moves toward his objective. The developer seeks primarily to protect his own personal assets (or those of his company) in his efforts to complete the project. The more he invests, the more he stands to lose if the project fails. With this consideration in mind, the developer may seek to minimize his own investments. That is, the developer tries to expand his own small initial asset input into a large amount of usable money. This is called *leverage*. He takes a small amount and levers or amplifies it into a large amount.

The amount of the mortgage loan should be a happy medium between too much and too little. If the mortgage is too small, there will not be enough money to cover the project. On the other hand, if the mortgage is too large, the developer will find that the individual mortgage payments will exceed his available revenue, and he may be unable to meet all of his obligations.

The amount the lender is willing to lend as long-term funding is derived from two concepts: the economic value of the project and the capitalization

Market Rent for Subject Property (Unfurnished)	
110 two-Bedroom A, B, or C units—1167 sq ft	
@ 20.5¢/sq ft = $239.24/mo or $240 × 110	$26,400.00
38 three-Bedroom A, B units—1555 sq ft	
@ 19.5¢/sq ft = $303.23/mo or $305.00 × 38	11,600.00
Total Estimated Monthly Income	$38,000.00
Other Income: Coin Laundry, Vending Machine	150.00
	$38,150.00
× 12 = Annual Total	457,800.00
Less Vacancy Factor of 5% (based on historical data)	-22,890.00
Adjusted Gross Annual Income	434,910.00
Less Estimated Expenses @ 29.45%	-128,080.00
NET INCOME BEFORE DEBT SERVICE	$306,830.00

Capitalized Value @ 9.5%	=	$\$3{,}229{,}789.00 = \frac{306{,}830}{0.095}$
Requested Loan Value	=	2,422,000.00
Loan/Value Ratio	=	75% (High) Governed by law
Long-Term Debt Service @ 9.75% Constant	=	236,145.00
Debt Service Coverage Ratio	=	$1.299 \simeq 1.3$
Loan per Unit	=	16,364.00
Loan per square foot	=	$12.92

FIGURE 7.2*a* Pro forma for 148 apartment units.

rate (cap rate). The economic value of the project is a measure of the project's ability to earn money. One method of predicting the economic value is called the *income approach* to value and is the method shown in figure 7.2*a*. Simply stated, it is the result of an estimated income statement of the project in operation. Like any income statement, it shows the various types of income and their sum. These are matched against the predicted sums of the different expenses. Although the predicted net income is a function of many estimated numbers, commonly a fairly reasonable degree of accuracy is achieved. The expected net income divided by the cap rate produces the economic value of the project. The cap rate used in Figure 7.2*a* is 9.5%. The capitalized economic value of the project is obtained by dividing the net income ($306 ,830) by the cap rate factor (0.095). This yields an economic value of $3,229,789.

How is the cap rate obtained? First, a lender generally provides a mortgage that is about 75% of the estimated economic value of the project. This is done because 25% value, or thereabouts, that must be invested by the developer will serve as an incentive for his making the project a success. That is, the lender furnishes 75% and the developer furnishes 25%. The lender must then decide what the interest rate will be, and takes up the developer's rate of return. The sum of these numbers, times their respective portions, gives the cap rate.

As an example, suppose that the lender decides that the interest rate will be 8½% and that the developer's planned rate of return will be 12%. Then, the cap rate is obtained as 8½% times 75% plus 12% times 25%, which gives 9.375% or 0.09375 as the cap rate factor. Obviously, the value of the cap rate can be adjusted by the values that the lender places on his interest rate and the developer's rate of return. These numbers are a function of the existing economic conditions and thus fluctuate with the state of the economy. The lender, therefore, cannot exert as much influence on their values as might at first be expected. In addition, the lender is in business to lend and wisely will not price himself out of the competition. He will attempt to establish a rate that is conservative but attractive. The expected income divided by the cap rate yields the economic value. The mortgage value may then be on the order of 75% of the calculated economic value. Not every lender will follow this type of formula approach; some, for example, may have a policy of lending a fixed proportion of their own assessed valuation, which may not be based on the economic value but instead on their estimate of the market value of the property.

Another factor affecting the economic value of the project is the percentage of space that is rented, since this number may largely determine the net income. Ideally, all the available space will be rented upon completion and, if the project is marketed properly and is of desirable quality, the chances should be strong that this will be the case. Of course, at different times and places, if there is insufficient advertising, or a poor market, or if there

1. Excavation and grading	$67,500
2. Storm sewers	48,000
3. Sanitary sewers	84,030
4. Water lines	28,000
5. Electric lines	14,000
6. Foundations	31,000
7. Slabs	96,273
8. Lumber and sheathing	185,000
9. Rough carpentry	185,000
10. Finish carpentry	81,362
11. Roofing and labor	20,035
12. Drywall and plaster	70,000
13. Insulation	28,888
14. Millwork	140,556
15. Hardware	8,813
16. Plumbing	165,000
17. Heating and air conditioning	95,025
18. Electrical	90,350
19. Linoleum and tile	17,752
20. Carpeting	101,881
21. Kitchen cabinets	62,075
22. Painting and decorating	107,000
23. Masonry—block	20,680
24. Masonry—brick	100,200
25. Ranges and hoods	29,638

	Item	Cost
	26. Disposals	3,139
	27. Exhaust fans	1,022
	28. Refrigerator	35,040
	29. Paving	20,915
	30. Walks and curbs	20,792
	31. Landscaping	30,000
	32. Fence and walls	36,792
	33. Fireplace	51,100
	34. Cleanup	29,200
NONCONSTRUCTION	35. Lender's fee	32,000
NONCONSTRUCTION	36. Surveyor's fee	1,000
NONCONSTRUCTION	37. Architect's fee	12,500
NONCONSTRUCTION	38. Land cost	80,000
NONCONSTRUCTION	39. Attorney's fee	7,500
NONCONSTRUCTION	40. Title insurance premium	5,762
NONCONSTRUCTION	41. Other closing costs	150
NONCONSTRUCTION	42. Hazard insurance premium	4,780
NONCONSTRUCTION	43. Construction loan interest	120,000
NONCONSTRUCTION	44. Appraisal	750
NONCONSTRUCTION	45. Building permit	1,500
NONCONSTRUCTION	46. Tax	50,000
	TOTAL	$2,422,000

FIGURE 7.2*b* Construction cost breakdown for 148 apartment units.

is a temporary surplus of available space, only a portion of the whole will be rented. A percentage of space rented at which the project will just be able to pay its way can be determined. If more space is rented, there will be a net profit, and if less is rented, there will be a net loss. This percentage is the break-even point. The lender may be willing to make a mortgage loan based on a net income using 100% occupancy even if only the break-even point is achieved. The amount of space rented that produces a balance between income and expense is called *rent roll achievement*. The amount of the loan if the break-even point is met or surpassed is called the *ceiling of the mortgage*. If less than the break-even point is achieved, the lender will consider that the project has less economic value and will, therefore, finance only a portion of the ceiling. This minimum amount is called the *floor of the mortgage*, and is normally 75% of the ceiling. If the break-even rent roll is not achieved by the completion of construction, the developer will usually be given an additional time period (generally only three to six months) to achieve the rent roll. If the rent roll is met during this period, the balance of the ceiling of the mortgage is made available.

The mortgage loan may be the critical financial foundation of the entire project, and may also involve protracted and complex negotiations. For this reason, the project developing company may exercise its right to hire a professional mortgage broker whose business it is to find a source of funds and service mortgage-loan dealings. The broker's reputation is based on his ability to obtain the correct size mortage at the best rate, which is also fair to his client. The broker acts as an advisor to his client, keeping him apprised of all details of the proposal financing in advance of actually entering into the commitment. For this service, the mortgage broker receives a fee of about 2% of the mortgage loan, although the rate and amount will vary with the size of the loan.

7.3 MORTGAGE LOAN COMMITMENT

Once the lending institution has reviewed the venture and the loan committee of the lender has approved the loan, a preliminary commitment is issued. Currently most institutions are reserving their final commitment approval until they have reviewed and approved the final construction plans and specifications.

The commitment issued is later embodied in a formal contract between the lender and borrower, with the borrower pledging to construct the project following the approved plans, and the lender agreeing that upon construction completion, and the achievement of target occupancy, he will provide the funds agreed upon at the stated interest rate for the stated period of time. As noted earlier, the actual amount of funds provided generally is less than the entire amount needed for the venture. This difference, called owner's equity, must be furnished from the entrepreneur's own funds or from some other source. The formal commitment will define the floor and ceiling amounts of the long-term loan.

During the construction period, no money flows from the long-term lender to the borrower. Funds necessary for construction must be provided by the entrepreneur or obtained from a short-term construction lender. Typically the lender of the long-term finance will pay off the short-term loan in full, at the time of construction completion, thereby cancelling the construction loan and leaving the borrower with a long-term debt to the mortgage lender.

7.4 CONSTRUCTION LOAN

Once the long-term financing commitment has been obtained, the negotiation of a construction loan is possible. Very often commercial banks make construction loans because they have some guarantee in knowing the loan will be repaid from the long-term financing. However, even in these situations, there are definite risks involved for the short-term lender. These risks relate to the possibility that the entrepreneur or contractor may, during construction, find themselves in financial difficulties. If this occurs it may not be possible for the entrepreneur/contractor to complete the project, in which case the construction leader may have to take over the job and initiate action for its completion. This risk is offset by a discount (1-2%), which is deducted from the loan before any money is disbursed. For example, if the amount of construction money desired is $1,000,000, the borrower signs a note that he will pay back $1,020,000. The borrower, in effect, pays immediately an interest of $20,000. This is referred to as a discount and may be viewed as an additional interest rate for the construction loan. The current trend to minimize these risks is to require the borrower to designate his intended contractor and design architect. Some commercial banks are evaluating and seeking to approve the owner's intended contractor, his prime subcontractors, and the owner's architect, as a prerequisite to approving the construction loan. This evaluation extends to an evaluation of their financial positions, technical capabilities, and current work loads.

To minimize the risks involved, the banks will also base their construction loans on the floor of the mortgage loan, and only 75 to 80% of this floor will be lent. Of course the developer may need additional funds to cover construction costs. One way to assure this is to finance the gap between the floor and ceiling of the long-term mortage loan. The entrepreneur goes to a lender specializing in this type of financing and obtains a standby commitment to cover the difference or gap between what the long-term lender provides and the ceiling of the long-term mortgage. Then, if the entrepreneur fails to achieve the break-even rent roll, he still is assured of the ceiling amount. In this situation, the construction lender will provide 75 to 80% of the ceiling rather than the floor. If the floor of the loan is $2,700,000 and the ceiling is $3,000,000, the financing of the gap can lead to an additional $240,000 for construction (i.e., 80% of $300,000). Financing of the gap is usually expensive, requiring a prepaid amount of as much as 5% to the gap lender. This would be $15,000 in the above example paid for money that may not be required if the rent roll is achieved. Nevertheless, the additional $240,000 of construction funidng may be critical to completion of the project and, therefore, the $15,000 is well spent in ensuring that the construction loan will include this gap funding.

Once the construction loan has been approved, the lender sets up a *draw schedule* for the builder or contractor. This draw schedule allows the release of funds in a defined pattern, depending on the size and length of the project. Smaller projects, such as single-unit residential housing, will be set up for partial payments based on completion of various stages of construction, (i.e., foundation, framing, roofing, and interior), corresponding to the work of the various subcontractors who must be paid (see Figure 7.3). For larger projects, the draw schedule is based on monthly payments. The contractor will invoice the owner each month for the work he has put in place that month. This request for funds is usually sent to the owner's representative or architect who certifies the quantities and value of work in place. Once approved by the architect and representative, the bank will issue payment for the invoice, less an owner's retainage (see Chapter Three).

The owner's retainage is a provision written into the contract as an incentive for the contractor to continue his efforts, as well as a reserve fund to cover defective work which must be made good by the contractor before the retainage will be released. Typically this retainage is 10%, although various decreasing formulas are also used. When the project is completed, approved, cleared, and taken over by the owner these retainages are released to the builder. The construction lender will not pay the builder any interest on these retained funds, nor does the builder have to pay interest on dollars he did not receive or use.

In addition to the funds mentioned, the developer should be aware that some front money is usually required. These funds are needed to make a good-faith deposit on the loan to cover architectural, legal, and surveying fees and for the typical closing costs. The entire topic of construction funding is discussed in detail in D. A. Halperin, *Construction Funding*, John Wiley & Sons, 1974.

7.5 OWNER FINANCING USING BONDS

Large corporations and public institutions commonly use the procedure of issuing bonds to raise money for construction projects. A bond is a kind of

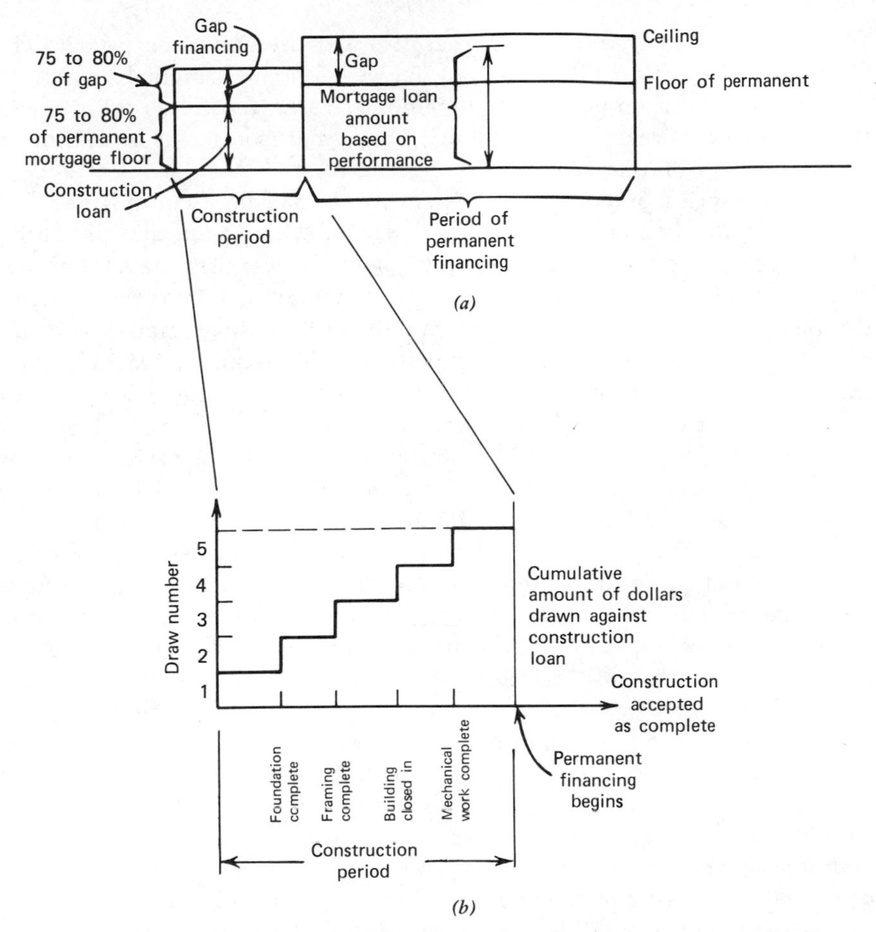

FIGURE 7.3 (*a*) Profile of project financing by the entrepreneur. (*b*) Draw schedule.

formal IOU issued by the borrower in which he has promised to pay back a sum of money at a future point of time. Sometimes this proviso is supported by the pledging of some form of property by way of security for the cover. A *series* of bonds or debentures, issued on the basis of a prospectus, are the general type of security issued by corporations, by cities, or by other institutions, but not by individual owner-borrowers. For convenience in exposition, owner-financing means financing arrangements made by those corporations or institutions that are the owners of the project property. In the illustrative material that follows, "Joe" stands as a surrogate for "any borrower" ("Joan" would have served as well!). During the period in which he has use of the money, the borrower promises to pay an amount of interest at regular intervals. For instance, Joe borrows $1000 and agrees to pay back the $1000 (referred to as the principal) in full at the end of

10 years. He pays an annual interest of 8% at the end of each year. That is, he, in effect, pays a rent of $80.00 per year on the principal sum of $1000 for 10 years and then pays back the amount borrowed. The rent is payable at the end of each year. The sequence of payments for this situation would be as appears in Figure 7.4.

Where a series of bonds are issued, there may be a commitment to pay the interest due in quarterly installments rather than in one amount at the end of the year. A bond, as a long-term promissory note, may take any one of a variety of forms depending on the circumstances; mortgage bonds involve the pledging of real property, such a land and buildings; debentures do not involve the pledging of specific property. Apart from the security offered, there is the question of interest rates and the arrangements to be made for the repayment of the principal sum. Sometimes a sinking fund may be set up to provide for the separate investment, at interest, of capital installments that will provide for the orderly retirement of the bond issue. Investors find this type of arrangement an attractive condition in a bond issue.

In preparing for a bond or debenture issue, financial statements must be drawn up and sometimes a special audit may be required. A prospectus for the issue may need to be drawn up and this will involve settling the terms of issue and of repayment, the interest rates payable, and the series of promises or conditions related to the issue, such as its relative status in terms of priority of repayment, limitations on borrowing, the relative value of the security, and the nomination of a trustee to watch the interests of bond or debenture holders. These details are usually settled with the aid of specialists—the CPA firm or the mortgage broker.

Public bodies may need the approval of some local regulatory authority, and corporations may have to file and have approved a prospectus for the proposed bond issue. Charters or other constitutional documents must of course confer on the public body or corporation the power to borrow money in this way; this power is exercised by the council or by the board of directors or governors. For public offerings that are particularly attractive, banks bid for the opportunity to handle the placement of the bonds. The banks recover their expense and profit by offering to provide a sum of money slightly less than the amount to be repaid. As noted above, this is called

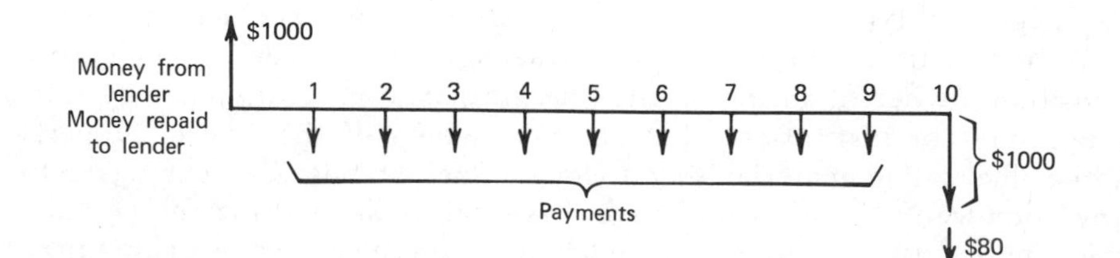

FIGURE 7.4 Sequence of payments for a bond.

discounting the loan. The fact that more will be repaid by the borrower than is lent by the lender leads to a change in the actual interest rate. This is established through competitive bidding by the banks to provide the amount of the bond issue. The bank that offers the lowest effective rate is normally selected; this represents the basic cost incurred for the use of the money.

Consider the following situation in which a city that has just received a major league baseball franchise decides to build a multipurpose sports stadium. The design has been completed and the architect's estimate of cost is \$40.5 million. The stadium building authority has been authorized to issue \$42 million in bonds to fund the construction and ancillary costs. The bonds will be redeemable at the end of 50 years with annual interest paid at 5% of the bond principal. Neither the term nor its rate purport to be representative of current market conditions. At this time the term for any bond issue would tend to be shorter, and its rate higher. In some commercial dealings "index number" escalation clauses are also occasionally seen. The banks bid the amounts for which they are willing to secure payment support. Suppose the highest bid received is \$41 million.

In order to determine the effective rate of interest a rate of return analysis may be used. The profile of income and expense is shown in Figure 7.5. The effective rate of interest is that rate for which the present worth of the expenses is equal to the present worth of the income (in this case, $\$41 \times 10^6$). That is,

$$\text{PW (Income)} = \text{PW (Expenses)}$$

Utilizing the information above, this expression for the bond issue problem becomes

$$\$41 \times 10^6 = \$2.1 \times 10^6(\text{PWUS},i,50) + \$42 \times 10^6(\text{PWUS},i,50)$$

The annual interest is \$2.1 million, and this is a uniform series of payments for 50 years. The \$42 million must be repaid as a single payment at the end of 50 years. The notation used in the equation is consistent with that used by Grant and Ireson, *Engineering Economy,* 4th edition.

In making this approach to a solution, a trial-and-error method must be used to solve the equation. That is, values of i must be assumed and the equation

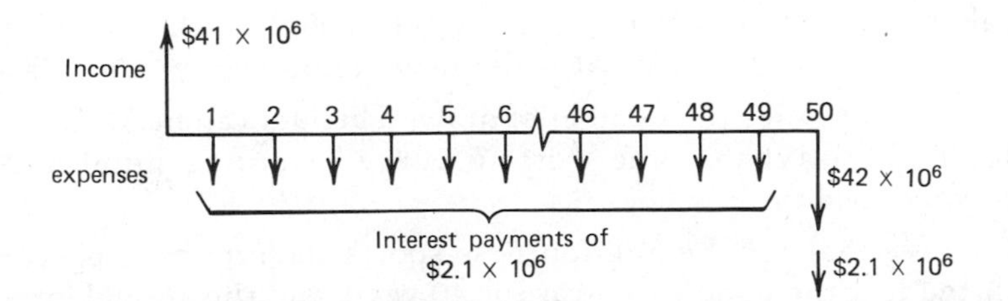

FIGURE 7.5 Profile of income-expense-form bond issue.

solved to see if the relationship (e.g., PW(Income) - PW(Expenses) = 0) is satisfied. In this case, two initial candidates for consideration are $i = 0.05$ and $i = 0.06$. Consulting appropriate tables for the present worth factors, the right side of the equation becomes

$$i = 0.05 \quad PW = \$2.1 \times 10^6 \, (18.256) + \$42 \times 10^6 \, (0.0872)$$
$$= \$42 \times 10^6 \qquad \text{difference} = + 1.0 \times 10^6$$

$$i = 0.06 \quad PW = \$2.1 \times 10^6 \, (15.762) + \$42 \times 10^6 \, (0.0543)$$
$$= \$35.38 \times 10^6 \qquad \text{difference} = -\$5.72 \times 10^6$$

Since the equation balance goes from plus to minus, the value satisfying the relationship is between 5% and 6%. Using linear interpolation, the effective interest rate is found to be

$$i = 0.05 + (0.06 - 0.05) \times \frac{1.0 \times 10^6}{(1.0 + 5.72) \times 10^6} = 0.515$$

or 5.15% as an approximation.

Alternatively, and practically, this kind of sum would be put through a hand-held calculator; some are already programmed for the task, others have the relevant program in a set or module, and others again may be programmed by the user to yield an accurate answer, almost as swiftly as the data can be keyed in (5.13%). Although short-cut iterative methods are employed in these programs, they customarily yield answers accurate to five places.

REVIEW QUESTIONS AND EXERCISES

7.1 What is the present level of the prime rate? How does this rate relate to the current financing and overdraft charges for new building construction in your locality? (How many points is this rate above the prime?) Does this overdraft rate vary with the magnitude of the monies involved?

7.2 Referring to the example in Figure 7.2*a*, suppose the market rent for a two-bedroom unit is $275 per month and for a three-bedroom unit is $330 per month. If the going cap rate is 10%, rework the pro forma calculations for the apartment project of Figure 7.2. Then determine lender's interest rate. What is the new break-even vacancy factor?

7.3 What determines the number of draws a builder can make in completing his facility? What is the existing policy regarding number of draws in your locality?

7.4 Suppose that for the multipurpose sports stadium example considered in the text the bond issue was for 40 years and the annual interest rate is 6% of the bond principal. Using the architect's estimate of cost of $41 million, determine the effective interest rate.

7.5 Suppose in the preceding problem, that the bonds are financed by a bank that discounts the bond issue to $40 million. What is the new effective interest rate?

CHAPTER EIGHT

Project Cash Flow

8.1 CASH FLOW PROJECTION

The projection of income and expense during the life of a project can be developed from several time-scheduling aids used by the contractor. The sophistication of the method adopted usually depends on the complexity of the project. One of the most commonly used and simplest aids is the so-called S-curve. In many contracts (e.g., public contracts such as those used by the Army Corps of Engineers), the owner requires the contractor to provide an S-curve of his estimated progress and costs across the life of the project. The contractor develops this by constructing a simple bar chart of the project, assigning costs to the bars and smoothly connecting the projected amounts of expenditures over time.

Consider the following highly simplified project (Figure 8.1) in which four major activities are scheduled across a four-month time span. Bars representing the activities are positioned along a time scale indicating start and finish times. The direct costs associated with each activity are shown above each bar. It is assumed that the monthly cost of indirect charges (i.e., site office costs, telephone, heat, light, and supervisory salaries, which cannot be charged directly to an activity) is $5000. Assuming for simplicity that the direct costs are evenly distributed across the duration of the activity, the monthly direct costs can be readily calculated and are shown below the time line. The direct charges in the second month, for example, derive from activities *A*, *B*, and *C*, all of which have a portion in the period. The direct charge is simply calculated based on the portion of the activity scheduled in the second month, as:

Activity A: 1/2 × 50,000 = \$25,000
Activity B: 1/2 × 40,000 = 20,000
Activity C: 1/3 × 60,000 = 20,000

\$65,000

The figure shows the total monthly and cumulative monthly expenditures across the life of the project. The S-curve is nothing more than a graphical presentation of the cumulative expenditures over time. A curve is plotted below the time-scaled bars through the points of cumulative expenditure. The name S-curve comes from the fact that the curve of cumulative expenditures has the appearance of a "lazy S." This general shape characteristic results because early in the project, activities are mobilizing and the expenditure curve is relatively flat. As many other activities come on-line, the level of expenditures increases and the curve has a steeper middle section. Toward the end of a project, activities are winding down and expenditures flatten again. The points

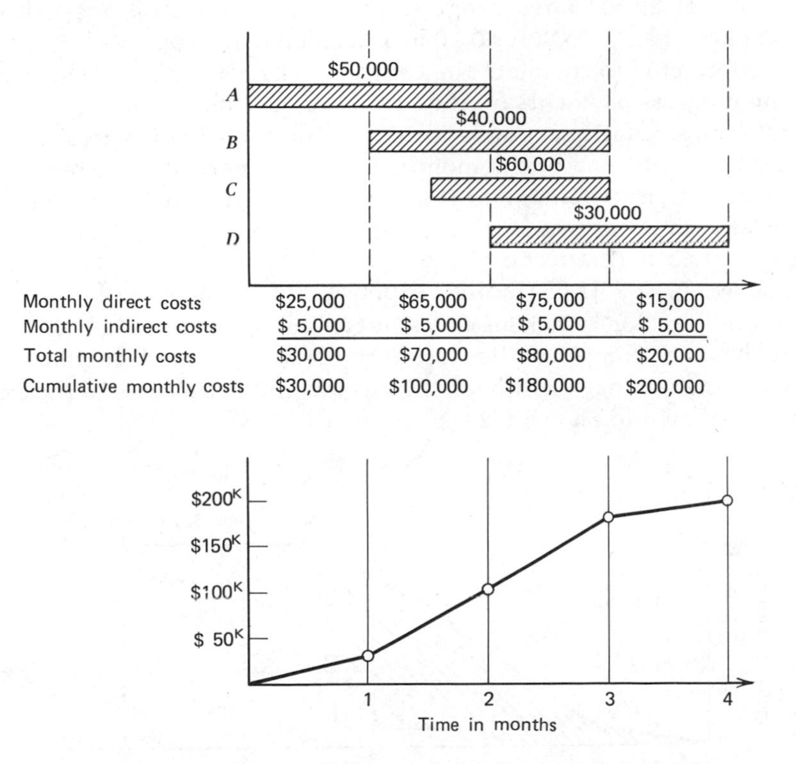

FIGURE 8.1 Development of S-curve.

are connected by a smooth curve, since the assumption is that the expenditures are relatively evenly distributed over each time period. This curve is essentially a graphical portrayal of the outflow of monies (i.e., expense flow) for both direct and indirect costs.

8.2 CASH FLOW TO THE CONTRACTOR

The flow of money from the owner to the contractor is in the form of progress payments. As noted above, estimates of work completed are made by the contractor periodically (usually monthly), and are verified by the owner's representative. Depending on the type of contract (e.g., lump sum, unit price, etc.), the estimates are based on evaluations of the percentage of total contract completion or actual field measurements of quantities placed. This process is best demonstrated by further consideration of the four-activity example just described. Assume that the contractor originally included a profit or markup in his bid of $10,000 (i.e., 5%) so that the total bid price was $210,000. The owner retains 10% of all validated progress payment claims until one-half of the contract value (i.e., $105,000 × 0.10) has been built and approved, as incentive for the contractor to complete the contract. The retainage will be deducted from the progress payments on the first $105,000, and eventually paid to the contractor on satisfactory completion of the contract. The progress payments will be billed at the end of the month, and the owner will transfer the billed amount minus any retainage to the contractor's account 30 days later. The amount of each progress payment payment can be calculated as

$$\begin{aligned}\text{Pay} &= 1.05 \times (\text{Indirect expense} + \text{direct expense})\\ &\quad -0.10 \times [1.05\,(\text{Indirect expense} + \text{direct expense})]\end{aligned}$$

The minus term for retainage drops out of the equation when 50% of the contract has been completed. Because of the delay in payment of billings by the owner and the retainage withheld, the income profile lags behind the expense S-curve as shown in Figure 8.2.

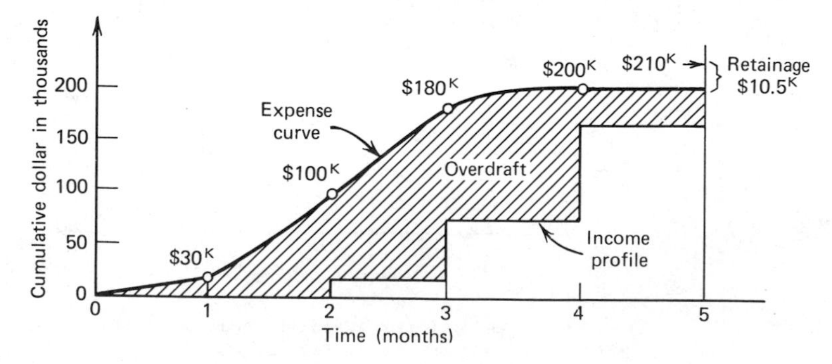

FIGURE 8.2 Expenses and income profiles.

The income profile has a stair-step appearance, since the progress payments are transferred in discrete amounts based on the above equation. The cross-hatched area between the income and expense profiles indicates the need on the part of the contractor to finance part of the construction until such time as he is reimbursed by the owner. This difference between income and expense makes it necessary for the contractor to obtain temporary financing. Usually, a bank extends a line of credit against which the contractor can draw to buy materials, make payments, and pay other expenses while waiting for reimbursement. This is similar to the procedure used by major credit card companies in which they allow credit card holders to charge expenses and carry an outstanding balance for payment. Interest is charged by the bank (or credit card company) on the amount of the outstanding balance or *overdraft.** It is, of course, good policy to try to minimize the amount of the overdraft and, therefore, the interest payments. The amount of the overdraft is influenced by a number of factors including the amount of markup or profit the contractor has in his bid, the amount of retainage withheld by the owner, and the delay between billing and payment by the owner.

Interest on this type of financing is usually quoted in relationship to the *prime rate*. The prime rate is the interest rate charged preferred customers who are rated as very reliable and who represent an extremely small risk of default (i.e., General Motors, Exxon, etc.). The amount of interest is quoted in the number of *points* (i.e., the number of percentage points) above the prime rate. The higher-risk customers must pay more points than more reliable borrowers. Construction contractors are normally considered high-risk borrowers; if they default the loan is secured only by some materials inventories and partially completed construction. In the event that a manufacturer of household appliances defaults, the inventory of appliances is available to cover part of the loss to the lender. Additionally, since construction contractors have an historically high rate of bankruptcy they are more liable to be charged additional interest rates in most of their financial borrowings.

Some contractors offset the overdraft borrowing requirement by requesting *front* or *mobilization* money from the owner. This shifts the position of the income profile so that no overdraft occurs (Figure 8.3). Since the owner is normally considered less of a risk than the contractor, he can borrow short-term money at a lower interest rate. If the owner agrees to this approach, he essentially takes on the interim financing requirement normally carried by the contractor. This can occur on cost-reimbursable contracts where

*Similar examples of this type of *inventory financing* can be found in many cyclic commercial undertakings. Automobile dealers, for instance, typically borrow money to finance the purchase of inventories of the new car models and then repay the lender as cars are sold. This is often achieved by a floor plan whereby a major distributor guarantees a specific-purpose loan or overdraft with the dealer's brokers. Clothing stores buy large inventories of spring or fall fashions with borrowed money and then repay the lender as sales are made.

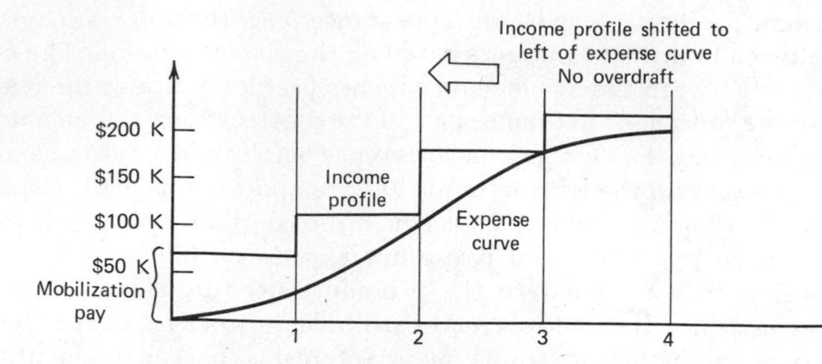

FIGURE 8.3 Influence of front or mobilization payment on expense and income profiles.

the owner has great confidence in the contractor's ability to complete the project. In such cases it represents an overall cost savings to the owner, since otherwise he will ultimately be back-billed for the contractor's higher financing rate if the contractor must carry the overdraft.

8.3 OVERDRAFT REQUIREMENTS

In order to know how much credit must be made available at the bank, the contractor needs to know what the maximum overdraft will be during the life of the project. With the information given regarding the four-activity project, the overdraft profile can be calculated and plotted. For purposes of illustration, the interest rate applied to the overdraft will be assumed to be *one percent* per month. That is, the contractor may pay the bank 1% per month for the amount of the overdraft at the end of the month. More commonly, daily interest factors may be employed for the purpose of calculating this interest service charge. Month-end balances might otherwise be manipulated by profitable short-term borrowings at the end of the month. The calculations required to define the overdraft profile are summarized in Table 8.1. The table indicates that the payment by the owner occurs at the end of a month based on the billing at the end of the previous month. It is assumed that the interest is calculated on the overdraft and added to obtain the amount financed. This amount is then reduced by the amount received from the owner for previous billings. To illustrate: The overdraft at the bank at the end of the second month is $100,300. The interest on this amount is $1003 and is added to the overdraft to obtain the total amount financed ($101,303). To obtain the overdraft at the end of the third month, the progress payment of $28,350 is applied to reduce the overdraft at the beginning of the third month to $72,953. The overdraft at the end of the period is, then, $72,953 plus the costs for the period. Therefore, the overdraft is $72,953 plus $80,000 or $152,953. The informa-

Table 8.1 OVERDRAFT CALCULATIONS

	Month 1	2	3	4	5	
Direct cost	$25,000	$65,000	$75,000	$15,000		
Indirect cost	5,000	5,000	5,000	5,000		
Subtotal	30,000	70,000	80,000	20,000		
Markup	1,500	3,500	4,000	1,000		
Total billed	31,500	73,500	84,000	21,000		
Retainage withheld	3,150	7,350	0	0		
Payment received			$28,350	$66,150	$84,000	$31,500
Total cost to date	30,000	100,000	180,000	200,000	200,000	
Total amount billed to date	31,500	105,000	189,000	210,000	210,000	
Total paid to date			28,350	94,500	178,500	210,000
Overdraft end of month	30,000	100,300	152,953	108,333	25,416	(-)5830
Interest on overdraft balance [a]	300	1,003	1,530	1,083	254	0
Total amount financed	$30,300	$101,303	$154,483	$109,416	$25,670	

[a] A simple illustration only. Most lenders would calculate interest charges more precisely on the amount/time involved employing daily interest factors.

tion in the table is plotted in Figure 8.4. The overdraft profile appears as a saw-tooth curve plotted below the base line. This profile shows that the maximum requirement is $154,483. Therefore, for this project the contractor must have a line of credit that will provide at least $155,000 at the bank plus a margin for safety, say $175,000 overall to cover expenses.

Requirements for other projects are added to the overdraft for this project to get a total overdraft or cash commitment profile. The timing of all projects presently under construction by the contractor leads to overlapping overdraft profiles that must be considered to find the maximum overdraft envelope for a given period of time.

Bids submitted that may be accepted must also be considered in the projection of total overdraft requirement. The plot of total overdraft requirements for a set of projects is shown in Figure 8.5

Cash flow management involves all of the techniques described in this chapter—and very much more. It is fairly true to say, for example, that you cannot budget the other fellow's payments! That is, cash inflows are affected by a significant degree of uncertainty. A cash flow management model of a relatively simple kind involves making provision for a set of at least 50 variables and requires a computer program to secure sufficiently timely and usable output cash management decision making and control.

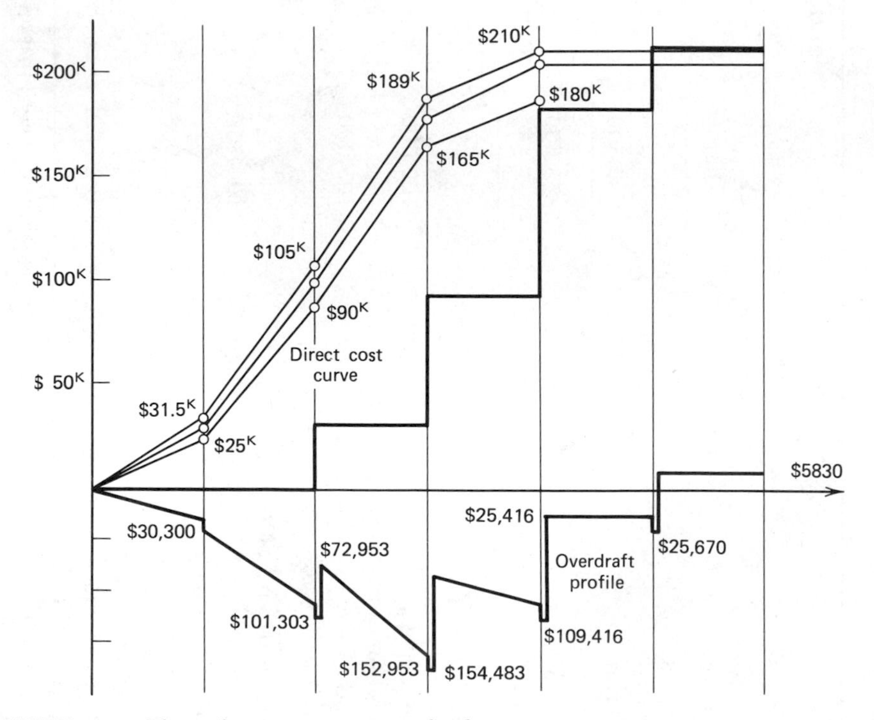

FIGURE 8.4 Plot of maximum overdraft.

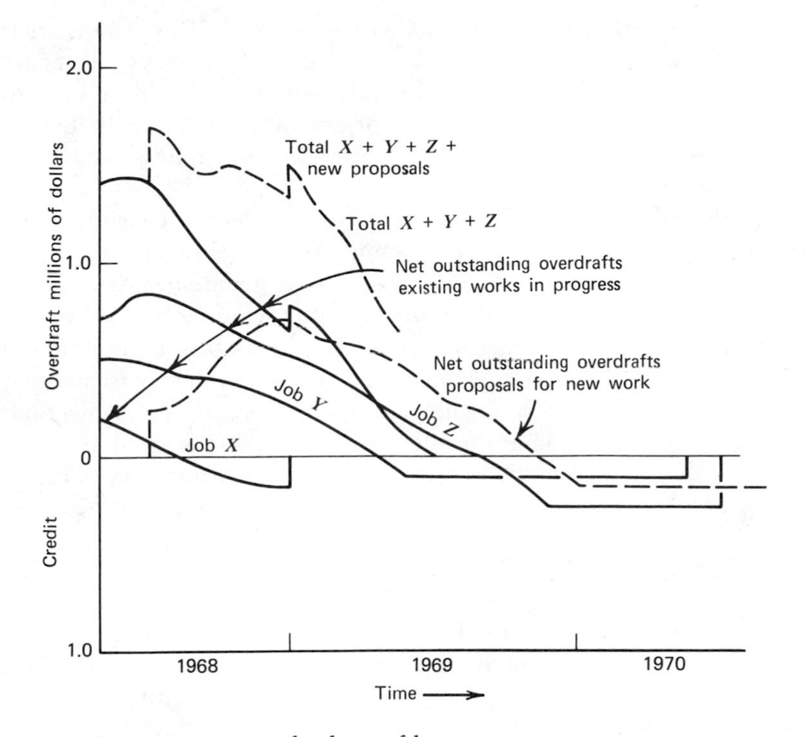

FIGURE 8.5 Composite overdraft profiles.

A simplified illustration of an estimate of bank overdraft requirements is given in Table 8.1. It is generally necessary in practice to prepare a much more detailed analysis that takes account of the expected levels of working capital and the extent of the support likely to be received by way of credit from suppliers and subcontractors. In addition it is necessary to estimate and allow for the variations that are likely to occur, especially lags and delays in the flow of moneys from progress-payment billings.

A more detailed system for gathering relevant data for meaningful cash flow forecasts will begin at the individual project level; a summary and a time schedule for the various elements and items making up total construction costs moving from time of incurrence to expected time of payment will be required. Similarly, a matching schedule for anticipated progress payments moving from the time of billing to the expected time of receipt must be prepared.

This data can be combined for all current projects and can incorporate capital outlays and general administrative cost schedules to yield a total organization-wide forecast of cash flow. It must be remembered that at any time it may become necessary to adjust this forecast in the light of actual events as they unfold, and to call up contingency plans to deal with unexpected shortfalls or an unexpected funds surplus. One of the primary purposes of the cash flow forecast is of course to make it possible to plan ahead, either to ensure that any additional funds requirements may be met by advance provisions or to ensure that a cash surplus does not remain idle when it should be earning some reasonable return. The importance of organization-wide cash-flow forecasting lies in the lead time provided by competent forecasting and the improved efficiency of cash management.

Most of the discussion and the illustrations reflect fairly simplistic manual system-based summaries, but increasing use is being made of computers, utilizing various modes of input related to programs of varying degrees of sophistication. Before these computer-financing models can be activated a great deal of detailed information is required, ranging from field details on individual projects to top management-level inputs relating to planned capital outlays and to a variety of financing packages involving, at various times, bond and debenture issues, share issues, mortgage loans, and bankers' loans including, of course, standby arrangements and contingency plans.

8.4 COMPARISON OF PAYMENT SCHEMES

Rate of return analysis is helpful in comparing various economic alternatives. The rate of return constitutes a parameter that can be used in evaluating one economic scheme against others. Particularly, with project cash-flow situations, it provides a vehicle for analyzing various proposals that have variations in retainage, payment delay, and amount of profit or markup. In order to illustrate this application, consider the following project payment situation.

A contractor is preparing to bid for a project. He has made his cost estimate together with the schedule of work. Table 8.2 gives his expected expenses and their time of occurrence. Other expenses such as insurance, bonds, and payroll taxes are included. For simplicity of analysis, it is assumed that all expenses are recognized at the end of the month in which they occur. The contractor is planning to add 10% to his estimated expenses to cover profits and office expenses. The total will be his bid price. He is also planning to submit for his progress payment at the end of each month. Upon approval the owner will subtract 10% for retainage and will pay the contractor one month later. The accumulated 10% for retainage will be paid to the contractor with the last payment at the end of the thirteenth month. The interest on the overdraft will not be considered in this analysis. The owner has offered an alternative to monthly progress payments. Under this payment scheme, the retainage will be 10%, but the contractor will be paid only twice. The first payment will be at the end of six months and the final payment will be received at the end of the thirteenth month. The question is, which of the two available payment schemes is most advantageous to the contractor? We do *not* really need to do our sums to be aware that a regular monetary payment schedule will be superior to a delayed payment schedule. We can also invariably assume that

Table 8.2 TABLE OF EXPENSES

Month	Mobilization Demobilization	Subcontractors	Materials	Payroll	Equipment	Field Overhead
0	$40,000	$0	$0	$0	$0	$0
1	0	10,000	10,000	10,000	20,000	1,000
2	0	30,000	20,000	15,000	10,000	5,000
3	0	30,000	30,000	20,000	20,000	6,000
4	0	40,000	30,000	20,000	30,000	6,000
5	0	50,000	40,000	40,000	20,000	6,000
6	0	50,000	40,000	40,000	15,000	6,000
7	0	40,000	30,000	40,000	10,000	6,000
8	0	40,000	10,000	20,000	10,000	6,000
9	0	70,000	10,000	10,000	10,000	6,000
10	0	30,000	5,000	5,000	10,000	6,000
11	0	30,000	5,000	5,000	5,000	6,000
12	20,000	50,000	0	5,000	5,000	5,000
Total	$60,000	$470,000	$230,000	$230,000	$165,000	$65,000

Total cost = $60,000 + $470,000 + $230,000 + $230,000 + $165,000 + $65,000
= $1,220,000

Profits +
overhead
@ 10% = $122,000

Bid price = $1,342,000

the person putting forward a proposal at least believes that it is to his or her advantage! We *do* need to do our sums, however, if we want to discover the "how much" the difference is. We also need to do very much more before we make a realistic decision; the *timing* of cash flow, for example, may be crucial. A plot of the two schemes is shown in Figure 8.6. The income profile is related to the expense profile by the equation:

$$\text{Income (Period } I + 1) = 1.1 \times [\text{Expense (Period I)}] - 0.10 \times 1.1\ [\text{Expense (Period I)}]$$

where $I = 0,12$

An intuitive indication of the economic desirability of these two proposals is given by their general structure, and by the area between the income and expense curves. As a general rule, the greater the amount that must be borrowed, the less desirable the proposal. Therefore, the closer the expense and income curves are, the more desirable the alternative. Based on this intuitive approach, it can be estimated that the first option will be more advantageous from an economic standpoint than the second (i.e., the two-payment alternative). The amount by which the first proposal is more desirable can be determined using rate of return analysis. Table 8.3 summarizes the calculations required to determine the rate of return under the monthly payment alternative (option A). These and subsequent calculations

Table 8.3 MONTHLY PAYMENT RATE OF RETURN

Month	Expenses	Payments	Net	Present Worth @ 6%	Present Worth @ 7%
0	-$40,000	$0	-$40,000	-$40,000	-$40,000
1	- 51,000	39,600	- 11,400	- 10,755	- 10,654
2	- 80,000	50,490	- 29,510	- 26,264	- 25,774
3	-106,000	79,200	- 26,800	- 22,501	- 21,877
4	-126,000	104,940	- 21,060	- 16,682	- 16,067
5	-156,000	124,740	- 31,260	- 23,361	- 22,288
6	-151,000	154,440	+ 3,440	+ 2,425	+ 2,292
7	-126,000	149,490	+ 23,490	+ 15,623	+ 14,627
8	- 86,000	124,740	+ 38,740	+ 24,305	+ 22,547
9	-106,000	85,140	- 20,860	- 12,347	- 11,346
10	- 56,000	104,940	+ 48,940	+ 27,328	+ 24,876
11	- 51,000	55,440	+ 4,440	+ 2,339	+ 2,109
12	- 85,000	50,490	- 34,510	- 17,151	- 15,322
13	$0	$218,350	+$218,350	+$102,362	+$90,615
				+$ 5,321	-$ 6,262

Rate of return = 6.46% (using linear interpolation)

Month	Monthly Payment Option A — Cumulative Monthly Expenses	Income	Net
0	- 40,000	0	- 40,000
1	- 91,000	39,600	- 51,400
2	-171,000	90,090	- 80,910
3	-277,000	169,290	-107,710
4	-403,000	274,230	-128,770
5	-559,000	398,970	-160,030
6	-710,000	553,410	-156,590
7	-836,000	702,900	-133,100
8	-922,000	827,640	- 94,360
9	-1028,000	912,780	-115,220
10	-1084,000	1017,720	- 66,280
11	-1135,000	1073,160	- 61,840
12	-1220,000	1123,650	96,350
13	-1220,000	1342,000	+ 122,000

Month	Two-Payment Option B — Cumulative Monthly Expenses	Income	Net
0	- 40,000	0	- 40,000
1	- 91,000	0	- 91,000
2	-171,000	0	-171,000
3	-277,000	0	-277,000
4	-403,000	0	-403,000
5	-559,000	0	-559,000
6	-710,000	702,900	- 7,100
7	-836,000	702,900	-133,100
8	-922,000	702,900	-219,100
9	-1028,000	702,900	-325,100
10	-1084,000	702,900	-381,100
11	-1135,000	702,900	-432,100
12	-1220,000	702,900	-517,100
13	-1220,000	1342,000	+ 122,000

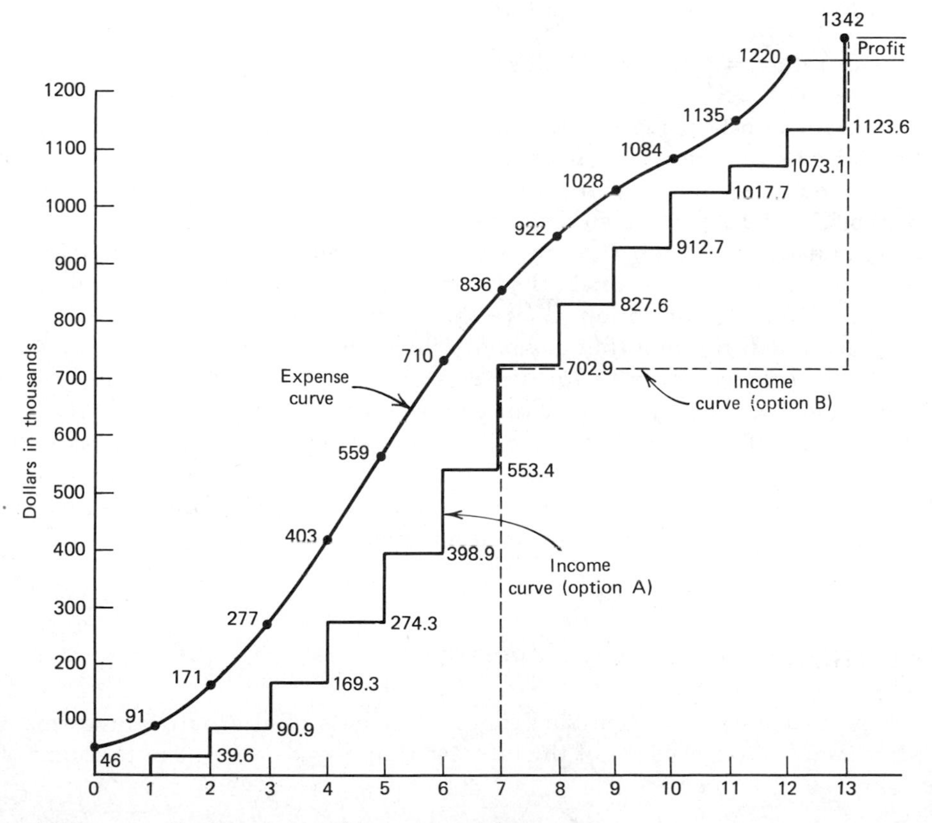

FIGURE 8.6 Comparison of payment schemes.

Table 8.4. TWO-PAYMENT RATE OF RETURN

Month	Expenses	Payments	Net	Present Worth @ 2.5%	Present Worth @ 3.0%
0	-$40,000	$0	-$40,000	-$40,000	-$40,000
1	- 51,000	0	- 51,000	- 49,756	- 49,515
2	- 80,000	0	- 80,000	- 76,145	- 75,408
3	-106,000	0	-106,000	- 98,431	- 97,005
4	-126,000	0	-126,000	-114,156	-111,949
5	-156,000	0	-156,000	-137,881	-134,566
6	-151,000	702,900	+551,900	+475,902	+462,208
7	-126,000	0	-126,000	-105,999	-102,450
8	86,000	0	- 86,000	- 70,584	- 67,889
9	-106,000	0	-106,000	- 84,877	- 81,240
10	- 56,000	0	- 56,000	- 43,747	- 41,670
11	- 51,000	0	- 51,000	- 38,869	- 36,843
12	- 85,000	0	- 85,000	- 63,202	- 59,617
13	$0	$639,100	639,100	+463,616	+435,196
			Summation	$15,871	- $ 748

Rate of return = 2.98% (using linear interpolation)

would generally be made with the aid of a program or programmable calculator or minicomputer. However, in order to understand what is involved, and later to manipulate such data with confidence, a number of these problems should be worked on a step-by-step basis as illustrated here. Checking the tabulation of cumulative net cash flow in Figure 8.6 establishes that the maximum overdraft ($160,030) occurs at the beginning of month 6. Similar calculations for the two-payment option (B) are shown in Table 8.4.

As expected, the monthly payment scheme has a rate of return of 6.46% versus slightly less than 3% for the two-payment alternative. In addition, the maximum overdraft at the bank for option B is $559,000 versus the $160,030 for option A. If interest on the overdraft had been included in this analysis, the disparity between the two options would have been even greater. Determination of the impact of interest and calculation of the revised rates of return are left as a recommended exercise for the reader.

8.5 THE DECISION-MAKING PROCESS

It is customary to present an economic analysis of alternative investment projects by taking account of the interest-weighted time value of the sums of money involved. Certainly these analyses are useful; in fact, where outlays and inflows of money are uneven and different, project by project, these analytical methods are useful in reducing the money elements to a common base or reference level. It would be misleading, however, to suggest that comparative investment-selection strategies were no more than a single-valued function of some rate of interest, as if business-project decisions were equivalent to examining a series of bond purchases and sales with comparative yields-to-maturity as the only relevant decision-making factor.

Apart from net rates of return, the pattern and timing of cash flows, of back receipts and payments, is itself a crucial element. A wide variety of 10-year cash flow patterns can be shown to be equivalent, including one where nothing is received for 9 years, 364 days, the whole inflow being received on the last day of the tenth year. Whether that particular project has an equivalent or a superior yield is not the point at all; the key question is: Could the contractor survive in this financial desert for 9 years, 364 days without bread or water, on the strength of the promise of a grand blowout on day 3652?

We must also look at the way in which a particular project combines or meshes with others that are in progress or planned. This obliges us to adopt an organization-wide, group-project approach.

There are also a variety of qualitative factors that may influence our selection or rejection of a project such as its size and type as well as the kinds of materials and skilled manpower it calls for.

Although the wide and complex variety of interacting business events that go toward the makeup of a successful or unsuccessful project are not bounded or fenced in by an interest rate, there is at least one more interest rate-related factor we should take into account. This is sometimes referred to as the reinvestment *assumption*—the idea that your rate of return result is only valid if you assume that you can reinvest the monies released by the project at the project rate of return. This is better regarded as a reinvestment *problem*—that, of course, you must usefully and profitably employ available funds at all times and not merely during those time intervals spanned by the project under analysis.

There are other more exotic and esoteric mathematical problems thrown up by some offbeat cash-flow patterns but no useful purpose would be served by pursuing that particular line of inquiry.

To conclude simply and practically, the decision-making process in any real life situation is highly complex; a full set of information is required for sensible decision making; every aspect of a project must be weighted and assessed, all of the quantitative factors and all of the qualitative factors—so far as these are known or may be estimated. Our information is always incomplete and we close or jump that information gap by exercising business judgement. It is this infinite variety of known and unknown factors that makes business activities so challenging and presents us with that incommensurable quality that we call business risk.

REVIEW QUESTIONS AND EXERCISES

8.1. Given the following cost expenditures for a small warehouse project (to include direct and indirect charges), calculate the peak financial requirement, the average overdraft, and the rate of return on invested money.
Assume 12% markup.
Retainage 10% throughout project.
Finance charge = 1½% month.
Payments are billed at end of month and received one month later.
Sketch a diagram of the overdraft profile.

Month	1	2	3	4
Indirect + Direct Cost ($)	$69,000	$21,800	$17,800	$40,900

8.2. During the negotiation between a contractor and an owner final agreement on the price of the contract was reached. The terms of progress payment were not resolved. Two proposals were under discussion as follows:

(a) Monthly progress payments with a 10% retainage on each to be paid at the end of the project (proposal A).
(b) Two payments, the first at the middle of the project minus a 10% retainage, the final payment including the retainage at project completion (proposal B).
(i) Which proposal has the highest present value to the owner?
(ii) Which proposal has the highest present value to the contractor?
(iii) A third proposal was also discussed. It involves a monthly payment with 20% retainage on the first 50% value of the progress payments with no retainage for the remaining 50%. From the contractor's point of view, rank the three proposals in order of highest rate of return. The third proposal is proposal C.

8.3. The table and graph below represent a contractor's overdraft requirements for a project. Complete the table shown below for costs, markup, total worth, retainage, and pay received. Retainage is 10%, markup is 10%, and interest is 1% per month. The client is billed at the end of the month. Payment is received the end of the next month, to be deposited in the bank the first of the following month.

Overdraft	-50,000	-120,500	-82,205	-13,727	+10,336
Interest	- 500	-1,205	- 822	- 137	---
Cumulative overdraft	-50,500	-121,705	-83,027	-13,864	+10,336

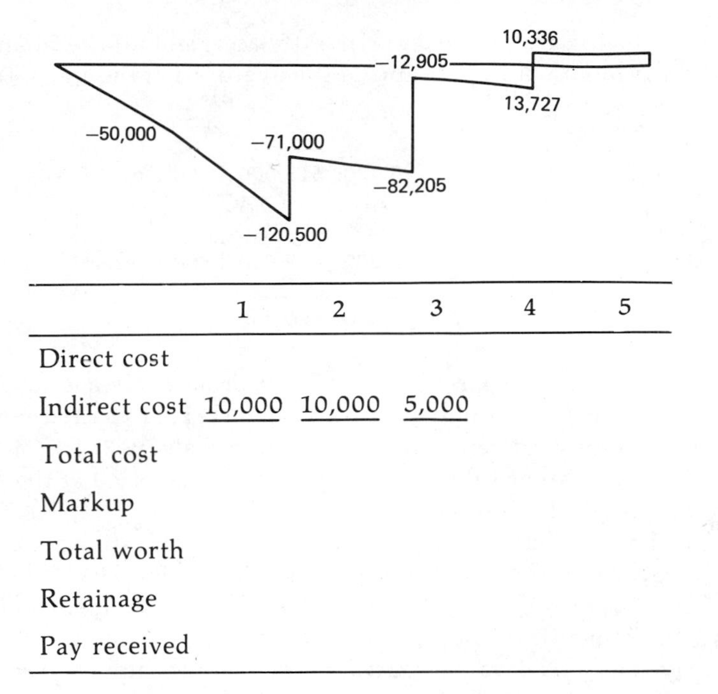

	1	2	3	4	5
Direct cost					
Indirect cost	10,000	10,000	5,000		
Total cost					
Markup					
Total worth					
Retainage					
Pay received					

8.4. The contract between Ajax Construction Co. and Mr. Jones specifies that the contractor will bill Mr. Jones at the end of each month for the amount of work finished that month. Mr. Jones will then pay Ajax a specified percentage of the bill the same day. The accumulated retainage was to be paid one month after project completion. The latest cumulative billing was $5 million, of which Ajax has actually received $4.5 million. The project is to be finished two months from now. Ajax estimates the bills for the remaining two months will be $100,000 and $50,000. Mr. Jones being short of cash at present proposed the following alternate:

Rather than follow the contract and make the three payments required, he will make one final payment (for the two months work plus the retainage) five months from now. Mr. Jones will also pay a 4% monthly interest rate because of the delay in payment.

(a) Find what would be the total final payment according to the actual contract and the new final payment according to the new proposal.

(b) Should the contractor accept the new proposal and why?

8.5. Given the bar chart in Figure 8.1 with the direct costs for each activity as shown calculate the rate of return of the contractor. Assume that (a) the markup is 5%, (b) retainage is 5% on the first 50% of worth, and 0% thereafter, (c) payment requests are submitted at the end of each month and payments are received one month later, and (d) the finance charge is 1% per month of the amount of the overdraft at the end of the month.

Timing and Allocation	$25,000	$65,000	$75,000	$15,000	Total Direct Costs	$180,000
Indirect Costs $5000/month	5,000	5,000	5,000	5,000	Total Indirect Costs	20,000
	$30,000	$70,000	$80,000	$20,000		$200,000

8.6. A contractor is preparing to bid for a project. He has made his cost estimate together with the schedule of work. His expected expenses and their time of occurrence are as shown in Table 8.2. For simplicity of analysis he assumed that all expenses are recognized at the end of the month in which they occur.

(a) The contractor is planning to add 10% to his estimated expenses to cover profits and office expenses. The total will be his bid price. He is also planning to submit for his progress payment at the end of each month. Upon approval the owner will subtract 5% for retainage and pays the contractor one month later. The accumulated retainage will be paid to the contractor with the last payment (i.e., end of month 13).

(i) What is the monthly rate of return?

(ii) What is the annual rate of return?

(iii) What is the peak financial requirements and when does it occur?

(b) Plot the S-curve against time.

8.7. Suppose that in problem 8.5, the 5% retainage holds for the full completion period of the project. What is the contractor's additional overdraft requirement?

8.8. Suppose for the project defined in problem 8.5, activities B, C, and D maintain their relative scheduling to each other, but activity B cannot start until activity A has been completed. Recalculate the maximum overdraft and rate of return of the contractor given the same markup, retainage, payment procedures, and financial charges of problem 8.5.

CHAPTER NINE

Equipment Costs

9.1 GENERAL

Equipment resources play a major role in any construction activity. Decisions regarding equipment type and combination can have a major impact on the profitability of a job. In this respect, the manager's object is to select the equipment combination that yields the maximum production at the best or most reasonable price. Quite obviously, the manager must have a basic understanding of the costs associated with a particular piece of equipment. He must also be capable of calculating the rate of production of the piece or combination of equipment. The cost and the rate of production combine to yield the cost per unit of production. For example, if it is estimated that the cost of a particular fleet of haulers and loaders is $250 per hour and the production rate is 750 cu yd/hr, the unit price can be easily calculated as $0.33 per cubic yard.

Construction equipment can be divided into two major categories. Productive equipment describes units that alone or in combination lead to an end product that is recognized as a unit for payment. Support equipment is required for operations related to the placement of construction such as movement of personnel and materials and activities that influence the placement environment. Typical production units are pavers, haulers, loaders, rollers, and entrenchers. Hoists, cranes, vibrators, scaffolds, and heaters represent typical classes of support equipment. In most cases, equipment units are involved either in handling construction materials at some point in the process of placing a definable piece of construction (e.g., crane lifting a boiler, pavers moving concrete or asphalt into lifts on a base course) or in controlling the environment in which a piece of construction is realized (e.g., heaters controlling ambient temperature, prefabricated forms controlling the location of concrete in a frame or floor slab).

In heavy construction, large quantities of fluid or semifluid materials such as earth, concrete, and asphalt are handled and placed, leading to the use of machines. The equipment mix in such cases has a major impact on production, and the labor component controls production rates only in terms of the skill required to operate machines. Therefore, heavy construction operations are referred to as being equipment-intensive. Heavy construction contractors normally have a considerable amount of money tied up in fixed equipment assets, since capitalizing a heavy construction firm is a relatively expensive operation.

Building and industrial construction require handwork on the part of skilled labor at the point of placement and are therefore normally less equipment-intensive. Equipment is required to move materials and man-power to the point of installation and to support the assembly process. Emphasis is on hand tools; and, although heavy equipment pieces are important, the building and industrial contractors tend to have less of their capital tied up in equipment. Also because of the variability of equipment needs from project to project, the building contractor relies heavily on the rental of equipment. The heavy-construction contractor, because of the repetitive use of many major equipment units, often finds it more cost-effective to own this equipment.

9.2 EQUIPMENT OWNING AND OPERATING COSTS

The costs associated with construction equipment can be broken down into two major categories. Certain costs (e.g., depreciation, insurance, and interest charges) accrue whether the piece of equipment is in a productive state or not. These costs are fixed and directly related to the length of time the equipment is owned. Therefore, these costs are called *fixed* or *ownership* costs. The term *fixed* indicates that these costs are time dependent and can be calculated based on a fixed formula or a constant rate basis. On the other hand the operation of a machine leads to operating costs that occur only during the period of operation. Some of these costs accrue because of the consumption of supplies, such as tires, gas, and oil, and the widespread practice of including the operator's wages in the operating costs. Other costs occur as a result of the need to set aside moneys for both routine and unscheduled maintenance. Thus operating costs are *variable costs.*

The total of owning and operating costs for items of equipment such as tractors, shovels, scrapers, dozers, loaders, and backhoes is invariably expressed on an hourly basis. These two categories of cost accrue in different ways. *Ownership costs* are usually arrived at by relating the estimated total service life in hours to the total of those costs. If the equipment is idle for some of those hours the relevant costs would be taken up as part of general operating overhead; when the equipment is in use, the hourly costs are charged to the job or project.

Operating costs are variable in total amount, being a function of the number of operating hours, but these hourly costs are found to be relatively constant.

The *hourly charge* for a piece of equipment is made up of four elements. To the ownership and operating costs is added an allowanced for estimated hourly overhead ocsts. The fourth element is a markup for income or profit. A schematic illustration of this break down of the hourly charge for a piece of equipment is shown in Figure 9.1.

Ownership costs are composed of two elements: first an estimate for depreciation on the cost of using the equipment itself. Each piece of equipment represents an estimated number of hours of useful service life and the depreciable value, the major part of its original cost, is divided by the hours total to yield a charging rate for this element of equipment costs. The second component of ownership costs consists of estimates of allowances for interest, insurance, and taxes.

Operating costs cover a broader range of items, the principal elements being: fuel, oils and lubricants, hydraulics fluid, grease, filters, and other supplies; maintenance, general overhauls, and repairs; and parts replacement (cutting edges, blades, buckets), tire replacements, and the like. Also included here are the direct labor costs—the operator's wages— including all of the expense loadings for holidays, sick leave, and insurance.

To the direct operating costs just enumerated are added allowances for general overhead expenses and the indirect costs of supervisory labor. This total establishes the total hourly cost of owning and operating a unit of equipment. A percentage markup is added to provide for an income or profit element, the whole yielding an hourly charging rate.

Some of these costs are incurred and paid for concurrently with the operation of the equipment, but the allowances or estimates included for items such as repairs and maintenance are provisions for costs that will have to be paid at some future time.

General administrative costs, including items such as telephones, stationery, postage, heat, light and power, and the costs of idle equipment in general, are aggregated together as general overhead expense, an allowance that forms part of the hourly charging rate.

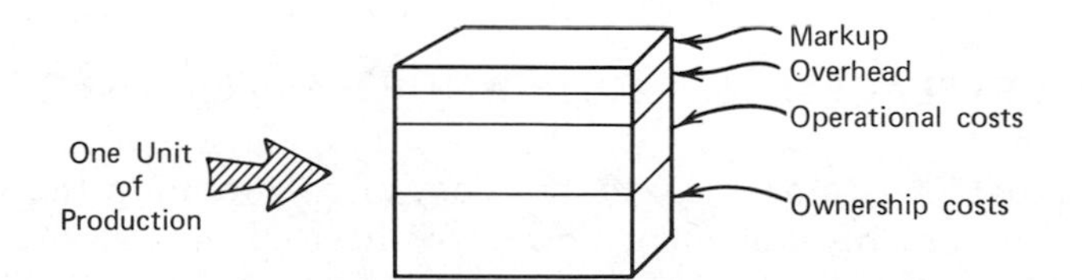

FIGURE 9.1 Cost components in a production unit.

The amount estimated for depreciation represents the recovery, as a component of the hourly charging rate, of the initial cost of the equipment. The depreciable amount, taken up in this way, is arrived at by deducting from the total delivered cost, including all attachments, the amount representing the cost of the tires and the estimated end-of-useful trade-in or resale value, sometimes referred to as the salvage value.

This net depreciable value, divided by the estimated hours of service life, yields the hourly cost element for depreciation. The cost of tires, and their replacement, is divided also by their estimated hours of service life to yield that hourly cost element.

In summary, we are saying that depreciation of equipment is a function of use. If idle hours and down time are included in the overhead, then depreciation based on use is combined with depreciation based on time. In either case, what is involved is a process of cost allocation and capital recovery.

The amount charged to a unit of production for depreciation is most commonly calculated on the basis of the service life of the equipment in terms of total hours. If the depreciable value to be recovered is, for example, $60,000 and the service life is estimated generally by the manufacturer as limited to five years, at least $12,000 must be recovered in each year. An estimate will be made of the number of hours in a normal working year for which the equipment might be operated. If a 40-hour week and a 50-week year are taken as a basis, yielding 2000 hours over each of five years, this is related to a 10,000-hour estimated service life and gives an hourly charging rate of $6 per hour for capital cost recovery, termed depreciation.

Dependent on the operating conditions, the actual service life might, in some instances, extend beyond the estimate used enabling the contractor to reduce the cost per unit quoted (if it might improve his competitive position). We have to remind ourselves, of course, that often more is involved than the recovery of some years-past capital outlay. If replacement costs are rising, the contractor has to recover as part of his costs the additional cost of replacement. Again if equipment becomes obsolete more rapidly than was expected, the costs of that obsolescence must also be recovered if the contractor is to preserve his level of income and stay in business. Manufacturers provide tables setting out their estimates of useful service life under varying conditions (see Table 9.1), and some also provide hourly owning- and operating-cost estimate sheets, with hourly cost guidelines.

9.3 HIGHER REPLACEMENT COSTS AND OBSOLESCENCE

The capital cost of equipment and the cost of maintaining the necessary operating inventory of equipment comprise three elements. The previous section considered the recovery of the base cost of the equipment by an hourly charge—taken up on various projects, or in general overhead with respect to

Table 9.1 ESTIMATED SERVICE LIFE TABLE (Caterpillar Tractor Co.)

Type of Equipment	Excellent Conditions: Hours	Average Conditions: Hours	Severe Conditions: Hours
Track-type tractors Traxcavators Wheeled loaders Wheeled tractors Scrapers	12,000	10,000	8,000
Motor graders	15,000	12,000	10,000

To determine the cost per hour due to depreciation, the above information may be used as follows:

$$\text{Depreciation cost per hour} = \frac{\text{Purchase price - Tire value}}{\text{Estimated service life in hours}}$$

downtime or idle hours. It is argued that operating costs should include an amount that makes provision for the expected higher costs of replacement.

It may be necessary to reestimate the spreading or prorating of these equipment costs from time to time as it becomes clear that a particular item has become technically obsolete by reason of being superseded by more modern, more efficient, or more economical units. Where technological changes are significant, the actual service life of some units may prove to be shorter than was originally estimated, with a consequent need for revision of cost estimates and hourly charging rates (so far as the competitive situation allows).

The manner or rate of writing down equipment costs, including them in hourly charging rates, is sometimes referred to as *amortization* while the expense item deducted from income or profit for tax purposes will retain the descriptive term *depreciation*.

Depreciation means, of course, a decline in or a loss of value and may include both the element of loss of value due to wear and tear and the ultimate loss of value associated with obsolescence and the final scrapping or downgrading of equipment. All of these elements of cost making up the total depreciable value of the equipment are recognized as legitimate costs of doing business, along with all other costs. Because depreciation is a deduction from income, and thus operates to reduce the amount of tax that otherwise would be payable, it becomes of importance to consider the incidence of tax and the various available (IRS) approved methods of depreciation.

9.4 THE TAX SIGNIFICANCE OF DEPRECIATION

From an accounting point of view, depreciation charges simply aim to distribute the net cost of the asset over its useful life by charges against income made in a systematic and rational manner. The hourly charges, which reflect the method adopted, are designed to recover the capital outlay made when the asset was acquired. The method of calculation of that hourly charge may not be the same as the method adopted for the preparation of the financial statements. The method adopted for the purpose of reporting income for taxation purposes may be different again.

The whole of the depreciable value of an asset may be offset against income as a deduction for tax purposes at rates and by methods approved by the Internal Revenue Service (IRS). There is a clear advantage in taking larger deductions in early years of the life of an asset, since this alters the incidence of the assessed tax over time.

Corporations pay approximately 50¢ in taxes out of every dollar of profit above $100,000. It follows that, if the firm can increase its depreciation deduction in year 1 from $40,000 to$73,000, or even to $80,000, it will be able to retain much larger amounts of its earnings for a longer period of time. The advantage of larger deductions in the early years is offset by smaller deductions in later years, and the total deduction is the same under all methods. The deduction of $40,000 refers to the result of adopting a straight-line method (say 10% per year on $400,000) over a 10 year period. The sum-of-the-years-digits method would permit a deduction of 10 fifty-fifths in the first year of a 10-year-lived asset. The increase of $33,000 represents a saving or deferment of tax liability which is worth $16,500. The double-declining-balance method would permit a deduction of 20% or $80,000. Compared to a $40,000 deduction this represents a saving or deferment of tax liability that is worth $20,000. These amounts are of the nature of short-term, interest-free loans.

Each of the depreciation methods will be discussed in more detail in the next chapter. The choice of method with the greatest tax advantage to the firm will include preserving the ability of the firm to switch to its straight-line method in about the sixth or seventh year of a 10-year-lived asset. The method or combination of methods that will yield the greatest advantage in terms of net present value will depend on the firm's cost of capital and the characteristics of the equipment in terms of service life and salvage value. Short service life, high cost of capital, and high salvage value would lead to a preference for the sum-of-the-years-digits method. The strategy adopted may also depend on the expected incidence of losses/profits over the firm's life.

9.5 A PRACTICAL EXAMPLE

A typical equipment record for a tractor dozer purchased in June 1970 is shown in Figure 9.2. The depreciation for this unit has been calculated using the double declining method and is show in the depreciation portion of the form. This example indicates one other aspect of depreciation that must be considered. The service life year of an asset normally does not coincide with the tax year. Therefore, the amount of depreciation that can be recovered in a given tax year must be calculated by proportioning the service life year to the calendar year. The depreciation of the unit is shown graphically in Figure 9.3. Inspection of the figure indicates that the service life of the unit is 4½ years. Therefore, the double-declining-balance rate to be applied is 44.44%. The service life year depreciation is calculated in Table 9.2. After using DDB for the first three service life years, it was decided to switch to the straight-line method for the last year and a half. Table 9.2 also shows the basis used for adapting the service life years to the calendar year. It was assumed that the unit was in use and eligible for depreciation between 1 July 1970 and 31 December 1974. The appropriate conversions are shown on the right side of Table 9.2.

9.6 INTEREST, INSURANCE, AND TAX—IIT COSTS

In addition to the amortization/depreciation component, the ownership costs include a charge for other fixed costs that must be recovered by the equipment owner. Throughout the life of the unit, the owner must pay for insurance, applicable taxes, and either pay interest on the note used to purchase the equipment or lose interest on the money invested in equipment if the unit was paid for in cash. These costs are considered together as what can be called the IIT costs. Recovery of these charges is based on percentages developed from accounting records that indicate the proper levels that must be provided during the year to offset these costs. The percentages for each cost is applied to the *average annual value* of the machine to determine the amount to be recovered each hour or year with respect of these cost items.

The average annual value is defined as

$$\text{AAV} = \frac{C(n+1)}{2n}$$

where AAV is the average annual value, C is the initial "new value" of the asset, and n is the number of service life years. This expression assumes that the salvage value is zero. What the formula does is to level the de-

HALCON, INC. EQUIPMENT RECORD 25

Kind: Crawler Tractor	Make: Caterpillar
BASIC MACHINE	ATTACHMENTS
Eq. Info.: Serial No.: 66A8320	Kind: Bulldozer
Year Mfg.: 1970 Model D9G 90″ Gauge	Make: Caterpillar
Color: Yellow Weight : 92, 928#	Model: 9U 6J8797
Other Data: Inc. Blower Fan, Power	Serial No.: 19K578
Shift Transmission, 7-roller track	Includes Tilt Cylinder
frame, rigid drawbar, hydraulic track	
adjuster. Arrangement 3S3953	
Motor Info. Serial No.:	Kind: Hydraulic Control
Make: Caterpillar	Make: Caterpillar
Model D353 Fuel: Diesel	Model: 193 9S6853
Other Data: 24 Volt direct electric	Serial No.: 28H9756
start and fuel priming pump	W/3 valves
Tires - Make	Kind: Canopy
Axle - No. - Size - Ply - Type	
1.	Make: Tube-Lok
2.	Model: COE 0401
3.	Serial No.:
4.	
5.	

FIGURE 9.2 Typical equipment record.

Other Data and Remarks: Equipped with 9S8366 Alternator, 5S8033 Fuel Tank Cap Lock, 5D1910 Hydraulic tank cap lock, 2S3025 Oil Filler Cap lock, 2S6096 Radiator cap lock, 7S7701 Fan — Reversible blade, 9M5934 Fan Blast deflector, Wiggins fuel tank filler adapter, 9S9466 Crankcase guard, 9S5646 Instrument panel guard, 7S8014 Track Roller guard, 2H9476 Front pull hook, 4S5014 Lighting system, 7S6285 Radiator Core protector grid, 8S1417 rain cap, 8S1742 24" extreme service tracks, 6D4709 seat belt & 6J9247 Push plate moldboard.

License: Year:

State and No.:

Class and Cost:

Purchased from: Wallace Machinery Company, Oxnard, California Date 6/10/70 New

Cost $88,279 Sales tax: $4,570.47 Freight: $3,133.00 Inv. No.:

Purchase Details: Included in Security Agreement & 8¾% Note with Wallace Machinery Company payable to Bank of America.

DEPRECIATION EXPENSE

Original Capitalized Amount	$88,279
Salvage Value	6,541.30
Est. Life in Months	54 months
Depr. Method	D. D. B.

Year of Life		Annual	
Year	Amt. per mo.	Year	Amt.
1	$3,269.17	1970	$19,615
2	2,542.75	1971	30,513
3	1,412.83	1972	16,954
4	743.50	1973	8,922
5	955.66	1974	5,734
6		19	
7		19	
8		19	
9		19	

ASSIGNMENT — MOVEMENT

Date	Project and Location	Job No.
6/16/70	Rincon Freeway, California	144

SOLD or DISPOSED, Details, to Whom, Price, etc.:

Table 9.2 CALCULATION OF DDB/SL DEPRECIATION FOR TRACTOR DOZER

SLY	RATE %	Book Value, End Previous Year	Depreciation	Book Value, End of This Year	Calendar Year	Expression for Calendar Year	Calendar Year Depreciation
		$	$	$			$
1	44.44	88,279.	39,231.	49,048.	1970	1/2 SLY (1)	19,615.
2	44.44	49,048.	21,797.	27,251.	1971	1/2 SLY (1)+ 1/2 SLY (2)	30,513.
3	44.44	27,251.	12,110.	15,141.	1972	1/2 SLY (2)+ 1/2 SLY (3)	16,954.
4	SL	15,141.	5,734.	9,407.	1973	1/2 SLY (3)+ 1/2 SLY (4)	8,922.
5	SL	9,407.	2,867.	6,541.	1974	1/2 SLY (4)+ 1/2 SLY (4-4.5)	5,734.
		Total	$81,738			Total	$81,738

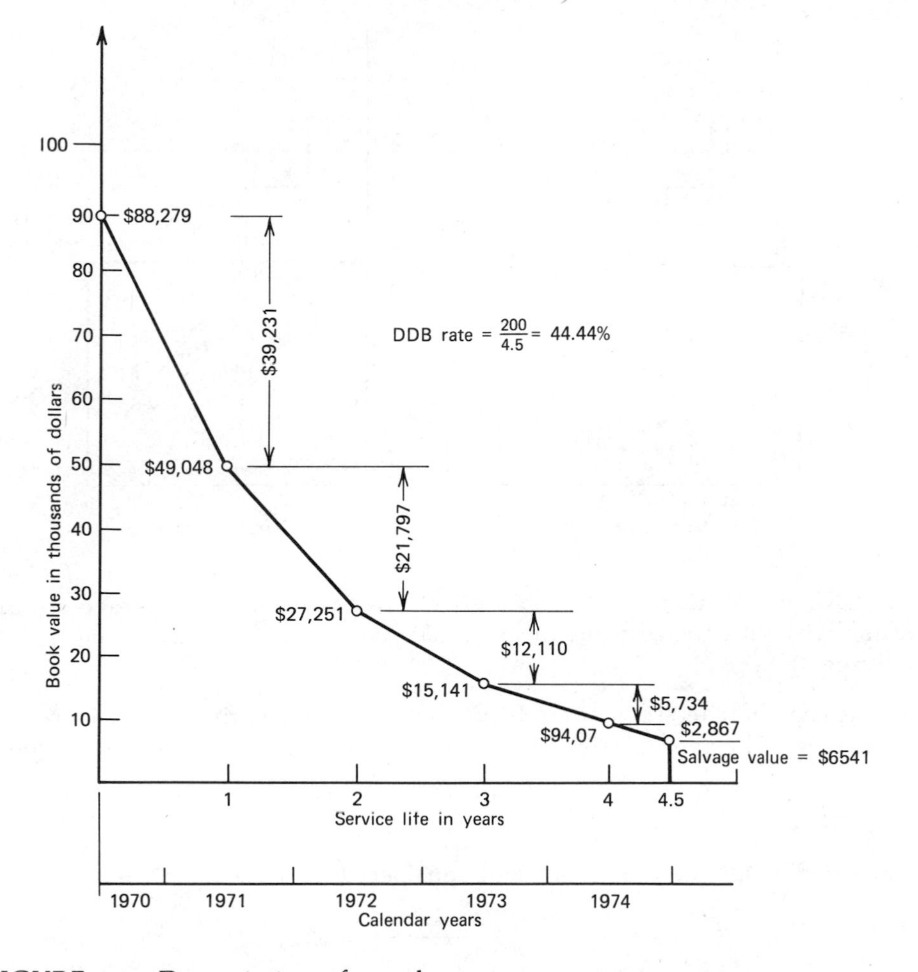

FIGURE 9.3 Depreciation of crawler tractor.

clining value of the asset over its service life so that a constant average value on a year basis is achieved. This is indicated graphically in Figure 9.4.

Applying this formula to a machine with initial capital value of $16,000 and a salvage value of $1000, the average annual value is calculated as

$$AAV = \frac{\$16{,}000 \times (6)}{10} = \$9600$$

The area under the rectangle representing the average annual value (AAV) equals the area under the stepwise figure representing the straight-line decline in value. Using this fact, the formula can be derived. If we consider the salvage value, the area under the stepped curve is increased by the area

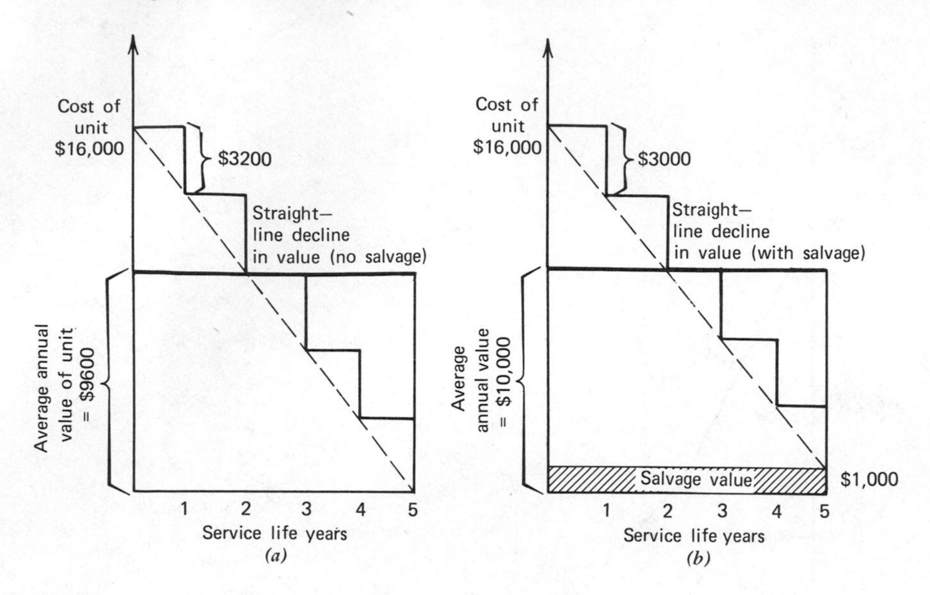

FIGURE 9.4 Interpretation of average annual value. (*a*) Average annual value without salvage value included. (*b*) Average annual value considering salvage value.

of the crosshatched segment in Figure 9.4*b*. Therefore, the AAV is increased somewhat. The appropriate expression for AAV including the salvage value is

$$\text{AAV} = \frac{C(n+1) + S(n-1)}{2n}$$

For the $16,000 piece of equipment considered, this yields

$$\text{AAV} = \frac{16000(6) + 1(4)}{10} = \$10,000$$

Verification of this expression is left as an exercise for the reader.

Assume that the proper levels of the annual provision to cover IIT costs for the unit are as follows:

Interest	=	8% of AAV
Insurance	=	3% of AAV
Taxes	=	2% of AAV
Total	=	13% of AAV

The amount to cover these ownership costs must be recovered on an hourly basis by backcharging the owner. Therefore, an estimate of the number of hours the unit will be operational each year must be made. Assume the number of hours of operation for the unit is 2000 hours/year. Then, the IIT cost per hour would be

$$\text{IIT} = \frac{0.13(\text{AAV})}{2000} = \frac{0.13(9600)}{2000} = \$0.624 \text{ or } \$0.62 \text{ per hour}$$

Manufacturers provide charts that simplify this calculation.

The interest component may be a nominal rate or an actual rate or, again, it may reflect some value of the cost of capital to the company. Some contractors also include here a charge for the protective housing or storage of the unit when it is not in use. These adjustments may raise the annual provision to cover IIT costs by from 1 to 5% with the following effect (in this case, each 1% charge may represent a 5¢ per hour increase in the charging rate).

Percentage of AAV (%)	General Provision for IIT Costs ($)	Hourly Rate (Base 2000 hours) ($)
13	1248	0.62
14	1344	0.67
15	1440	0.72
16	1536	0.77
17	1632	0.82
18	1728	0.87

Ultimately it is the competitive situation that sets practical limits to what may be recovered. That is why the tax limitation strategies discussed earlier are of such importance.

Figure 9.5 shows a chart for calculating the hourly cost of IIT. To use the chart the total percent of AAV and the estimated number of annual operating hours are required. Entering the y axis with the percent (use 13% from above) and reading down from the intersection of the 13% line with the 2000-hr slant line, the multiplier factor is 0.039. The hourly charge for IIT is calculated as

$$\text{IIT/hr} = \frac{\text{Factor} \times \text{Delivered price}}{1000}$$

In the example discussed,

$$\text{IIT/hr} = \frac{0.039 \times 16{,}000}{1000} = \$0.624 \quad \$0.62$$

If the amortization/depreciation costs using the SL method for the $16,000 unit in the example is $1.50 per hour, the owner must recover $1.50 plus $0.62 or $2.12 per hour for fixed costs.

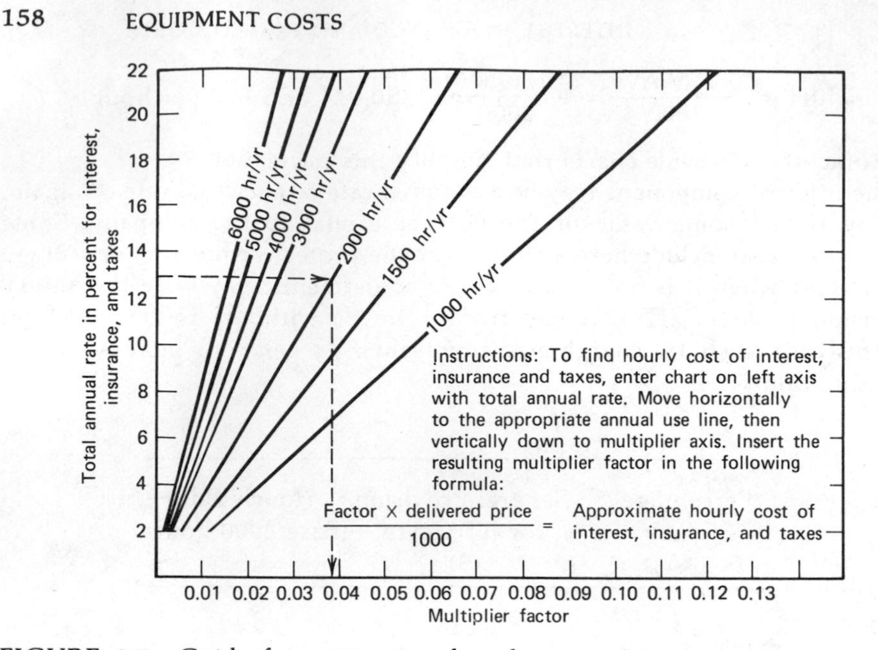

FIGURE 9.5 Guide for estimating hourly cost of interest, insurance, and taxes (Caterpillar Tractor Co.).

9.7 OPERATING COSTS

The major components contributing to the operating or variable costs are fuel, oil, grease (FOG), tire replacement (on rubber-wheeled vehicles), and normal repairs. Normally, historical records (purchase vouchers, etc.) are available that help in establishing the rate of use of consumables such as fuel, oil, and tires. Maintenance records indicate the frequency of repair. The function that best represents the repair costs to be anticipated on a unit starts low and increases over the life of the equipment. Since repairs come in discrete amounts, the function has a stepwise appearance (see Figure 9.6).

The following guidelines for establishing the amount to set aside for repairs are taken from Caterpillar Tractor material:

Guide for Estimating Hourly Repair Reserve*

To estimate hourly repair costs, select the appropriate multiplier factor from the table below and apply it in the following formula:

$$\frac{\text{Repair factor} \times (\text{Delivered price} - \text{tires})}{1000} = \text{Estimated hourly repair reserve}$$

*As given in *Fundamentals of Earthmoving*, Caterpillar Tractor Co., 1968.

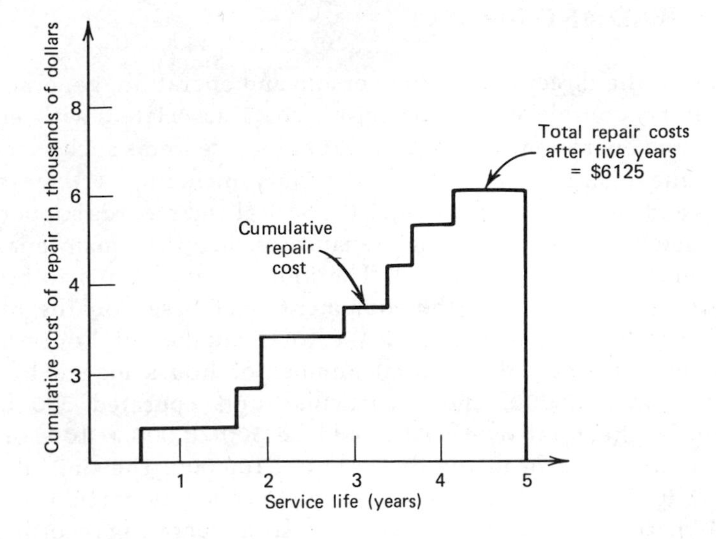

FIGURE 9.6 Repair cost profile.

	Operating Conditions		
	Excellent	Average	Severe
Track-type tractors	0.07	0.09	0.13
Wheel tractor scrapers	0.07	0.09	0.13
Off-highway trucks	0.06	0.08	0.11
Wheel-type tractors	0.04	0.06	0.09
Track-type loaders	0.07	0.09	0.13
Wheel loaders	0.04	0.06	0.09
Motor graders	0.03	0.05	0.07

The cost of tires on rubber-wheeled vehicles is prorated over a service life expressed in years or hours. Therefore, if a set of tires has an initial cost of $15,000 and a service life of 5000 hours, the hourly cost of tires set aside for replacement is

$$\text{Hourly cost of tires} = \frac{\$15{,}000}{5{,}000} = \$3.00$$

9.8 OVERHEAD AND MARKUP

In addition to the direct costs of ownership and operation, general overhead costs must be considered in recovering costs associated with equipment ownership and operation. Overhead charges include items such as the costs of operating the maintenance force and facility including: (1) wages of the mechanics and supervisory personnel, (2) clerical and records support, and (3) rental or amortization of the maintenance facility (i.e., maintenance bays, lifts, machinery, and instruments). The industry practice is to prorate the total charge to each unit in the equipment fleet based on the number of hours it operates as a fraction of the total number of hours logged by the fleet. For instance, if the total number of hours logged by all units in the fleet was 20,000, and a particular unit operated 500 hours, its proportion of the total overhead would be 500/20,000 × 100, or 2.5%. If the total cost of overhead for the year is $100,000, the unit above must recover $2,500 in backcharge to the client to cover its portion of the overhead. Overhead rates are updated annually from operating records to ensure adequate coverage. If overhead costs overrun projections, the coverage will be inadequate and the overrun will reduce profits.

The last component of the total charge associated with a unit of production is the profit or markup expressed as a percentage of total hourly operating costs which, in turn, may be expressed in cubic yards of material moved or in some other bid-relevant measure. The amount of markup per cubic yard, square foot, or linear foot is a judgment that contractors must make based on their desire to win the contract and the nature of the competition. In a "tight" market where competition is strong, the allowable margin of profit that still allows the bidder to be competitive may be only 1 or 2%. In a "fat" market, where a lot of jobs are available, the demand is greater and the client is ready to pay a higher markup to get the work under way. Competition is also bidding higher profit so the markup can be adjusted upward.

Bidding strategy will include attention to the concept of marginal costs, which may permit the acceptance of jobs yielding less than the desired rate of return. In general, bidding, based on margins as low as 1 or 2%, bearing in mind that the whole bid is a matter of estimates, is uncomfortably close to what one might call the disaster area; the area of operating *losses*.

REVIEW QUESTIONS AND EXERCISES

9.1 What are the major cost components that must be considered when costing out a piece of equipment? How can a contractor manipulate amortization for a piece of equipment in order to increase or reduce direct costs charged per unit of production? Why are tires on a rubber-tired vehicle not considered for depreciation?

9.2 You have just bought a new pusher dozer for your equipment fleet. Its cost is $100,000. It has an estimated service life of 4 years. Its salvage value is $8000.

(a) Calculate the depreciation for the first and second year using the straight-line, double-declining-balance, and the sum-of-the-years-digits methods.

(b) The tax, interest, and insurance components of ownership cost based on average annual value are:

Tax:	2%
Insurance:	2%
Interest:	7%

What cost per hour of operation would you charge to cover interest, tax, and insurance?

9.3 You have just bought a used track-type tractor to add to your production fleet. The initial capitalized value of the tractor is $80,000. The estimated service life remaining on the tractor is 9000 hours, and the anticipated operating conditions across the remainder of its life are normal. The salvage value of the tractor is $8,000.00.

You have decided to use a composite method of depreciation. During the first three years, you use the declining balance method of depreciation. Since the tractor is used, the maximum rate of depreciation that can be used is 1½ times the straight-line rate. After the first three years of depreciation, the straight-line method will be used. The tractor was purchased on 1 July 1972.

(a) What amount of depreciation will you claim for each calendar year during the period 1972-1975?

(b) What percent of the total depreciable amount is taken in the first year?

(c) The tax, interest, and insurance components of ownership cost based on average annual value are:

Tax:	3%
Insurance:	2%
Interest:	8%

What cost per hour of operation would you charge to cover interest, tax, and insurance?

(d) If the total average operating cost for the tractor is $13.50 per hour and the amount of overhead cost prorated to this tractor for the year is $4000.00, what would your total hourly cost for the operation of the tractor be (during the first year of its service life)?

9.4 You have purchased a scraper with the following characteristics:

(a) Initial cost = $70,000
(b) Operating cost = $20,000/year
(c) Tire cost @ 2 years = $13,000
(d) Major overhaul in third year = $15,000
(e) Salvage value = $5,000

The unit has a service life of four years. How much must you recover each year if you require i = 10% on your investment?

9.5 On the form shown below, determine the hourly operating and owning costs for a track-type tractor fitted with blade and rippers operating under severe conditions. The necessary information has been filled in (problem as described in *Fundamentals of Earthmoving*, Caterpillar Tractor Co., 1968).

9.6 Determine appropriate values for the owning and operating costs of your automobile.

9.7 How would you decide whether a piece of equipment has become obsolescent? If possible give several examples of equipment items that have been superseded.

9.8 Determine the current rental rates and salvage values of two equipment items in regular use in your vicinity. *Hint.* Consult the Green Book or that issued by the Associated Equipment Dealers.

9.9 Visit a local contractor and ascertain the extent of the maintenance and repair work on a major equipment item such as a bulldozer, crane. or material unit. What sort of records are being kept on each plant item?

HOURLY OWNING AND OPERATING COST ESTIMATE

Machine Designation—Track Type Tractor

DEPRECIATION VALUE

1.	Delivered Price (including attachments)	$45,543
2.	Less Tire Replacement Costs: Front ________ Drive ________ Rear ________	-----
3.	Delivered Price Less Tires	-----
4.	(Optional) Less Resale Value or Trade-In	$ 8,000
5.	NET VALUE FOR DEPRECIATION	

OWNING COSTS

6. Depreciation: $\frac{\text{NET DEPRECIATION VALUE (ITEM 5)}}{\text{DEPRECIATION PERIOD IN HOURS}}$

$\frac{\text{VALUE}}{\text{HOURS}}$: __________ __________

7. Interest, Insurance, Taxes:
   Annual Rates: Int. 7% Ins. 2% Taxes 2%
   Estimated Annual Use in Hours 2000 hr

$$\frac{\text{Factor} \times \text{Delivered Price (Item 1)}}{1000}$$

$$\frac{\quad \times \quad}{1000}$$ ______

8. TOTAL HOURLY OWNING COSTS ______

OPERATING COSTS

		Unit Price × Consumption	
9.	Fuel:	$0.17/gal × 6.7 gal/hr	______

10. Lubricants, Filters, Grease:

	Unit Price	×	Consumption	
Engine	1.25/gal	×	0.04 gal/hr	______
Transmission	1.25/gal	×	0.03 gal/hr	______
Final Drives	1.40/gal	×	0.02 gal/hr	______
Hydraulics	1.25/gal	×	0.02 gal/hr	______
Grease	0.25/lb	×	0.05 lb/hr	______
Filters	______	×	______	$ 0.17

11. Tires: $\frac{\text{REPLACEMENT COST}}{\text{ESTIMATED LIFE IN HOURS}}$

$\frac{\text{COST}}{\text{LIFE}}$: ______ ______

12. Repairs: $\frac{\text{Factor} \times \text{Del. Price Less Tires}}{1000}$

$$\frac{\quad \times \quad}{1000}$$ ______

13. Special Items:
   Ripper Tips (3 @ $21.00) Life 80 hr
   Shank Protectors (3 @ $40.00) Life 400 hr ______

14. TOTAL HOURLY OPERATING COSTS $ ______

15. Operator's Hourly Wage $ 3.90

16. TOTAL HOURLY OWNING AND OPERATING COSTS ______

CHAPTER TEN

Depreciation

10.1 METHODS OF DEPRECIATION

The method by which depreciation is calculated for tax purposes must conform to standards established by the Internal Revenue Service (IRS). The four most commonly used methods are:

1. Straight line.
2. Declining balance.
3. Sum of years digits.
4. Production.

All of these methods are approved by the IRS. Declining balance and sum of years digits are referred to as accelerated methods, since they allow larger amounts of depreciation to be taken in the early years of the life of the asset. The contractor usually selects a method that offsets or reduces the reported profit for tax purposes as much as possible. If taxable profits are anticipated to be large in a given year, then the company will try to take as much depreciation as possible and will use an accelerated method on as many asset items as possible. Most heavy construction contractors assume that each machine in the fleet is a small "profit center" and therefore attempt to utilize the depreciation available on each piece to reduce or alter the incidence of the taxable profit or income it will generate. Each individual asset may be depreciated separately under IRS rules, which support industry practice in this area. The contractor may establish as many accounts as desired. He can use a separate account for each item, or two or more items may be included in one account. Asset items may also be grouped, by common useful lives or common uses. In a similar approach, a composite account may be used without special regard for the character of the depreciable assets involved.

For a variety of reasons as we will see later, the production method of calculating depreciation is commonly used by heavy construction contractors. One thing that limits the number of depreciation strategies available to the firm is the IRS stipulation affecting any change in methods applied to a given asset during its life. It is possible to make one change from an accelerated to a straight-line method. It is not possible to then change from a straight-line to an accelerated method.

The major factors to be considered in calculating the depreciation of an asset are shown in Figure 10.1 The three major factors form the three sides of the depreciation "box" that are linked by the method of depreciation selected. They are:

1. Initial cost or basis in dollars.
2. Service life in years or hours.
3. Salvage value in dollars.

The amount that can be depreciated or claimed by way of a tax deduction is the difference between the initial net value of the asset and its residual or salvage value. This is referred to as the *depreciable amount* and establishes the maximum number of depreciation dollars available in the asset during its service life.

The declared initial cost of the asset must be acceptable in terms of the IRS definition of depreciable cost. For instance, suppose a $75,000 scraper is purchased. The tires on the scraper cost $15,000. These tires are considered a current period expense and therefore are not depreciable. That is, they are

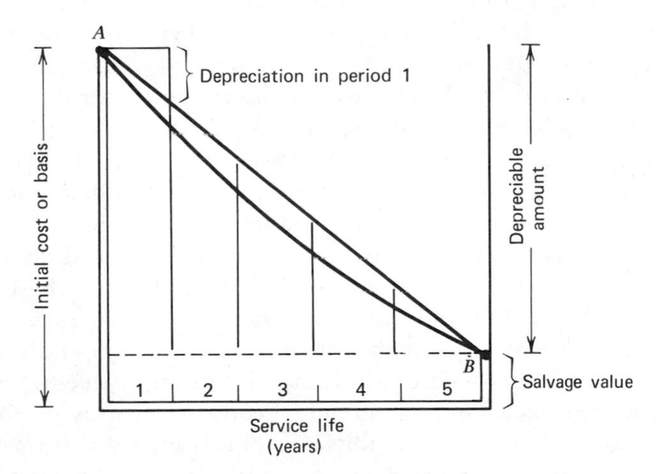

FIGURE 10.1 Factors in depreciation.

not part of the capital asset for purposes of depreciation just as a typewriter ribbon is not a depreciable portion of a typewriter but is a current period expense. The initial cost of the scraper for depreciation purposes is $60,000. The initial depreciable cost or basis is often referred to as the net first cost. In addition to the purchase price minus major expense items such as tires, items such as freight costs and taxes *are* included in the net first cost and are part of the amount depreciable. If we have purchased a Caterpillar 824B Wheel Tractor (300 hp) Power Shift, the net first cost for purposes of depreciation would be arrived at as follows:

Purchase price:		$54,800	(FOB Peoria, Illinois)
Less tires	−	9,400	
		$45,400	
Tax @ 5% of asset		2,270	
Freight	+	1,330	
Net first cost		$49,000	

The depreciable basis for the calculation of depreciation allowances is this first cost of $49,000. In the application of some methods of depreciation, an estimated salvage value must be deducted where this exceeds 10% of the base value. In the most general case, the depreciation allowance for any tax year is limited to so much as is necessary to recover the remaining cost or other basis, less salvage value, during the remaining useful life of the property. However calculated, by whatever method, an asset may not be depreciated below a reasonable salvage value (with the exception already noted).

As mentioned previously the major equipment manufacturers publish tables (see, for example, Table 9.1) indicating the typical service life of a unit given defined operating conditions. These tables are available from manufacturers in Equipment Handbooks, which provide operational and performance characteristics relating to equipment units. Typically, equipment units have a service life of five years or 10,000 hours. However, units operating under severe conditions such as those encountered in continuous subzero temperature zones (the construction of the Alaska pipeline) or in the tropics may have service lives of only two or three years. Obviously, the more compressed the service life, the larger the depreciation deduction in those years as an offset to the income generated by that equipment.

Each asset has a residual or scrap value that is referred to as the salvage value. The concept is that, toward the end of its useful life, the value of the asset will not decline below this residual value. In the case of machines, the residual value is normally assumed to be the sale value or trade-in value at the end of its normal service life or, ultimately, a scrap value, if the machine is sold as scrap for recycling. Equipment vendors can provide information concerning a reasonable and acceptable level of salvage value for a given piece of equipment. There is a tendency to use a zero value for salvage because of the provision in the Internal Revenue Code that salvage value up to a 10% level may be ignored for tax depreciation purposes. A reasonable salvage value if it exists must be recognized.

The various methods of depreciation allow recovery of the depreciable amount over the service life of the asset. That is, the method selected allows transit along one of the paths from *A* to *B* as shown in Figure 10.1.

10.2 THE STRAIGHT-LINE METHOD

The accountant (and the IRS) would describe the straight-line method of calculating allowable depreciation as being based on the assumption that the depreciation, or the loss in value through use, is uniform during the useful life of the property. In other words, the net first cost or other basis for the calculation, less the estimated salvage value, is deductible in equal annual amounts over the estimated useful life of the equipment. The engineer would call this a linear method. This simply means that the depreciable amount is linearly prorated or distributed over the service life of the asset. Let us assume that we have a piece of equipment that has an initial cost or base value of $16,000 and a salvage value of $1,000. The service life is five years and the depreciable amount is $15,000 (initial cost minus salvage value). If we linearly distribute the $15,000 over the five-year service life (i.e., take equal amounts each year), we are using the straight-line method of depreciation. The amount of depreciation claimed each year is $3000. This is illustrated in Figure 10.2.

The remaining value of the piece of equipment for depreciation purposes can be determined by consulting the stepwise curve of declining value. During the

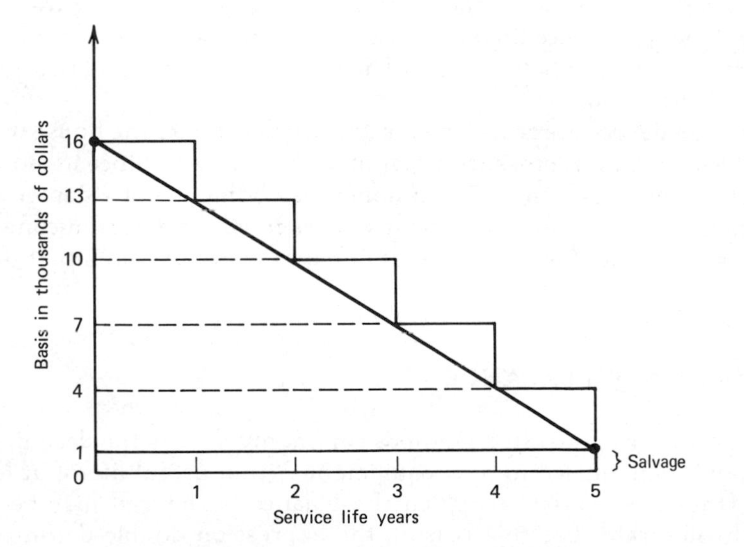

FIGURE 10.2 Straight-line depreciation.

third year of the asset's service life, for example, the remaining base value or *book value* of the asset is $10,000. If we connect the points representing the book value at the end of each year (following subtraction of the depreciation), we have the "straightline."

The concept of the base value or book value has further tax implications. For instance, if we sell this asset in the third year for $13,000 we are receiving more from the buyer than the book value of $10,000. We are gaining $3000 more than the depreciated book value of the asset. The $3000 constitutes a *capital gain*. The reasoning is that we have claimed depreciation up to this point of $6000 and we have declared that as part of the cost of doing business. Now the market has allowed us to sell at $3000 over the previously declared value, demonstrating that the depreciation was actually less than was claimed. We have profited and, therefore, have received taxable income. Since this is treated as a capital gain rather than as a business profit, we are taxed, not at the full rate (17 up to 46% depending on total net taxable income), but at the alternative rate of 30%. Current corporation income tax rates include a basic rate of 17% on taxable incomes up to $25,000 and an increase up to 46% for taxable incomes above $100,000. The alternative capital gain tax rate on corporations is 30%, which compares with the overall rate on taxable incomes over $100,000 of 46%, a difference which represents a substantial saving.

The base value for depreciation is affected if we modify substantially the piece of equipment. Assume in the above example, that in the third year we perform a major overhaul ($2000) and modification ($3000) on the engine of the machine at a total cost of $5000. Since this is in part a capital improvement, the term *basis* is used to refer to the depreciation base. The modification increases the base value of the unit by $3000 as shown in Figure 10.3. It also may extend the service life of the asset. Something similar occurs if we make some improvements to a building. The value is increased and this added value can be depreciated.

If we can depreciate real property, can we depreciate the house in which we live? Depreciation represents a cost of doing business. Since in most cases we do not "do business" in our own home, our home is not a depreciable asset. You can, however, think of some instances in which a person conducts some business at home. Special depreciation rules apply to that situation.

10.3 DECLINING BALANCE

The first of the accelerated methods commonly used is the declining-balance method. When applied to new equipment with a useful life of at least three years, the effective *rate* at which the balance is reduced may be twice the straight-line rate. For this reason, the expression double-declining balance (DDB) is used when this IRS option is applied to new assets. For assets that

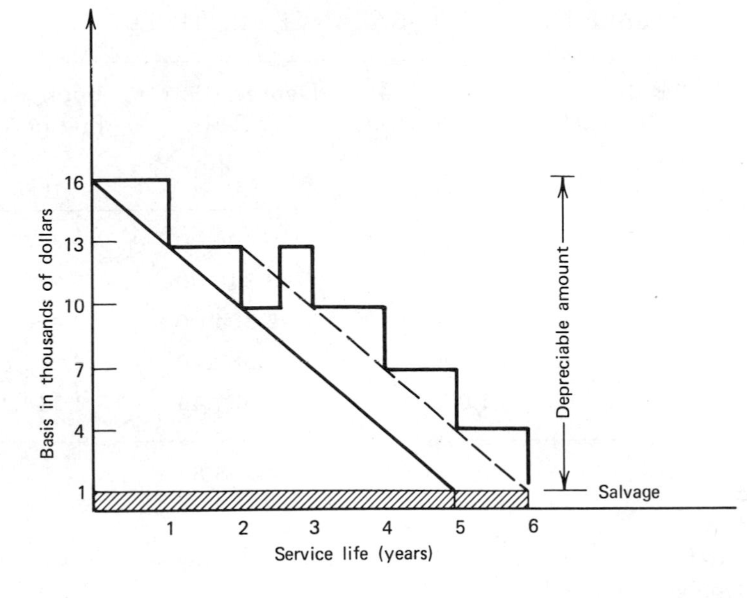

FIGURE 10.3 Adjustment of basis.

are not purchased new but are secondhand, the optional rate is 150% of the straight-line rate.* In this method, it is the rate that is important, since it remains constant throughout the calculations. Formally stated, in the declining-balance method, the amount of depreciation claimed in the previous year is subtracted from the book value (base value) at the beginning of the previous year before computing the next year's depreciation. Once the balance of the property's value has been reduced in this way, a constant rate, as described above, is applied to the reduced book value to get the depreciation for the next year. For new equipment the rate is calculated by dividing 200% by the number of service years — 200/SLY. For used equipment the rate is 150% divided by the service life years.

To illustrate, consider the $16,000 piece of equipment used in discussing the straight-line method. We will assume the piece is purchased new at this price. Since the service life of the unit is five years, the constant rate to be applied will be 200%/5 = 40%. The calculations for this example are summarized in Table 10.1.

A repetitive process of calculation can be detected. The constant rate of 40% (column 2) is applied to the book value at the end of the previous year (column 3) to obtain the depreciation (column 4). The reduced value of the property is column 3 - column 4, as shown in column 5. The "Book Value End of This Year" for year N is the "Book Value end of Previous Year" for year $N+1$.

*The assumption is that the straight-line rate represents 100%.

Table 10.1 DOUBLE-DECLINING-BALANCE METHOD

SLY	Rate Applied to Balance (%)	Book Value End of Previous Year ($)	Depreciation for This Year ($)	Book Value End of This Year ($)
1	40	16,000	6,400.00	9,600
2	40	9,600	3,840.00	5,760
3	40	5,760	2,304.00	3,456
4	40	3,456	1,382.40	2,073.60
5	40	2,073.60	829.44	1,244.16
		Total:	$14,755.84	

It follows that the value in column 3 for year 2 will be the same as the value in column 5 for year 1.

Another interesting fact will be noted. The amount of depreciation taken over the five-year service life is less than the depreciable amount. The book value at the end of five years is $1244.16 and the salvage is $1000. Therefore, $244.16 has not been recovered. Typically, the method is changed to the straight-line approach in the fourth or fifth year to ensure closure on the salvage value. This underlines the fact that the only role played by the depreciable value in the declining-balance method is to set an upper limit on the amount of depreciation that can be recovered, that is, an asset may not be depreciated below a reasonable salvage value. A common mistake is to apply the rate to the *depreciable value* in the first year. The rate is always applied to the *total* remaining book value, which in this example during the first year is $16,000.

If the piece of equipment had been purchased used, for $16,000, the procedure would be the same but the rate would be reduced. In this case, the rate would be 150%/5 or 30%. The 150% calculations are summarized in Table 10.2. In this situation, since $1,689.12 in unclaimed depreciation would remain at the end of year 5, the method could be changed to the straight-line approach in the fourth or fifth year with some advantage. A comparison of the double-declining-balance methods and the straight-line method is shown in Figure 10.4.

The proportionately higher rate of recovery in the early service life years is revealed by this figure. More depreciation is available in the first year using the double-declining-balance method ($6400) than in the first two years using the

Table 10.2 150 DECLINING-BALANCE METHOD

SLY	Rate (%)	Book Value End of Previous Year ($)	Depreciation for This Year ($)	Book Value End of This Year ($)
1	30	16,000	4,800	11,200
2	30	11,200	3,360	7,840
3	30	7,840	2,352	5,488
4	30	5,488	1,646.40	3,841.60
5	30	3,841.60	1,152.48	2,689.12
		Total	$13,310.88	

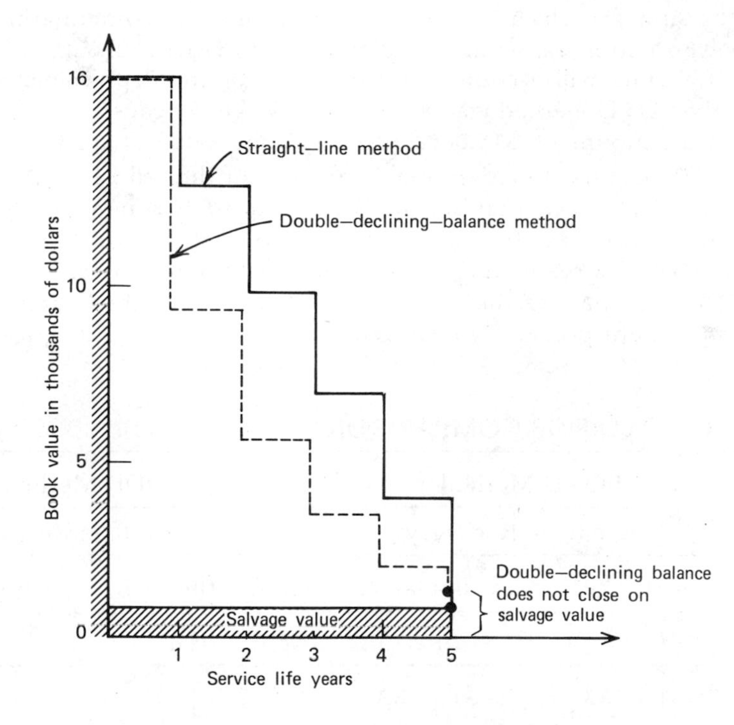

FIGURE 10.4 Comparison of double-declining-balance and straight-line methods.

straight-line method ($6000). Equipment rental firms that intend to sell the equipment after the first two years of ownership are in a good position to capitalize on this feature of the accelerated methods. Of course, if they sell at a price well above the book value, they must consider the impact of the capital gains tax.

10.4 SUM-OF-THE-YEARS-DIGITS (SOYD) METHOD

The sum-of-the-years-digits method is an alternate accelerated method of capital recovery through depreciation where the rate of recovery differs from the DDB method. Under the DDB method the first year's claim is larger than under the SOYD method, but in subsequent or later years not only does it fall behind the SOYD rate of recovery but also it fails to close on the salvage value. This is best illustrated in Table 10.3.

For the application of the SOYD method, the years' digits, 1 to 5, are added to give the sum 15, which then becomes the constant denominator for the fractions whose numerators are the digits of each year: 1, 2, 3,..., n. The sum of these fractions will account for the whole of the depreciable amount.

Under the SOYD method we have shown two percengages: %(*i*) deals with a depreciable amount of $15,000 with a salvage value of 6.25% or $1000 excluded; %(*ii*) deals with a depreciable amount of the full $16,000. The IRS code provides, in effect, that salvage values up to 10% may be ignored in depreciation calculations for tax purposes, although an asset may not be depreciated below a reasonable salvage value for tax purposes.

Table 10.4 summarizes the calculations for SOYD method applied to the piece of equipment previously analyzed.

Table 10.3 SOYD/DDB COMPARISON

Year	SOYD Method				DDB Method	
	Percentage Recovery				Percentage Recovery	
	For the Year		Cumulative		For the Year	Cumulative
	% (*i*)	% (*ii*)	% (*i*)	% (*ii*)	(%)	(%)
1	31	33	31	33	40	40
2	25	27	56	60	24	64
3	19	20	75	80	14	78
4	13	13	88	93	9	87
5	6	7	94	100	5	92

Table 10.4 SOYD METHOD CALCULATIONS

$$\sum_{i=1}^{n} i = \frac{n(n+1)}{2} = \frac{5(6)}{2} = 15$$

Denominator = 5 + 4 + 3 + 2 + 1 = 15

SLY	Fraction	Depreciable Amount ($)	Depreciation ($)	Book Value Beginning of Year ($)	Book Value End of Year ($)
1	5/15	15,000	5,000	16,000	11,000
1	4/15	15,000	4,000	11,000	7,000
3	3/15	15,000	3,000	7,000	4,000
4	2/15	15,000	2,000	4,000	2,000
5	1/15	15,000	1,000	2,000	1,000

Total $15,000

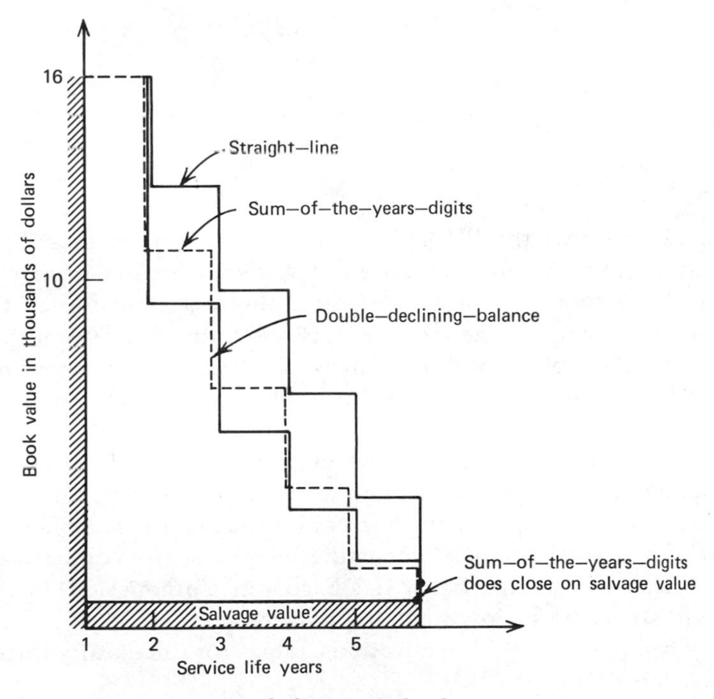

FIGURE 10.5 Comparison of three methods.

In Table 10.4, the calculations have been made as if the salvage value of $1000, 6.25% the base value of $16,000, had to be deducted; this would have resulted if it exceeded $1600 or 10% of the base value. This approach serves to illustrate more clearly the difference between these methods.

Figure 10.5 shows a comparison of SL, DDB, and SOYD methods. It is apparent that where salvage values have to be allowed for, the SOYD method is slightly "less accelerated" than DDB, but offers a much quicker rate of recovery than the SL method.

As long as the equipment unit or item of property has a service life in years that can be expressed as an integer (e.g. 3, 4, 5), the calculation of the fraction to be used with the depreciable amount is very straightforward. If the service life is 3.5 or 4.5 years, the determination of the fraction is more difficult. If the service life is 3.5 years, the depreciation for each service life year is as follows:

SLY	Fraction	Depreciation ($)
1	3.5/8 = 7/16	6562.50
2	5/16	4687.50
3	3/16	2812.50
3.0 - 3.5	1/16	937.50
	Total	$15000.00

We should note that the IRS-approved methods for depreciating building as real estate are not the same as those applicable to personal property such as equipment. In general, the only accelerated method allowing for these real estate additions or improvements is the 150% declining-balance method or any other consistently applied method, approved by the IRS, that does not give a greater allowance in the first two-thirds of its useful life than the 150% DB method.

Once again, however, it should be emphasized that IRS rules are not controlling: outside the tax assessment arena, the contractor may apply any method he chooses in any way which suits him and makes good business sense. Of course, for formal financial reporting purposes, the contractor may be bound by certain rules (e.g., Securities Exchange Commission) but he is still quite free to structure his own bidding policies.

For more complex service life durations, tables for calculating the depreciation are available from the IRS.

10.5 PRODUCTION METHOD

It was stated earlier that the contractor tries to claim depreciation on a given unit of equipment at the same time the equipment is generating profit in order to reduce the tax that might otherwise be payable. The production method allows this, since the depreciation is taken based on the number of hours the unit was in production or use for a given year. The asset's cost is prorated and recovered on a per-unit-of-output basis. If the $16,000 equipment unit we have been discussing has a 10,000-hour operation time, the $15,000 depreciable amount is prorated over 10,000 hours of productive service life. This method is popular with smaller contractors, since it is easy to calculate and ensures that the depreciation available from the asset will be recovered at the same time the unit is generating profit. A reasonable estimate of the total operating hours for a piece of equipment may be obtained by referring to the odometer on the unit together with the logbook or job cards.

In some cases, unless this method is used, the units may be depreciated during a period when they are not generating income and, consequently, the full benefit of the depreciation deduction may be lost. The objective of the contractor is to have a depreciation deduction available in years in which it can be more effectively applied to reduce taxable income. It may not be possible to defer depreciation and to take it in years in which it can be applied with more advantage. Therefore, the strategy should be to have it available in the years in which profits are likely to be high. The production method ensures that depreciation deduction is available when the machine is productive, and theoretically profitable or income-producing.

At the time of construction of the Alaska pipeline, contractors with contracts for the access road to parallel the pipeline purchased large equipment fleets in anticipation of project start-up. Then, environmental groups delayed the project several years during which time the contractors were forced to put their equipment fleets in mothballs. Since these units were not in use and were not productive and profitable during this delay period, the contractors claimed no depreciation. Nevertheless, the production method allowed them to apply the depreciation at the proper time when the job mobilized and the units were put into production.

In some situations, the production method might be less desirable. If we own an entrenching machine (service life of 10,000 hours), but only operate it 500 hours per year, using the production method would stretch the period of recovery out over 20 years. If the machine is sold after five years, we would have claimed only one-quarter of the available depreciation. One advantage offsetting this apparent disadvantage is that we might have a smaller adjustment to make by way of capital gain on the sale. In such a case, clearly a method other than the production method would be more appropriate and more balanced way of dealing with the matter.

10.6 PERSPECTIVE ON DEPRECIATION

The capital sum expended on the purchase of equipment involves the purchase of a finite number of equipment service hours. When a contractor *hires* equipment the cost of those service hours, consumed within the accounting period, is an allowable deduction from gross income along with other legitimate business expenses. The provision made for the deduction of depreciation charges, by whatever approved method, places the lump sum purchase of equipment service hours on a similar footing and allows the capital sum to be replaced. It also allows the in-period hiring charges to be offset against income derived from the billing of service charges to customers or owners.

There is some degree of flexibility in the manner in which these claims may be made for allowable deductions from gross income in arriving at any year's taxable income or profit. It has been noted that some economic and monetary advantage can be derived from the way in which we choose to organize these deductions both in amount and timing. There are, however, practical limitations. The total amount that may be claimed is limited to the cost or other approved basis less the reasonable salvage value, and of course an asset may not be depreciated below the salvage value. There are also limitations on the degree of acceleration or the earlier timing of these deductions. First, however, there is a bonus first-year depreciation allowance applicable to either new or used property, which is calculated at the rate of 20% of the net first cost, up to a total allowance of $10,000 for any one taxpayer, with respect to personal property having a remaining life of six years. Then there is the provision that while the straight-line method is applicable to *any* depreciable property, the other accelerated methods are applicable only to equipment having a useful life of three years or more and being new in use. For used or secondhand property, 150% of the straight-line rate may be used in the manner illustrated.

In addition to the methods described here, for tax purposes a taxpayer may use any other consistent method such as the set of compound interest methods (including sinking fund and annuity methods) *provided* that the total deductions during the first two-thirds of the useful life of the asset do not exceed the total allowable under the declining-balance method. There is a further general concession applicable to depreciable personal property having a useful life of three years or more. Salvage values of less than 10% of the cost or other basis for depreciation may be ignored and need not be taken into account in the calculation of depreciation. That is, a taxpayer may reduce the salvage value taken into account by up to 10% of the basis or base value of the property.

The general drift of the whole range of provisions made in the IRS code relating to depreciation is to achieve a rough equality between taxpayers who hire equipment and are allowed a full period-by-period allowance of that expense and other taxpayers who, for a variety of reasons, prefer to own equipment and in the process purchase (for example) five years of service hours in the one year. Similar, but less successful attempts, are made in relation to real property; to buildings owned, as opposed to those that are leased or rented. However, the failure of the law to achieve equity in this area is matched by the ingenuity of taxpayers in devising appropriate accounting and tax strategies.

Leaving the world of tax deductions and returning to the reality of the construction industry, it should be noted that for the purposes of accounting and reporting to shareholders or bondholders, consistent and generally accepted accounting principles should be followed. These may well differ from some aspects of the IRS code provisions and also from what the contractor may consider acceptable industry practice. It makes no difference to the accountant or the IRS what the contractor may charge to and recover from customers or owners. He may bill additional amounts to cover higher replacement costs, or more rapid obsolescence. It is of no concern to the accountant or to the IRS that the contractor may continue to include hourly charges in his bids for equipment long since written down fully for tax purposes. If the contractor can make additional income from the use of owned equipment beyond the recovery of the original capital outlay plus profit margins, that is simply a normal part of business income or profit-making and ought to be seen as quite unremarkable in a profit-oriented private enterprise system. The marketplace has its own way of cutting back excess gains. The forces of normal competition within the industry place a break on excessive recovery.

Throughout this discussion of depreciation methods we have been using the concept of the estimated useful life. It is worth noting that, since 1970, taxpayers have had the option of adopting the Class Life ADR system; ADR stands for asset depreciation range. There is much less likelihood of agent-taxpayer disputation of whether the reasonableness of the estimated life of a class of assets being depreciated is in line with the ADR guidelines. The ADR guidelines for depreciable assets used in general contract construction suggest a useful life of five years with lower and upper limits of four and six years, respectively, and with a guideline repair allowance of 12.5%.

10.7 CURRENT VALUE ON THE DEPRECIATION BASE

In discussions thus far, the net first cost, on an historical basis, has been related to the concept of the estimated useful life. It has been noted that some attention must be given to adjusting an estimate of service life once it becomes clear that equipment has been superseded by technologically and economically superior later models. It is for this reason that attention has been turned to employing current values or current replacement costs as the depreciation bases in place of the lower first cost on an historical basis.

This involves a shift from the concept of cost recovery based on original capital outlays to a concept of capital maintenance, whereby sufficient value is recovered to maintain the quality of equipment at current prices. This might involve an increase in the hourly charge on the order of 20 to 25 cents or a 10 to 12% surcharge. Unless it became an industry-wide practice, an increase of this magnitude might not be acceptable to either owners or contractors. The cost of a unit of equipment can thus be recovered two or three times over. It is a case of what the market will bear! If at the end of five years the first cost of a unit has been recovered, a backcharge for its use can still be charged. A charge for the use of the unit can still be included and achieve what amounts to an additional component of profit. The machine may remain in service for 10 or even 15 years and, depending on the competition, the backcharge can be left in the bid and in the unit price of production. As a practical matter, this backcharge may be manipulated throughout the life of the unit to allow the contractor to maintain a competitive edge in tight bidding situations. In a market with heavy job volume, the contractor inserts the backcharge. In a tight market, he can omit it.

REVIEW QUESTIONS AND EXERCISES

10.1 What is the difference between depreciation and amortization?

10.2 Under what circumstances would you prefer the production method of depreciation?

10.3 A forklift has an initial cost of $40,000. It is estimated that its useful life is four years. It is also estimated that it will have a salvage value of $10,000. Calculate the depreciation for the first and second year using the straight-line method, the sum-of-years-digits-method, and the double-declining-balance method.

10.4 You have just purchased a 100-ton crane unit for $65,000. The crane will have a resale value of $15,000 five years from now. The annual operating and maintenance cost for the crane will be $16,000. A major overhaul requiring $8000 will be required at the end of three years. You will sell the crane after five years. How much would you have to recover annually to have a 15% return on the investment?

CHAPTER ELEVEN

Equipment Productivity

11.1 PRODUCTIVITY CONCEPTS

Now that a basis for charging each unit of production has been established, the rate of production or the number of productive units that can be generated per hour, per day, or over another period of time must be considered. Our discussions here will be limited primarily to heavy construction units such as haulers, graders, and dozers. The concepts developed, however, are applicable to all types of construction equipment performing basically repetitive or cyclic operations. The cycle of an equipment piece is the sequence of tasks, which is repeated to produce a unit of output (e.g., a cubic yard, a trip load, etc.).

There are two characteristics of the machine and the cycle that dictate the rate of output. The first of these is the cyclic *capacity* of the machine or equipment, which establishes the number of units produced per cycle. The second is the cyclic *rate* or *speed* of an equipment piece. A truck, for instance, with a capacity of 16 cu yd, can be viewed as producing 16 yd each time it hauls. The question of capacity is a function of the size of the machine, the state of the material that is to be processed, and the unit to be used in measurement. A hauler such as a scraper pan usually has a rated capacity, "struck," versus its "heaped" capacity. The bowl of the scraper can be filled level (struck), yielding one capacity, or can be filled above the top to a heaped capacity. In both cases, the earth hauled tends to take on air voids and bulks, yielding a different weight per unit volume than it had in the ground when excavated (i.e., its *in situ* location). The material has a third weight-to-volume ratio when it is placed in its construction location (e.g., a road fill or, an airport runway) and is compacted to its final density. This leads to three types of measure: (1) bank cubic yards cu yd (bank) (in-situ volume), (2) loose cubic yards cu yd (loose), and (3) compacted cubic yards. Payment in the contract is usually based on the placed earth construction, so that the "pay" unit is the final compacted cubic yard. The relationship between these three measures is shown in Figure 11.1.

The relationship between the bulk or loose volume and the bank volume is defined by the percent swell. In Figure 11.1, the percent swell is 30%. Percent swell is given as

$$\text{Percent swell} = \left(\frac{1}{\text{Load factor}} - 1\right) \times 100$$

where

$$\text{Load factor} = \frac{\text{Pounds per cubic yard—loose}}{\text{Pounds per cubic yard—bank}}$$

Tables such as Table 11.1 give the load factor for various types of materials indicating their propensity for taking on air voids in the loose state. Each material has its own characteristic load factor. In the example above, the material has a load factor of 0.77. Therefore,

$$\text{Percent swell} = \left(\frac{1}{0.77} - 1\right) \times 100 = 30\%$$

Therefore, we would expect 10 yd of bank material to expand to 13 yd during transport. The shrinkage factor relates the volume of the compacted material to the volume of the bank material. In the example, the shrinkage factor is 10%, since the bank cubic yard is reduced by 10% in volume in the compacted state.

In order to understand the importance of capacity, consider the following situation. A front-end loader has an output of 200 *bank* cu yd of common earth per hour. It loads a fleet of 4 trucks (capacity 18 *loose* cu yd each), which haul the earth to a fill where it is compacted with a shrinkage factor of 10%. Each truck has a total cycle time of 15 minutes, assuming it does not have to wait in line to be loaded. The earth has a percent swell of 20%. The job requires a volume of 18,000 compacted cu yd. How many hours will be re-

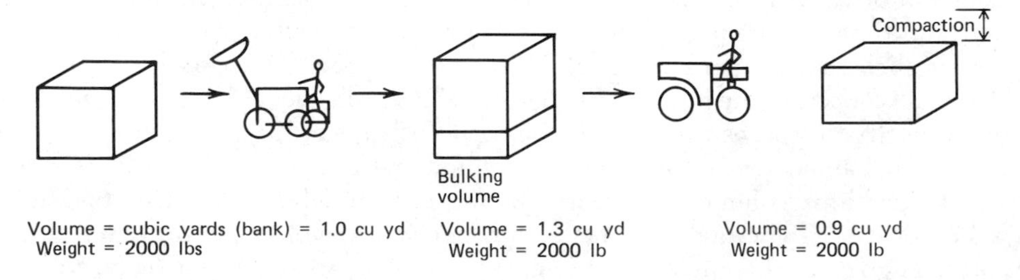

FIGURE 11.1 Volume relationships.

Table 11.1 APPROXIMATE MATERIAL CHARACTERISTICS [a]

Material	Pounds per Cubic Yard—Bank	Percent of Swell	Load Factor	Pounds per Cubic Yard—Loose
Clay, natural bed	2960	40	0.72	2130
Clay and gravel, Dry	2960	40	0.72	2130
wet	2620	40	0.72	2220
Clay, natural bed, Anthracite	2700	35	0.74	2000
bituminous	2160	35	0.74	1600
Earth, loam, Dry	2620	25	0.80	2100
wet	3380	25	0.80	2700
Gravel, ¼-2 in., Dry	3180	12	0.89	2840
wet	3790	12	0.89	3380
Gypsum	4720	74	0.57	2700
Iron ore, Magnetite	5520	33	0.75	4680
Pyrite	5120	33	0.75	4340
Hematite	4900	33	0.75	4150
Limestone	4400	67	0.60	2620
Sand, dry, loose	2690	12	0.89	2400
wet, packed	3490	12	0.89	3120
Sandstone	4300	54	0.65	2550
Trap rock	4420	65	0.61	2590

[a] The weight and load factor will vary with factors such as grain size, moisture content, and degree of compaction. A test must be made to determine an *exact* material characteristic.

quired to excavate and haul the material to the fill? Two types of productive machines are involved: four trucks and a front-end loader. We must see which unit or set of units is most productive. Reference all calculations to the loose cubic yard production per hour. Then the loader productivity (given 20% swell) is

$$200 \text{ cu yd (bank)/hr} = 1.2(200) \text{ or } 240 \text{ cu yd (loose)/hr}$$

The truck fleet production is

$$4 \text{ trucks} \times \frac{60 \text{ minute/hr}}{15 \text{ minute/cycle}} \times 18 \text{ cu yd (loose) truck}$$

$$= 72 \text{ cu yd (loose)} \times 4 \text{ cycle/hr} = 288 \text{ cu yd (loose)/hr for 4 trucks}$$

Because the loader production is lower, it constrains the system to a maximum output of 240 cu yd (loose)/hr. We must now determine how many loose cubic yards are represented by 18,000 cu yd (compacted).

$$18{,}000 \text{ cu yd compacted} = \frac{18{,}000}{0.9} \text{ or } 20{,}000 \text{ cu yd (bank)}$$

$$20{,}000 \text{ cu yd (bank)} = 24{,}000 \text{ cu yd (loose) required}$$

Therefore, the number of hours required is

$$\text{Hours} = \frac{24{,}000 \text{ cu yd (loose)}}{240 \text{ cu yd (loose)/hr}} = 100$$

This problem illustrates the interplay between volumes and the fact that machines that interact with other machine cycles may be constrained or constraining.

11.2 CYCLE TIME AND POWER REQUIREMENTS

The second factor affecting the rate of output of a machine or machine combination is the time required to complete a cycle which determines the cyclic rate. This is a function of the speed of the machine and, in the case of heavy equipment, is governed by (1) the power required, (2) the power available, and (3) the usable portion of the power available that can be developed to propel the equipment unit.

The power required is related to the rolling resistance inherent in the machine due to internal friction and friction developed between the wheels or tracks and the traveled surface. The power required is also a function of the grade resistance inherent in the slope of the traveled way. Rolling resistance in tracked vehicles is considered to be zero, since the track acts as its own roadbed, being laid in place as the unit advances. The friction between track and support idlers is too small to be considered. Rolling resistance for rubber-

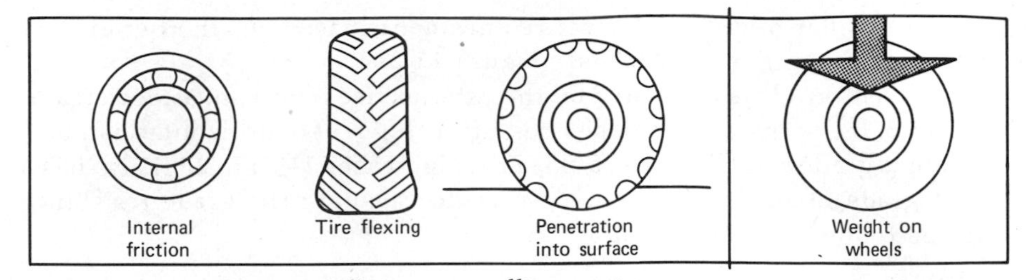

FIGURE 11.2 Factors influencing rolling resistance.

wheeled vehicles is a function of the road surface and the total weight on the wheels. Tables such as Table 11.2 are available in Equipment Handbooks giving the rolling resistance in pounds per tons of weight. Figure 11.2 indicates visually the factors influencing rolling resistance and therefore contributing to the required power that must be developed to move the machine.

If tables are not available, a rule of thumb can be used. The rule states that the rolling resistance is approximately 40 lb/ton plus 30 lb/ton for each inch of penetration of the surface under wheeled traffic. If the estimated deflection is 2 in. and the weight on the wheels of a hauler is 70 tons, we can calculate the approximate rolling resistance as

$$RR = (40 + 2(30))\ \text{lb/ton} \times 70\ \text{tons} = 7{,}000\ \text{lb}$$

The second factor involved in establishing the power required is the grade resistance. In some cases, the haul road across which a hauler must operate will be level and, therefore, the slope of the road will not be a consideration. In most cases, however, slopes (both uphill and downhill) will be encountered

Table 11.2 TYPICAL ROLLING RESISTANCE FACTORS (Caterpillar Tractor Co.)

Surface	Factor
A hard, smooth stabilized roadway without penetration under load (concrete or blacktop)	40 lb/ton
A firm, smooth-rolling roadway flexing slightly under load (macadam or gravel-topped road)	65 lb/ton
Snow-packed	50 lb/ton
Loose	90 lb/ton
A rutted dirt roadway, flexing considerably under load; little maintenance, no water (hard clay road, 1 in. or more tire penetration)	100 lb/ton
Rutted dirt roadway, no stabilization, somewhat soft under travel (4 to 6 in. tire penetration)	150 lb/ton
Soft, muddy, rutted roadway, or in sand	200-400 lb/ton

and lead to higher or reduced power requirements based on whether gravity is aiding or resisting movement (see Figure 11.3).

The percent grade is calculated by the ratio of rise over run, as depicted in Figure 11.3. If, for instance, a slope rises 6 ft in 100 feet of horizontal distance, the percent grade is 6. Similarly, a slope that increases 1½ ft in 25 ft also has a percent grade of 6. Percent grade is used to calculate the grade resistance using the following relationship:

$$\text{GR} = \text{Percent grade} \times 20\ \text{lb/ton/\% Grade} \times \text{Weight on wheels (tons)}$$

If the 70-ton piece of equipment referred to previously is ascending a 6% grade, the grade resistance is

$$\text{GR} = 6\%\ \text{Grade} \times 20\ \text{lb/ton/\% Grade} \times 70\ \text{tons} = 8400\ \text{lb}$$

Assuming the rolling resistance calculated above holds for the road surface of the slope and assuming the equipment is wheeled, the total power required to climb the slope will be

$$\text{Power required} = \text{RR} + \text{GR} = 7000\ \text{lb} + 8400\ \text{lb} = 15{,}400\ \text{lb}$$

If the slope is downward, an aiding force is developed, and the total power required becomes

$$\text{Power required} = \text{RR} - \text{GR} = 7000\ \text{lb} - 8400\ \text{lb} = -1400\ \text{lb}$$

The sign of the grade resistance becomes negative, since it is now aiding and helping to overcome the rolling resistance. Since a negative rolling resistance has no meaning, the power required on a downward 6% grade is zero. In fact, the 1400 lb represents a downhill thrust that will accelerate the machine, and lead to a braking requirement.

Traveled ways or haul roads normally consist of a combination of uphill, downhill, and level sections. Therefore, the power requirement varies and must be calculated for each section. Knowing the power required for each haul road section, a gear range that will provide the required power can be selected. The gear range allows a speed to be developed and, given the speed, we can

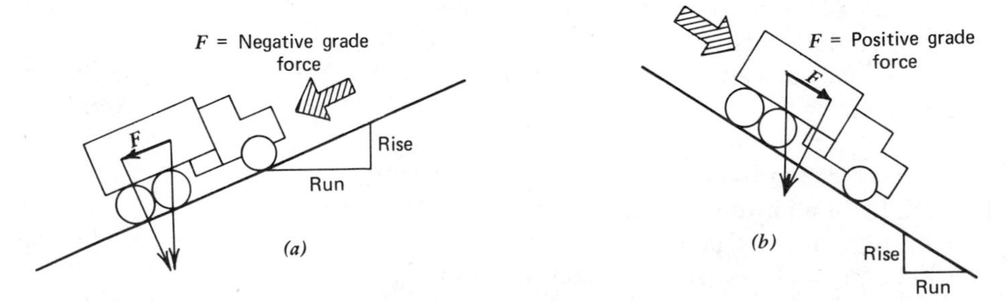

FIGURE 11.3 Grade resistance. (*a*) Negative (resisting) force. (*b*) Positive (aiding) force.

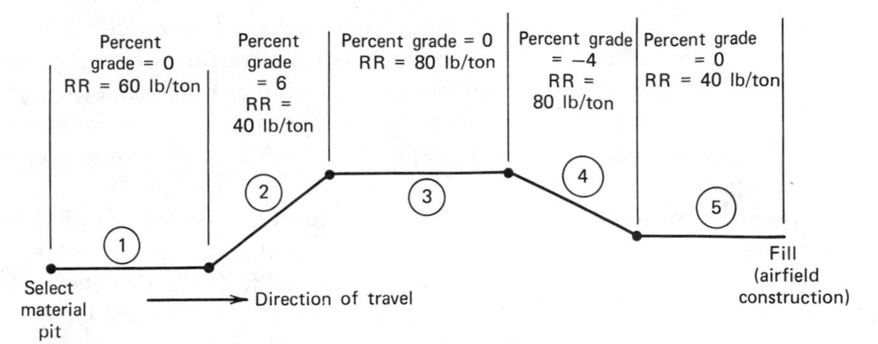

FIGURE 11.4 Typical haul road profile.

develop the time required to transit each section and the total time required for a cycle.

Consider the haul road profile shown in Figure 11.4 with rolling resistance and percent grade values as shown. The calculation of power required for each section of the road based on a 70-ton machine is shown in Table 11.3. Given the power requirements, the next section indicates how a gear range is selected. As noted above, this allows determination of the the speed across each section and the time required.

11.3 POWER AVAILABLE

The power available is controlled by the engine size of the equipment and the drive train, which allows transfer of that power to the driving wheels or power take-off point. The amount of power transferred is a function of the gear being used. Most automobile drivers realize that lower gears transfer more power to overcome hills and rough surfaces. Lower gears sacrifice speed

Table 11.3 CALCULATIONS FOR HAUL ROAD SECTIONS[a]

Section	Percent Grade	Grade Resistance	Rolling Resistance	Power Required
1	0	0	4200 lb	4200 lb
2	6	8400 lb	2800 lb	11,200 lb
3	0	0	5600 lb	5600 lb
4	-4	-5600 lb	5600 lb	0 lb
5	0	0	2800 lb	2800 lb

[a] All calculations assume travel from pit to fill.

in order to provide more power. Higher gears deliver less power, but allow higher speeds. Manufacturers publish figures regarding the power available in each gear for individual equipment pieces in Equipment Handbooks that are updated annually. This information can be presented in a tabular format such as that shown in Table 11.4 or in graphical format such as the nomograph shown in Figure 11.5.

For tracked vehicles, the power available is quoted in drawbar pull. This is the force that can be delivered at the pulling point (i.e., pulling hitch) in a given gear for a given tractor type. Power available for a wheeled vehicle is stated in pounds of rimpull. This is the force that can be developed by the wheel at its point of contact with the road surface. Manufacturers also provide information regarding *rated* power and *maximum* power. Rated power is the level of power that is developed in a given gear under normal load and over extended work periods. It is the base or reference level of power that is available for continuous operation. The maximum power is just what it indicates. It is the peak power that can be developed in a gear for a short period of time to meet extraordinary power requirements. For instance, if a bulldozer is used to pull a truck out of a ditch, a quick surge of power would be used to dislodge the truck. This short-term peak power could be developed in a gear using the maximum power available.

Most calculations are carried out using rated power. If, for example, the power required for a particular haul road section is 25,000 lb based on the procedures described in Section 11.2, the proper gear for the 270-HP Track-Type

Table 11.4 SPEED AND DRAWBAR PULL (270-HP TRACK-TYPE TRACTOR)

					Drawbar Pull Forward [a]			
	Forward		Reverse		At Rated RPM		Maximum at Lug	
Gear	mph	km/h	mph	km/h	lb	kg	lb	kg
1	1.6	(2.6)	1.6	(2.6)	52,410	(23790)	63,860	(28990)
2	2.1	(3.4)	2.1	(3.4)	39,130	(17760)	47,930	(21760)
3	2.9	(4.7)	2.9	(4.7)	26,870	(12200)	33,210	(15080)
4	3.7	(6.0)	3.8	(6.1)	19,490	(8850)	24,360	(11060)
5	4.9	(7.9)	4.9	(7.9)	13,840	(6280)	17,580	(7980)
6	6.7	(10.8)	6.8	(10.9)	8,660	(3930)	11,360	(5160)

[a] Usable pull will depend on traction and weight of equipped tractor.

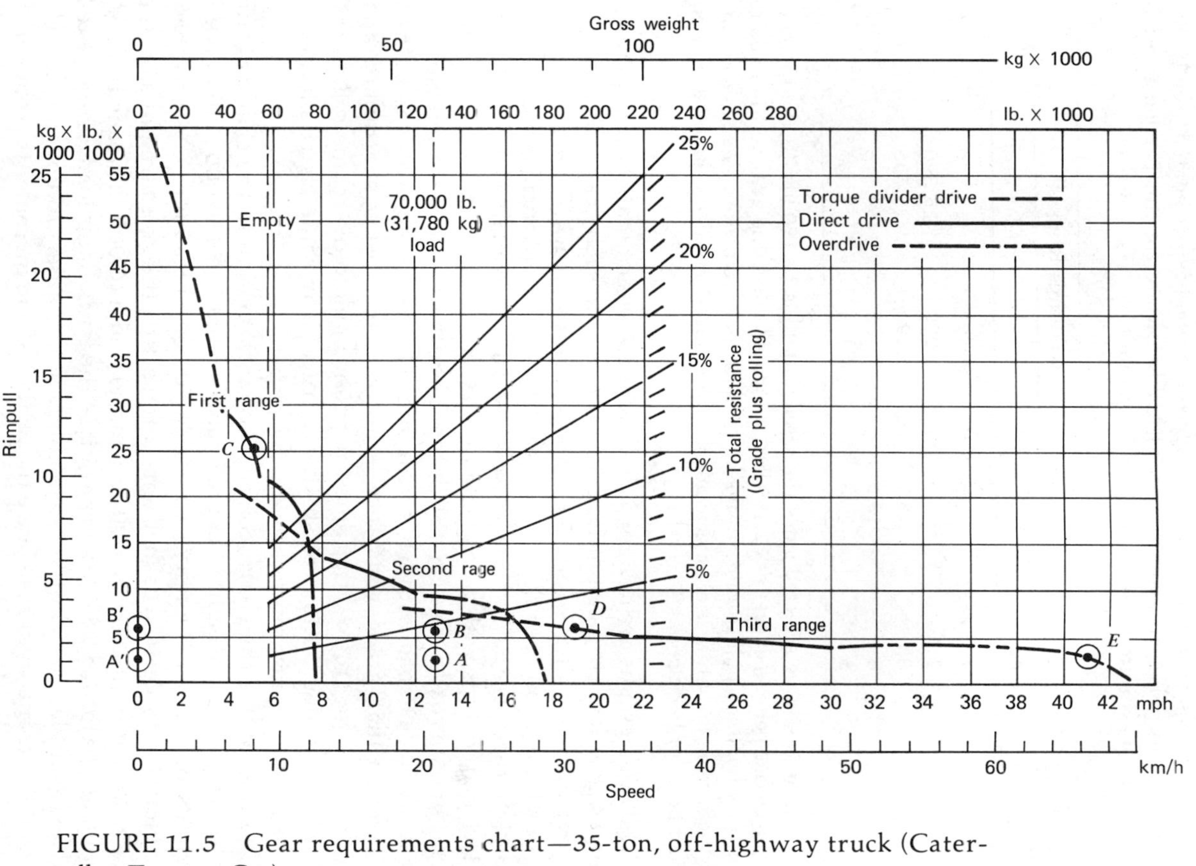

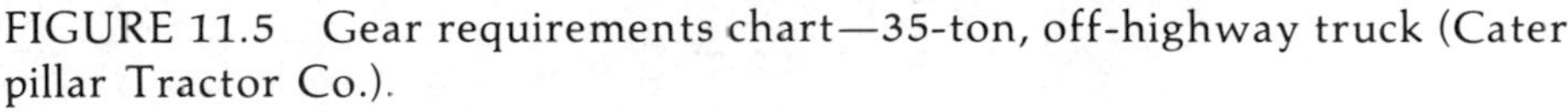
FIGURE 11.5 Gear requirements chart—35-ton, off-highway truck (Caterpillar Tractor Co.).

tractor is third gear. This is determined by entering Table 11.4 and comparing power required with rated power. Consider the following example*:

The sum of the rolling resistance and grade resistance that a particular wheel-type tractor and scraper must overcome on a specific job has been estimated to be 10,000 lbs. If the 'pounds pull-speed' combinations listed below are for this particular machine, what is the maximum reasonable speed of the unit?

		Pounds Rimpull	
Gear	Speed	Rated	Maximum
1	2.6	38,670	49,100
2	5.0	20,000	25,390
3	8.1	12,190	15,465
4	13.8	7,185	9,115
5	22.6	4,375	5,550

Third gear would be selected since the rated rimpull is 12,190 lbs. (If the total power required had been in excess of 12,190 lbs., we would select 2nd gear because you recall that rated pounds pull should always be used for gear selection. The reserve rimpull of the maximum rating is always available—at reduced speed—to pull the unit out of small holes or bad spots.)

The nomographs are designed to allow quick determination of required gear ranges as well as the maximum speed attainable in each gear. The nomograph shown in Figure 11.5 is for a 35-ton, off-highway truck. To illustrate the use of this figure, consider the following problem. On a particular road construction job, the operator has to choose between two available routes linking the select material pit with a road site fill. One route is 4.6 miles (one-way) on a firm, smooth road with a RR = 50 lb/ton. The other route is 2.8 miles (one-way) on a rutted dirt road with RR = 90 lb/ton. The haul road profile in both cases is level so that grade resistance is not a factor. Using the nomograph of Figure 11.5, we are to determine the pounds pull to overcome rolling resistance for a loaded 35-ton, off-highway truck. The same chart allows determination of the maximum speed.

In order to use the chart, consider the information in the chart regarding gross weight. The weight in pounds ranges from 0 to 280,000 (140 tons). The weights of the truck empty and with a 70,000-lb load (i.e., 35-ton capacity) are indicated by vertical dashed lines intersecting the gross weight axis (top of chart) at approximately 56,000 lb (empty) and 126,000 lb (loaded). For this problem, the loaded line is relevant, since the truck hauls loads from pit to fill.

Next consider the slant lines sloping from lower left to upper right. These lines indicate the total resistance (i.e., RR + GR) in increments from 0 to 25% grade. In the problem, there is no grade resistance. In dealing with rolling

*Example as given in *Fundamentals of Earthmoving*, Caterpillar Tractor Co., 1968

resistance, it is common to convert it to an equivalent percent grade. Then, the total resistance can be stated in percent grade by adding the equivalent percent grade for rolling resistance to the slope percent grade. To convert rolling resistance to equivalent percent grade, the following expression is used:

$$\text{Equivalent percent grade} = \frac{\text{RR}}{20 \text{ lb/ton/\% grade}}$$

For the rolling resistance values given in the problem, the equivalent percent grades become

Route	Distance	RR	Equivalent Percent Grade
1	4.6 miles	50 lb/ton	2.5
2	2.8 miles	90 lb/ton	4.5

In order to determine the required pounds pull, the intersection of the slant line representing the equivalent percent grade with the load vertical line is located. This intersection for route one is designated point *A* in Figure 11.5. The corresponding point for route two is labeled *B*.

The "pounds required" value is found using these points by reading horizontally across to the *y* axis, which gives the rimpull in pounds. For route 1, the approximate power requirement is 2500 lb. The requirement for route 2 is 5500 lb. Points *A* and *B* are also used to determine the maximum speed along each route.

Consider the curves descending from the upper left-hand corner of the chart to the lower right side. As labeled, these curves indicate the deliverable power available in FIRST, SECOND, and THIRD RANGES as well as the speed that can be developed. At 25,000 lb of rimpull, for example, on the *y* axis, reading horizontally to the right the only range delivering this much power is first range (see point *C*). Reading vertically down to the *y* axis, the speed that can be achieved at this power level is approximately 5 mi/hr.

Proceeding in a similar manner, it can be determined that two ranges, second and third, will provide the power necessary for route 2 (i.e., 5500 lb). Reading horizontally to the right from point *B*, the maximum speed is developed in third range at point *D*. Referencing this point to the *x* axis, the maximum speed on route 2 is found to be approximately 19 mi/hr. Route 1 requires considerably less power. Again reading to the right, this time from point *A*, the third range provides a maximum speed of approximately 41 mi/hr (see point *E*).

Now, having established the maximum speeds along each route and knowing the distances involved, it should be simple to determine the travel times required. Knowledge of the speeds, however, is not sufficient to determine the travel times since the requirements to accelerate and decelerate lower the effective speed between pit and fill. Knowing the mass of the truck and the horsepower of the engine, the classic equation, Force = mass × acceleration ($F = ma$), would allow determination of the time required to accelerate to and decelerate from maximum speed. This is not necessary, however, since the Equipment Handbooks provide travel time charts that allow direct readout of the travel time for a route and piece of equipment, given the equivalent percent grade and the distance. These charts for loaded and empty 35-ton trucks are given in Figure 11.6. Inspection of the chart for the loaded truck indicates that the distance to be traveled in feet is shown along the x axis. The equivalent percent grade is again shown as slant lines sloped from lower left to upper right. Converting the mileages given to feet

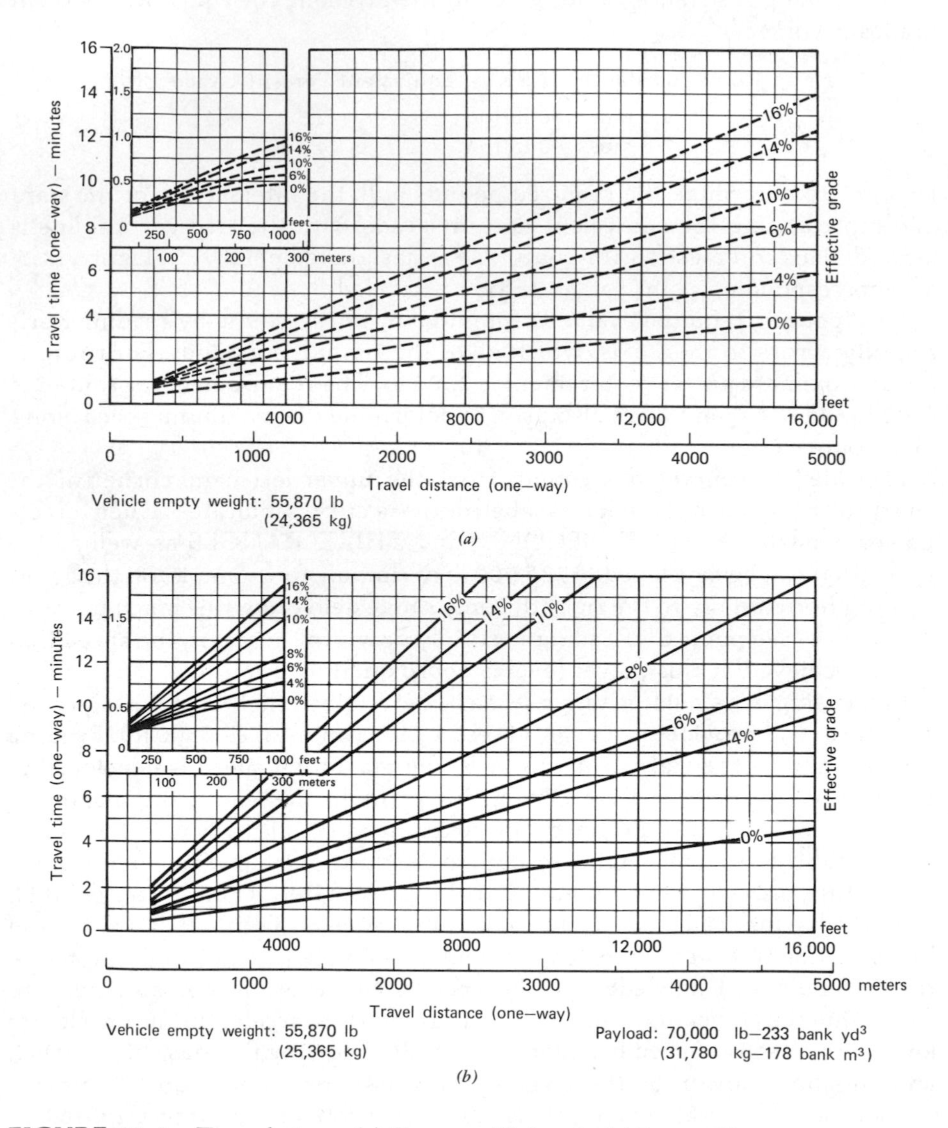

FIGURE 11.6 Travel time. (*a*) Empty. (*b*) Loaded (Caterpillar Tractor Co.).

yields the values 24,344 ft for route 1 and 14,784 ft for route 2. Entering the chart, with an equivalent percent grade of 4.5 for route 2, the travel time can be read (on the y axis) as 9.5 minutes.

A problem develops in reading the travel time for route 1, since the maximum distance shown on the chart is 16,000. One way of reconciling this problem is to break the 24,344 into two segments: (1) 16,000 ft as shown and (2) the 8344 ft remaining. The assumption is made that the 8,344 ft is traveled at the maximum speed determined previously to be 41 mi/hr. At this speed, the travel time for this segment is

$$T_2 = \frac{8344 \text{ ft } (60 \text{ minutes/hour})}{(41 \text{ mile/hr}) \ (5280 \text{ ft/mile})} = 2.31 \text{ minutes}$$

The time for the remaining 16,000 ft is read from the chart as 7.2 minutes. It is assumed that acceleration and deceleration effects are included in this time. Therefore, the required time for route 1 is also 9.5 minutes ($T_1 + T_2 = 7.2 + 2.31$). Therefore, the decision as to which route to use is based on wear and tear on the machines, driver skill, and other considerations. This problem illustrates the development of time, given information affecting power required and power available. Using the same procedure, the travel time empty returning to the pit from the fill can be determined and total cycle time can be determined.

11.4 USABLE POWER

To this point, it has been assumed that all of the available power is usable and can be developed. Environmental conditions play a major role in determining whether the power available can be utilized under operating conditions. The two primary constraints in using the available power are the road surface traction characteristics (for wheeled vehicles) and the altitude at which operations are conducted. Most people have watched the tires of a powerful car spin on a wet or slippery pavement. Although the engine and gears are delivering a certain horsepower, the traction available is not sufficient to develop this power into the ground as a driving force. Combustion engines operating at high altitudes experience a reduction in oxygen available within the engine cylinders. This also leads to reduced power.

Consider first the problem of traction. The factors that influence the usable power that can be developed through the tires of wheeled vehicles are the coefficient of traction of the surface being traveled and the weight of the vehicle on the driving wheels.

The coefficient of traction is a measure of the ability of a particular surface to receive and develop the power being delivered to the driving wheels, and has been determined by experiment. The coefficient of traction obviously varies based on the surface being traversed and the delivery mechanism (i.e., wheels, tracks, etc.). Table 11.5 gives typical values for rubber-tired and tracked vehicles on an assortment of surface materials.

Table 11.5 COEFFICIENTS OF TRACTION

Materials	Rubber Tires	Tracks
Concrete	.90	.45
Clay Loam, Dry	.55	.90
Clay Loam, Wet	.45	.70
Rutted Clay Loam	.40	.70
Dry Sand	.20	.30
Wet Sand	.40	.50
Quarry Pit	.65	.55
Gravel Road (loose, not hard)	.36	.50
Packed Snow	.20	.25
Ice	.12	.12
Firm Earth	.55	.90
Loose Earth	.45	.60
Coal, Stockpiled	.45	.60

The power that can be developed on a given surface is given by the expression:

Usable pounds pull = (Coefficient of traction) × (Weight on drivers)

In the consideration of rolling resistance and grade resistance, the entire weight of the vehicle or combination was used. In calculating the usable power, *only the weight on the driving wheels* is used, since it is the weight pressing the driving mechanism (e.g., wheels) and surface together. Equipment Handbooks specify the distribution of load to all wheels for both empty and loaded vehicles and combinations. The weight to be considered in the calculation of usable power for several types of combinations is shown in Figure 11.7. To illustrate the constraint imposed by usable power, consider the following situation. A 30-yd-capacity, two-wheel tractor-scraper is operating in sand and carrying 26-ton loads. The job superintendent is concerned about the high rolling resistance of the sand (RR = 400 lb/ton) and the low traction available in sand. The question is: Will the tractors have a problem with 26-ton loads under these conditions? The weight distribution characteristics of the 30-yd tractor-scraper are given below:

	Empty Weight (lb)	Percent	Loaded Weight (lb)	Percent
Drive wheels	50,800	67	76,900	52
Scraper wheels	25,000	33	70,900	48
Total Weight	75,800	100	147,800	100

The difference between the total weight empty and loaded is 72,000 lb or 36 tons. The loaded weight with 26-ton loads would be 127,800 lb. Assuming the same weight distribution given above for fully loaded vehicles, the wheel loads would be as follows:

	Percent	Weight in Pounds
Drive wheels	52	66,456
Scraper wheels	48	61,344
Total	100	127,800

The resisting force (assuming a level haul site) would be

$$\text{Pounds required} = 400 \text{ lb/ton} \times 63.9 \text{ tons} = 25{,}560 \text{ lb}$$

The deliverable or usable power is

$$\text{Usable power} = 0.20 \times 66{,}456 \text{ lb} = 13{,}291.20 \text{ lb}$$

Quite obviously, there will be a problem with traction, since the required power is almost twice the power that can be developed. The "underfoot"

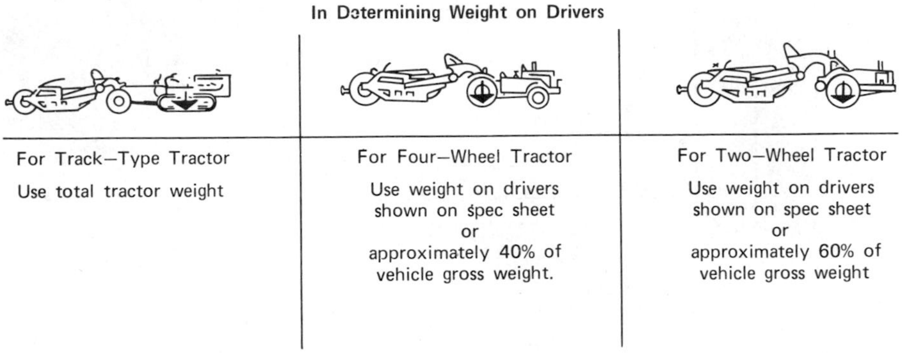

In Determining Weight on Drivers

For Track–Type Tractor	For Four–Wheel Tractor	For Two–Wheel Tractor
Use total tractor weight	Use weight on drivers shown on spec sheet or approximately 40% of vehicle gross weight.	Use weight on drivers shown on spec sheet or approximately 60% of vehicle gross weight

FIGURE 11.7 Determination of driver weights.

condition must be improved. A temporary surface (e.g., wood or steel planking) could be installed to improve traction. One simple solution would be to simply wet the sand. This yields an increased usable power:

$$\begin{aligned}\text{Usable power} &= 0.40 \times 66{,}456 \text{ lb} = 26{,}582.4 \text{ lb}\\ &> 25{,}560 \text{ lb}\end{aligned}$$

The impact of usable power constraints can be shown graphically (see Figure 11.8). Now, if the total resistance of the unit (rolling resistance plus grade resistance) is 10,000 lb, then an operating range for the machine is indicated in Figure 11.8*b*.

The altitude at which a piece of equipment operates also imposes a constraint on the usable power. As noted previously, the oxygen content decreases as elevation increases, so that a tractor operating in Bogota, Colombia (elevation 8600 ft), cannot develop the same power as one operating in Atlanta, Georgia (elevation 1050 ft). A good rule of thumb to correct for this effort is as follows.

Decrease pounds pull three percent (3%) for each 1000 ft (above 3000 ft)

Therefore, if a tractor is operating at 5000 ft above sea level, its power will be decreased by 6%.

11.5 EQUIPMENT BALANCE

In situations where two types of equipment work together to accomplish a task, it is important that a balance in the productivity of the units be achieved. This is desirable so that one unit is not continually idle waiting for the other unit to "catch up." Consider the problem of balancing productivity within the context of a pusher dozer loading a tractor scraper. A simple model of this

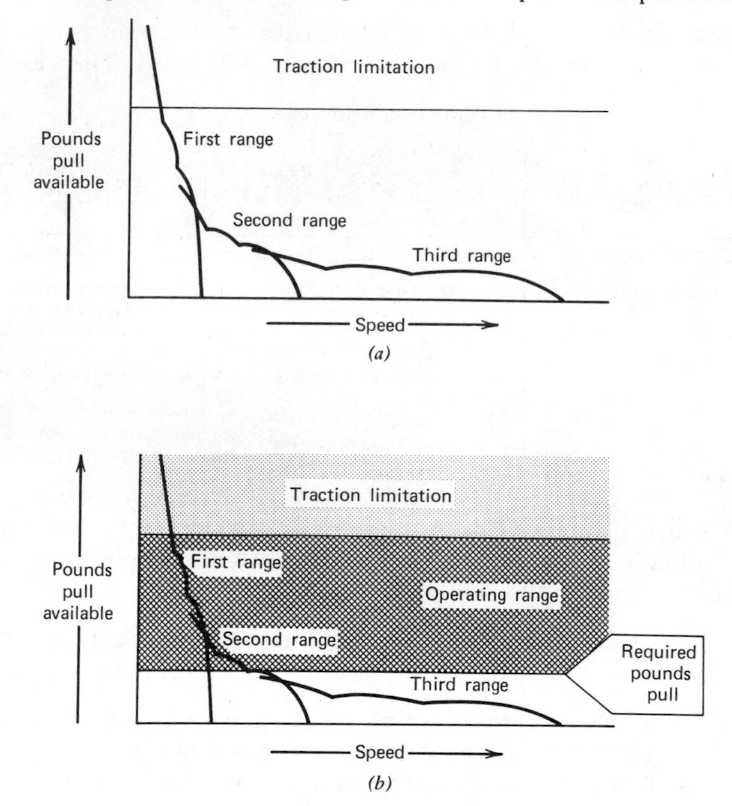

FIGURE 11.8 Impact of usable power constraints.

process is shown in Figure 11.9. The circles represent delay or waiting states, while the squares designate active work activities with associated times that can be estimated. The haul unit is a 30 yd^3 scraper and it is loaded in the cut area with the aid of a 385-hp pusher dozer. The system consists of two interacting cycles.

Assume that in this case the 30-yd^3 tractor scraper is carrying rated capacity and operating on a 3000-ft level haul where the rolling resistance (RR) developed by the road surface is 40 lb/ton.* Using the standard formula, this converts to

$$\text{Effective grade} = \frac{\text{RR}}{20 \text{ lb/ton/\% grade}} = \frac{40 \text{ lb/ton}}{20 \text{ lb/ton/\% grade}} = 2\% \text{ grade}$$

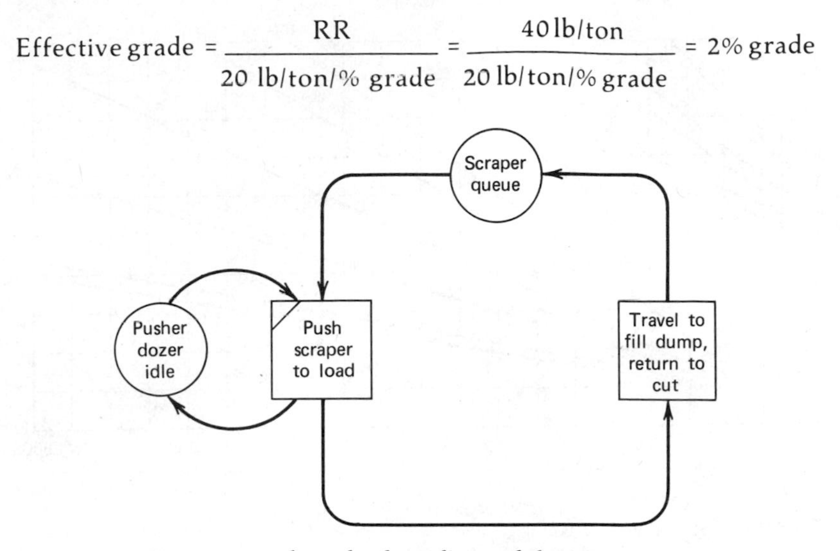

FIGURE 11.9 Scraper-pusher dual-cycle model.

By consulting the charts given in Figure 11.10, the following travel times can be established.

1. Time loaded to fill: 1.4 minutes.
2. Time empty to return: 1.2 minutes.

Assume further that the dump time for the scraper is 0.5 minutes and the push time using a 385-hp, track-type pusher tractor is 1.23 minutes, developed as follows:†

Load time	= 0.70
Boost time	= 0.15
Transfer time	= 0.10
Return time	= 0.28
Total	= 1.23 minutes

Using these deterministic times for the two types of flow units in this system (i.e., the pusher and the scrapers), the scraper and pusher cycle times can be developed, as shown in Figure 11.11, as shown on p. 197.

*See, for instance, *Fundamentals of Earthmoving*, p. 66, published by the Caterpillar Tractor Company, 1968.

†Example taken from Halpin and Woodhead, *Design of Construction and Process Operations*, John Wiley and Sons, Inc., New York, 1976.

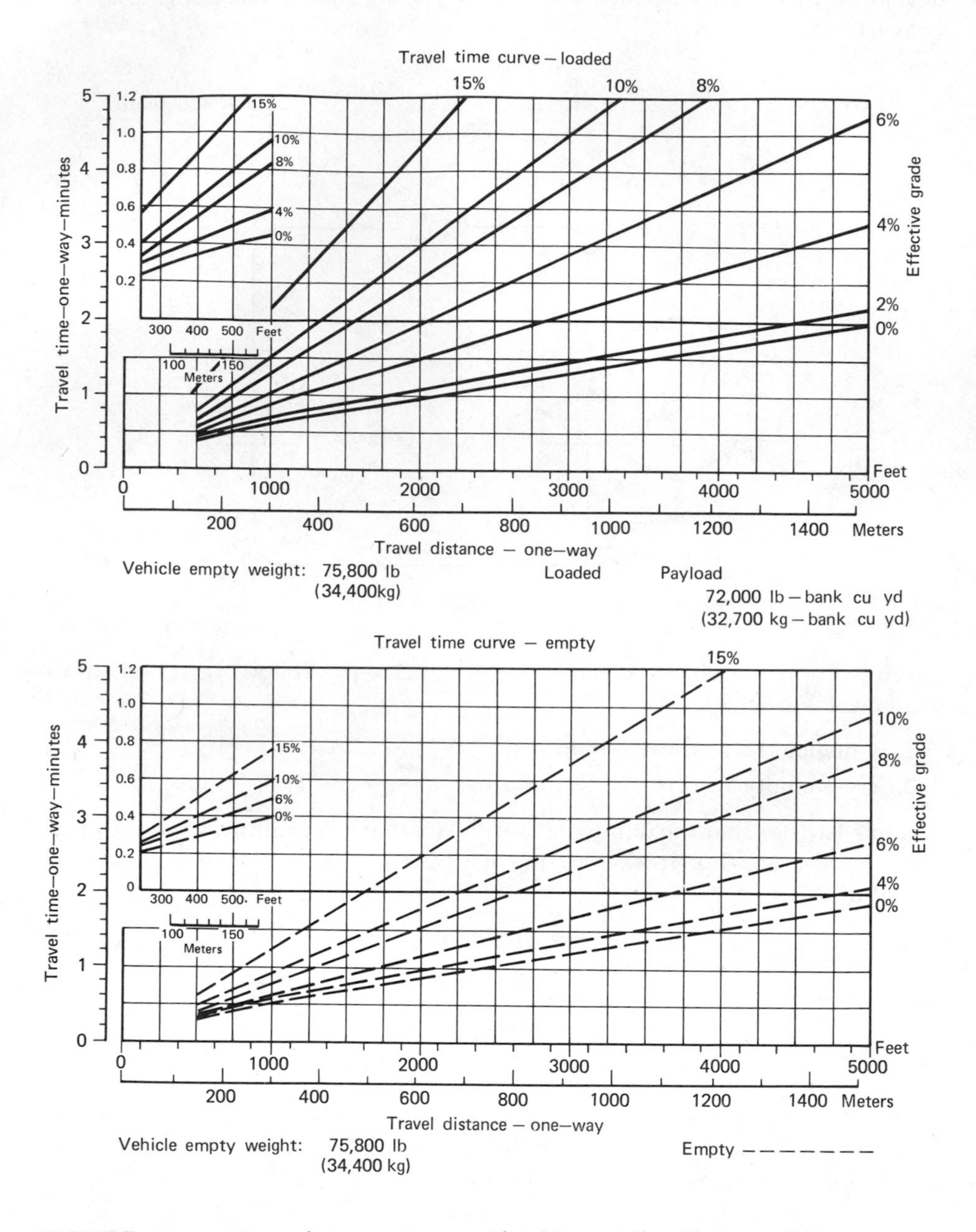

FIGURE 11.10 Travel time nomographs (Caterpillar Tractor Co.).

Pusher cycle = 1.23 minutes
Scraper cycle = 0.95 + 1.2 + 1.4 + 0.5 = 4.05 minutes

These figures can be used to develop the maximum hourly production for the pusher unit and for each scraper unit as follows.

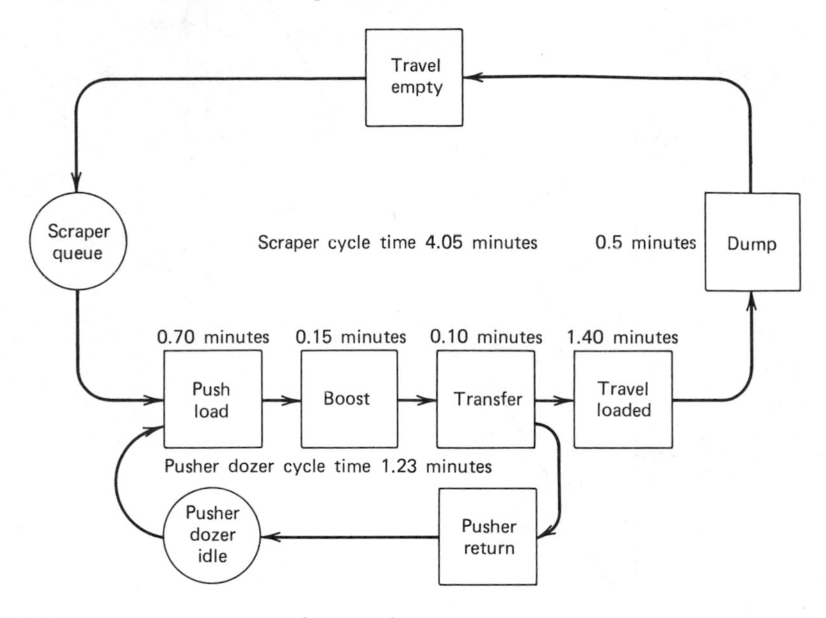

FIGURE 11.11 Scraper-pusher cycle times.

Maximum System Productivity (Assuming a 60-Minute Working Hour)

1. Per Scraper

$$\text{Prod (scraper)} = \frac{60 \text{ min/hr}}{4.05 \text{ min}} \times 30 \text{ cu yd (loose)}$$

$$= 444.4 \text{ cu yd (loose)/hr}$$

2. Based on Single Pusher

$$\text{Prod (pusher)} = \frac{60}{1.23} \times 30 \text{ cu yd (loose)}$$

$$= 1463.4 \text{ cu yd (loose)/hr}$$

Using these productivities based on a 60-minute working hour, it can be seen that the pusher is much more productive than a single scraper and would be idle most of the time if matched to only one scraper. By using a graphical plot, the number of scrapers that are needed to keep the pusher busy at all times can be determined.

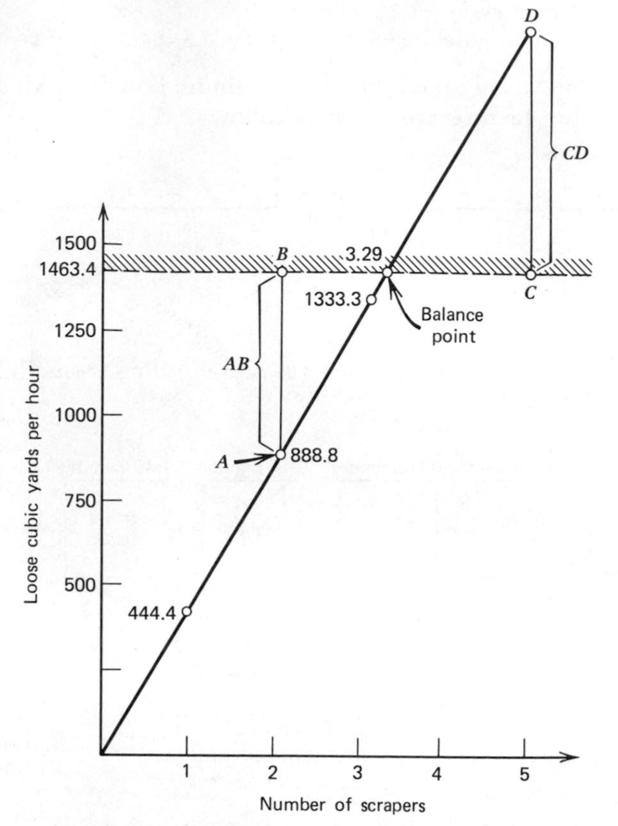

FIGURE 11.12 Productivity plot.

The linear plot of Figure 11.12 shows the increasing productivity of the system as the number of scrapers is increased. The productivity of the single pusher constrains the total productivity of the system to 1463.4 cu yd. This is shown by the dotted horizontal line parallel to the x axis of the plot. The point at which the horizontal line and the linear plot of scraper productivity intersect is called the *balance point*. The balance point is the point at which the number of haul units (i.e., scrapers) is sufficient to keep the pusher unit busy 100% of the time. To the left of the balance point, there is an imbalance in system productivity between the two interacting cycles; this leaves the pusher idle. This idleness results in lost productivity. The amount of lost productivity is indicated by the difference between the horizontal line and the scraper productivity line. For example, with two scrapers operating in the system, the ordinate AB of Figure 11.12 indicates that 574.6 cu yd, or a little less than half of the pusher productivity, is lost because of the mismatch between pusher and scraper productivities. As scrapers are added, this mismatch is reduced until, with four scrapers in the system, the pusher is fully utilized. Now the mismatch results in a slight loss of productivity caused by idleness of the scrapers. This results because, in certain instances, a scraper will have to wait to be loaded until the pusher is free from loading a preceding unit. If five scraper units operate in the system, the ordinate *CD* indicates that the loss in the productive capacity of the scraper because of delay in being push loaded is:

$$\text{Productive loss} = 5(444.4) - 1463.4 = 758.6 \text{ cu yd}$$

This results because the greater number of scrapers causes delays in the scraper queue of Figures 11.9 and 11.11 for longer periods of time. The imbalance or mismatch between units in dual-cycle systems resulting from deterministic times associated with unit activities is called *interference*. It is due only to the time imbalance between the interacting cycles. It does not consider idleness or loss of productivity because of random variations in the system activity durations. In most cases, only a deterministic analysis of system productivity is undertaken because it is sufficiently accurate for the purpose of the analyst.

11.6 RANDOM WORK TASK DURATIONS

The influence of mismatches in equipment fleets and crew mixes on system productivity was discussed in the last section in terms of deterministic work task durations and cycle times. In systems where the randomness of cycle times is considered, system productivity is reduced further. The influence of random durations on the movement of resource flow is to ensure that the various units eventually get bunched together and thus arrive at and swamp work tasks. Resulting delays impact the productivity of cycles by increasing the time that resource units spend in idle states pending release to productive work tasks.

Consider the scraper-pusher problem and assume that the effect of random variation in cycle activity duration is to be included in the analysis.

In simple cases such as the two-cycle system model of Figure 11.9, mathematical techniques based on *queueing theory* can be used to develop solutions for situations where the random arrival of scrapers to the dozer can be postulated. In order to make the system amenable to mathematical solution, however, it is necessary to make certain assumptions about the characteristics of the system that are not typical of field construction operations.*

Figure 11.13 indicates the influence of random durations on the scraper fleet production. The curved line of Figure 11.13 slightly below the linear plot of production based on deterministic work task times shows the reduction in production caused by the addition of random variation of cycle activity times. This randomness leads to bunching of the haul units on their cycle. With

*See Appendix A, Halpin and Woodhead, *Design of Construction and Process Operations*, John Wiley and Sons, Inc., New York, 1976.

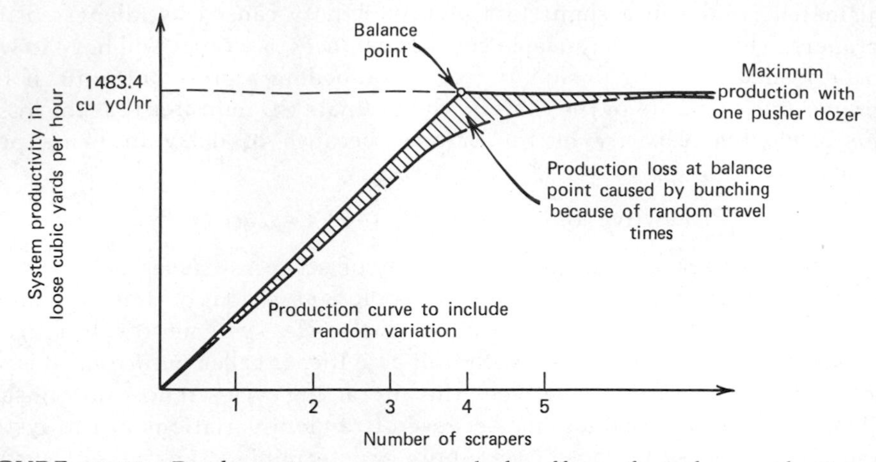

FIGURE 11.13 Productivity curve to include effect of random cycle times.

deterministic work task times, the haul units are assumed to be equidistant in time from one another within their cycle.

In deterministic calculations, all three of the haul units shown in Figure 11.14*a* are assumed to be exactly 1.35 minutes apart. In this system, there are three units, and the hauler cycle time is taken as a deterministic value of 4.05 minutes. In systems that include the effect of random variation of cycle times, "bunching" eventually occurs between the units on the haul cycle. That is, the units do not stay equidistant from one another but are continuously varying the distances between one another. Therefore, as shown in Figure 11.14*b*, a situation often occurs in which the units on the haul are unequally spaced apart in time from one another. This bunching effect leads to increased idleness and reduced productivity. It is intuitively clear that the three units that are bunched as shown in Figure 11.14*b* will be delayed for a longer period at the scraper queue, since the first unit will arrive to load only 1.05 minutes instead of 1.35 minutes in advance of the second unit. The bunching causes units to "get into each other's way." The reduction in productivity caused by bunching is shown as the shaded area in Figure 11.13 and occurs in addition to the reduction in productivity caused by mismatched equipment capacities.

This bunching effect is most detrimental to the production of dual-cycle systems such as the scraper-pusher process at the balance point. Several studies have been conducted to determine the magnitude of the productivity reduction at the balance point because of bunching. Simulation studies conducted by Morgan and Peterson of the research department of the Caterpillar Tractor Company indicate that the impact of random time variation is the

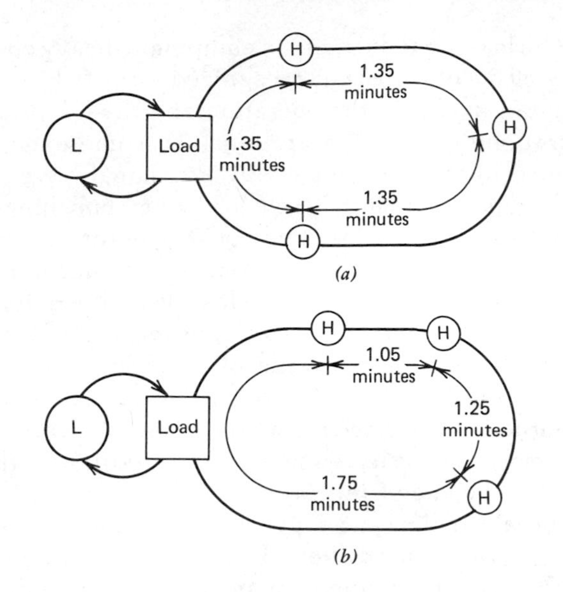

FIGURE 11.14 Comparison of haul unit cycles.

standard deviation of the cycle time distribution divided by the average cycle time. Figure 11.15 illustrates this relationship graphically.

As shown in the figure, the loss in deterministic productivity at the balance point is approximately 10% due to the bunching; this results in a system with a cycle coefficient of variation equal to 0.10. The probability distribution used in this analysis was lognormal. Other distributions would yield slightly dif-

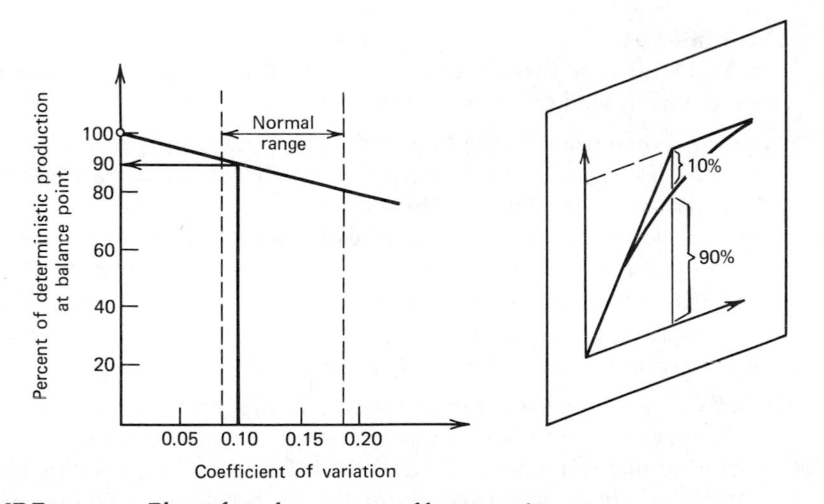

FIGURE 11.15 Plot of cycle time coefficient of variation.

ferent results. The loss in productivity in equipment-heavy operations such as earthmoving is well documented and recognized in the field, mainly because of the capital-intensive nature of the operation and the use of scrapers in both single-unit operations and fleet operations. To some extent, field policies have emerged to counteract this effect by occasionally breaking the queue discipline of the scrapers so that they self-load when bunching effects become severe. The resulting increased load and boost time for the scraper adds little to the system productivity, but it does break down the bunching of the scrapers.

Many cases exist in construction of the loss of productivity because of the interaction of randomly perturbed cycles. For example, in masonry operations, initial conditions relating to the status of scaffold stacks of bricks affect mason productivity until a transient "workup" phase has elapsed. A common solution to this situation is to have the masons' laborers workday begin earlier than that of the masons or to have some laborers work later than the masons to restock bricks at the end of the day.

Often the material handling and supply routes on construction sites provide situations where serious interaction develops between apparently totally independent activities. Competition for material hoists and transport space on congested sites reduces productivity, introduces random perturbations in activity durations, and sometimes reaches crisis magnitude. Very little information exists for estimators on the magnitude, influence, and cost of these interactions on construction sites.

REVIEW QUESTIONS AND EXERCISES

11.1 A customer estimates that he is getting 30 yd (loose) of gypsum in his scraper. Determine the percent overload if the load estimate is correct. The maximum load capacity of the scraper is 84,000 lb.

11.2 Stripping overburden in the Illinois coal belt, the Dusty Coal Company uses 270-HP Track-Type Tractors (with direct drive transmissions) and drawn scrapers. The overburden is a very soft loam that weighs 2800 lb/yd (loose). Estimated rolling resistance factor for the haul road is 300 lb/ton. If the scraper weighs 35,000 lb (empty) and carries 25 loose cubic yards per trip, what is the rolling resistance of the loaded unit? What operating gear and speed do you estimate for the loaded machines on level ground? (See Table 11.4)

11.3 The ABC Company is planning to start a new operation hauling sand to a ready-mix concrete plant. The equipment superintendent estimates that the company-owned 30-yd Wheel Tractor Scrapers can obtain 26-ton loads. He is concerned about the high rolling resistance of the units in the sand (RR factor 250 lb/ton) and the low tractive ability of the tractors on this job. Will traction be a problem? If so, what do you suggest to help?

11.4 Estimate the cycle time and production of a 30-yd wheel tractor scraper carrying rated capacity, operating on a 4500-ft level haul. The road flexes under load, has little maintenance, and is rutted. Material is 3000 lb/BCY. The scrapers are push-loaded by one 385-hp, track-type pusher tractor. How many scrapers can be served by this one pusher?

11.5 How many trips would one rubber-tired Herrywampus have to make to backfill a space with a geometrical volume of 5400 cu yd? The maximum capacity of the machine is 30 cu yd (heaped) or 40 tons. The material is to be compacted with a shrinkage of 25% (relative to bank measure) and has a swell factor of 20% (relative to bank measure). The material weighs 3000 lb/cu yd (bank). Assume that the machine carries its maximum load on each trip. Check by both weight and volume limitations.

11.6 You own a fleet of 30 cu yd tractor-scrapers and have them hauling between the pit and a road construction job. The haul road is clayey and deflects slightly under the load of the scraper. There is a slight grade (3%) from pit at the fill location. The return road is level. The haul distance to the dump location is 0.5 miles and the return distance is 0.67 miles. Four scrapers are being used.

(a) What is the rimpull required when the scraper is full and on the haul to the fill?
(b) What are the travel times to and from the dump location (see Figure 11.10)?
(c) The scrapers are push loaded in the pit. The load time is 0.6 minutes. What is the cycle time of the pusher dozer?
(d) Is the system working at, above, or below the balance point? Explain.
(e) What is the production of the system?

11.7 You are excavating a location for the vault shown below. The top of the walls shown are 1 ft below grade. All slopes of the excavation are ¾ to 1 to a toe 1 ft outside the base of the walls. The walls sit on a slab 1 ft in depth. Draw a sketch of the volume to be excavated, break it into components, and calculate the volume. The material from the excavation is to be used in a compacted fill. The front-end loader excavating the vault has an output of 200 *bank* cubic yards of common earth per hour. It loads a fleet of four trucks (capacity 18 *loose* cubic yards each) that haul the earth to a fill where it is compacted with a shrinkage factor of 10%. Each truck has a total cycle time of 15 minutes, assuming it does not have to wait in line to be loaded. The earth has a swell factor of 20%. How many hours will it take to excavate and haul the material to the fill?

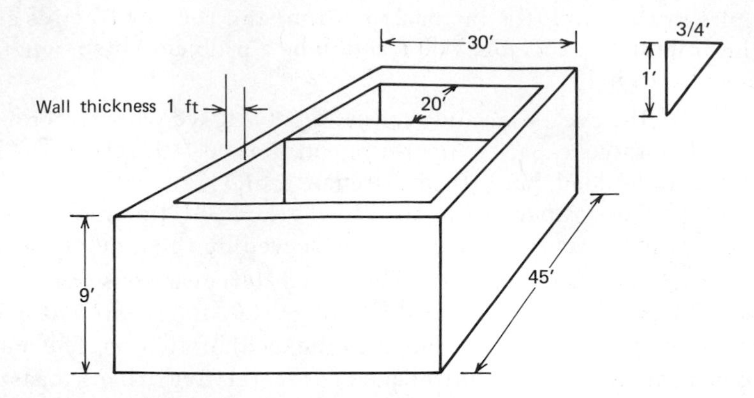

11.8 You have four 35-ton, off-highway trucks hauling from a pit to an airfield job. The haul road is maintained by a patrol grader and has a rolling resistance of 80 lb/ton. The road is essentially level and the distance one-way is 2.1 miles. The gross weight of the truck when loaded is 70 tons.

(a) What is the power required on the haul to the fill location?
(b) What is the maximum speed when hauling to the fill?
(c) What are the travel times to and from the fill (i.e., loaded and empty)?

The trucks are being loaded by a shovel with a 5-yd bucket (assume 7-load cycles per truck load). The cycle time for the shovel is 0.5 minutes.

(d) What is the total truck cycle time?
(e) What is the production of this system in cubic yards per hour assuming the trucks carry 35 cubic yards per load?
(f) Is the system working at, above, or below its balance point?
(g) If there is a probability of major delay on the travel elements to and from the fill of 7% and the mean value of delay is 5 minutes, what is the new system production?
(h) Is the new system above or below the balance point?

11.9 You are given the following information about a dry-batch paving operation. You are going to use one mixer that has a service rate of 30 services per hour. The dry-batch trucks you use for bringing concrete to the paver have an arrival rate of 7.5 arrivals per hour. Each truck carries 6 cu yd of concrete. You have a total amount of 13,500 cu yd of concrete to pour. You rent a truck at $15 per hour, and the paver at $60 per hour. If the job takes more than 80 hours, you pay a penalty of $140 per hour owing to delays in job completion. On the basis of least cost, determine the number of trucks you should use. Plot the cost versus the number of trucks used.

CHAPTER TWELVE
Materials Management

12.1 THE MATERIAL MANAGEMENT PROCESS

In the traditional contractual relationship, the owner contracts with a general contractor or construction manager to build his facilities and with an architect to perform the design. The general contractor, through this contract with the owner, is obligated to perform the work in accordance with the architect's instructions, specifications, and drawings. Thus, the architect is the owner's agent during the design and construction of a project. The lines of communication between the three parties are established as shown in Figure 12.1.

The materials that comprise facilities in building construction are subject to the obligation the contractor has to the architect. The contractor usually delegates this obligation to subcontractors and suppliers for the various

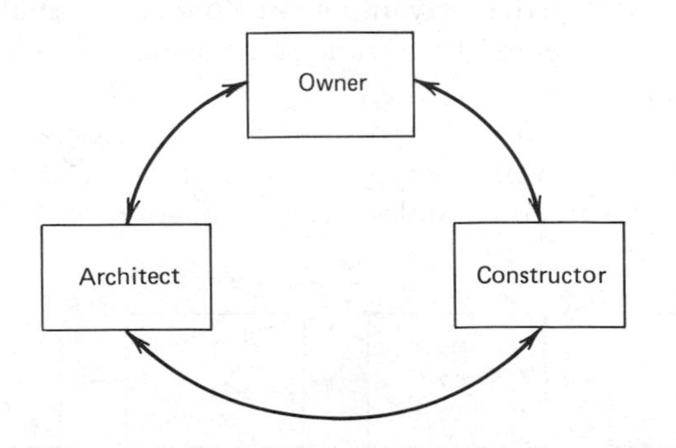

FIGURE 12.1 The Owner-architect-constructor relationship.

categories of work involved in the project. This delegation is accomplished through subcontracts and purchase orders. As a result of this delegation, a distinct life cycle evolves for the materials that make up the project. The four main phases of this cycle are depicted in Figure 12.2.

12.2 THE ORDER

When the contract for construction is awarded, the contractor immediately begins awarding subcontracts and purchase orders for the various parts of the work. How much of the work is contracted depends on the individual contractor. Some contractors subcontract virtually all of the work in an effort to reduce the risk of cost overruns and to have every cost item assured through stipulated-sum subcontract quotations. Others perform almost all the work with their own field forces.

The subcontract agreement defines the specialized portion of the work to be performed and binds the contrator and subcontractor to certain obligations. The subcontractor, through the agreement, must provide all materials and perform all work described in the agreement. The Associated General Contractors (A.G.C.) of America publish the *Standard Subcontract Agreement* for use by their members.

A Sample of this agreement can be found in Appendix I. Most contractors either adopt a standard agreement, such as that provided by the A.G.C., or implement their own agreement. In most cases, a well-defined and well-prepared subcontract is used for subcontracting work.

All provisions of the agreement between the owner and contractor are made part of the subcontract agreement by reference. The most important referenced document in the subcontract agreement is the general conditions. As previously noted in Chapter 2, the AIA Document 201, "General Conditions of the Contract for Construction," published by the American Institute of Architects, is the most widely used set of General Conditions. Article 4 of the General Conditions is particularly important concerning materials supplied by a subcontractor or supplier. Paragraph 4.2.1 provides that the "contractor shall perform no portion of the work at any time without Contract Documents, or where required, approved Shop Drawings, Product Data, or Samples for such portion of the work." Paragraph 4.12.8 reemphasizes that no work shall be performed unless required submittals are approved by the architects.

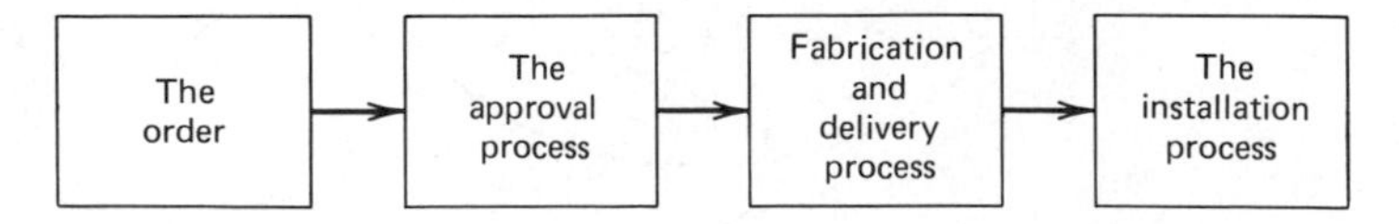

FIGURE 12.2 Material life cycle.

The purchase order is a purchase contract between the contractor and the supplier. This document depicts the materials to be supplied, their quantities, and the amount of the purchase order.

Purchase orders vary in complexity and can be as simple as a mail order house (e.g., Sears) order form or *almost* as complex as the construction contract itself. When complex and specially fabricated items are to be included in the construction, very detailed specifications and drawings become part of the purchase order. Some typical purchase order forms are shown in Figures 12.3 and 12.4. Figure 12.3 shows a form for field-purchased items procured from locally available sources. These items are usually purchased on a cash and carry basis. The purchase order in this case is used primarily to document the purchase for record-keeping and cost-accounting purposes (rather than as a contractual document). A more formal purchase order used in a contractual sense is shown in Figure 12.4. It is used in the purchase of more complex items from sources that are remote to the site.

Special Purchase Order

HCB HENRY C BECK COMPANY

VENDOR:

MAIL INVOICE TO:
HENRY C. BECK COMPANY
1210 S. Old Dixie Highway
Jupiter, Florida 33458

DATE:

CHG. TO JOB #21330

SHIP TO: 1210 S. Old Dixie Highway / Jupiter, Florida 33458

QUANTITY	ARTICLE	U.P.	AMOUNT	COST CODE

STATE AND LOCAL SALES TAXES MUST BE SET OUT SEPARATELY ON INVOICE

Invoice in Triplicate
To Above Address
No Later Than 25th of Month — Vendor's Acceptance (when required) SUPT. OR PROJECT MGR.
Show S.P.O. Number On Invoice
WHITE (ORIGINAL) - VENDOR'S COPY
CANARY -JOB OFFICE COPY
(MAIL TO DALLAS WITH INVOICE)
PINK - SUPERINTENDENT'S COPY
GOLDENROD - PROJECT MANAGER'S COPY

FIGURE 12.3 Field purchase order (Courtesy Henry C. Beck Co.).

Letter or transmittal form accompanying this order when mailed to Vendor should show the number of shop drawings and/or samples to be furnished and the address to which they must be sent; also the address to which Vendor is to mail correspondence relating to this order.

PURCHASE ORDER

HCB HENRY C BECK COMPANY

No.

VENDOR
ADDRESS

_______________19____

JOB:
Job Mailing Address:

Please ship the following to HENRY C. BECK COMPANY, at

SHIP VIA:

It is agreed that shipment will be made on or before or right is reserved to cancel order.

IMPORTANT NOTE: It is IMPERATIVE in the interest of prompt payment that all invoices be rendered in the original with two (2) copies. Mail together with two (2) copies of bills of lading and/or other papers to JOB at address above.

ITEM NO.	QUANTITY	DESCRIPTION	UNIT	AMOUNT

SALES or USE TAX (is) (is not) included in amounts shown above.

F. O. B.

TERMS:

HENRY C. BECK COMPANY

By_______________

See above IMPORTANT NOTE for invoicing instructions. They MUST be complied with.

Show above order number on invoices, and on the outside of each package containing Shipment.

Accepted:_______________

By_______________

FIGURE 12.4 Formal purchase order (Courtesy Henry C. Beck Co.).

Regardless of the complexity of the transaction, certain basic elements are present in any purchase order. Five items can be identified as follows:

1. Quantity or number of items required.
2. Item description. This may be a standard description and stock number from a catalogue or a complex set of drawings and specifications.
3. Unit price.
4. Special instructions.
5. Signatures of agents empowered to enter into a contractual agreement.

For simple purchase orders, the buyer normally prepares the order. If the vendor is dissatisfied with some element of the order, he may prepare his own purchase order document as a counterproposal.

The special instructions normally establish any special conditions surrounding the sale. In particular, they provide for shipping and invoicing procedures. An invoice is a billing document that states the billed price of shipped goods. When included with the shipped goods it also constitutes an inventory of the contents of the shipment. One item of importance in the order is the basis of the price quotation and responsibility for shipment. Price quotations normally establish an FOB location at which point the vendor will make the goods available to the purchaser. FOB means Free On Board and defines the fact that the vendor will be responsible for presenting the goods free on board at some mutually agreed on point such as the vendor's sales location, factory, or the purchaser's yard or job site. This is important because if the FOB location is other than the vendor's location, the vendor is indicating that the price includes shipment. The vendor may quote the price as cost, insurance, and freight (CIF). This indicates that the quoted price includes item cost plus the shipment cost to include freight and insurance expenses to the FOB location.

In the event the vendor ships the goods, it is of interest to establish at what point in time title of ownership passes from the vendor to the purchaser. This is established by the *bill of lading*. The bill of lading is a contractual agreement between a common carrier and a shipper to move a specified group of goods from point *A* to point *B* at a contracted price. If ownership passes to the purchaser at the vendor's location, the contract for shipment is made out between the purchaser and the common carrier. In cases in which the vendor has quoted a CIF price, he acts as the agent of the purchaser in retaining a carrier and establishing the agreement on behalf of the purchaser. The bill of lading is written to pass title of ownership at the time of pickup of the goods by the common carrier at the vendor's location. In such cases, if the common carrier has an accident and damages the goods during transfer, the purchaser must seek satisfaction for the damage since he is owner.

If goods are to be paid for cash on delivery (i.e., COD), the title of ownership passes at the time of payment. In such cases, the bill of lading is between vendor and common carrier. If damage should occur during shipment, recovery of loss falls to the vendor as owner.

The sequence of events in CIF and COD transactions is shown in Figure 12.5. This figure also indicates the relationship between order, bill of lading, and invoice. A typical bill of lading memorandum and invoice are shown in Figures 12.6 and 12.7

The invoice normally states the payment procedures and establishes trade discounts that are available to the purchaser if he pays in a timely fashion. Trade discounts are incentives offered by the vendor for early payment. If the purchaser pays within a specified period, he must pay the stated price minus a discount. Failure to pay within the discount period means that the full price is due and payable. Terminology relating to trade discounts is as follows:

1. *ROG/AOG.* The discount period begins upon receipt of goods (ROG) or arrival of goods (AOG).

2. *2/10 NET 30 ROG.* This expression appearing on the invoice means 2% can be deducted from the face amount if the contractor pays within 10 days of AOG/ROG.

3. *2/10 PROX NET 30.* A 2% cash discount is available if invoice is paid not later than the 10th of the month following ROG. Payment is due in full by the end of the following month.

4. *2/10 E.O.M.* The discount (2%) is available to the 11th of the month

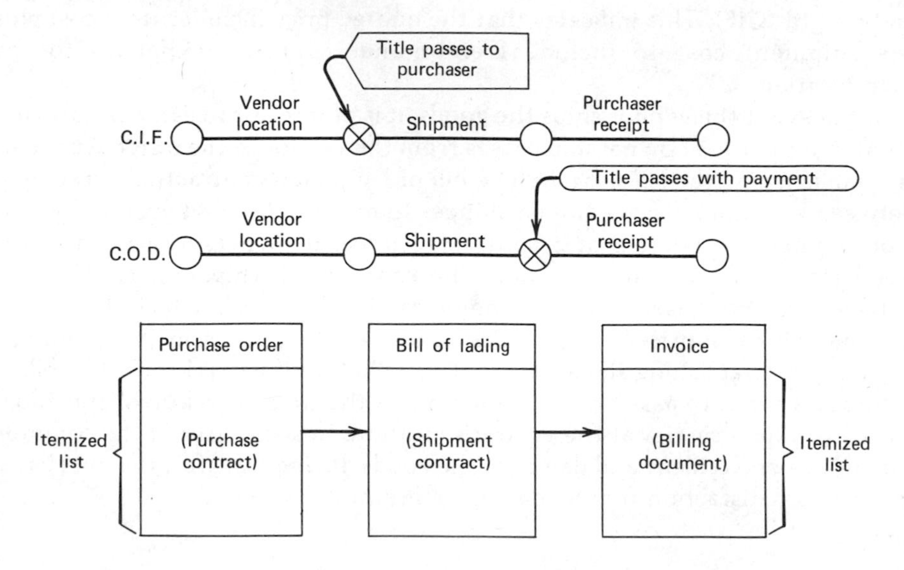

FIGURE 12.5 Procurement documents and title transfer sequence.

following ROG. Payment in full is due thereafter.

Trade discounts received are treated as earned income in financial statements.

The special conditions of the purchase order may include a "hold harmless" clause. Such clauses protect one of the parties to the purchase order from liability arising out of damages resulting from the conditions of the purchase order. A transit concrete mix company, for instance, may have the contractor submit his orders on their forms holding the vendor harmless for damages arising out of delivery of the concrete to the site. Thus, if the transit mix truck should back across a gas main on the site rupturing it during normal delivery, liability for repair costs will accrue to the contractor since the concrete vendor is "held harmless." The converse could, of course, occur if the contractor uses his own purchase order form, which holds him harmless in such event. These situations are not covered by normal liability insurance since such "contractually accruing" liability is considered to be outside the realm of normal liability. If the language of the order is prepared by the contractor, the hold harmless clause will operate to protect him. If the vendor's language is used, the special conditions will hold him harmless in these damage situations.

For the contractor's protection, reference is made in complex purchase orders (requiring special fabrication) to the contract specifications and other documents that define the materials to be supplied. Specifications detail the required *shop drawings, product data,* and *samples* that must be submitted for approval prior to fabrication and delivery. The provisions of the purchase order and the subcontract agreement require the subcontractor and supplier to obtain approval for their materials.

12.3 THE APPROVAL PROCESS

The contract drawings prepared by the architect are generally not specific enough to facilitate accurate fabrication of the materials involved. Therefore, to produce the necessary materials for a project, subcontractors and suppliers must provide details that further amplify the contract drawings. These details can be classified into three groups: (1) shop drawings, (2) product data, and (3) samples.

Shop drawings are defined in Paragraph 4.12.1 of the General Conditions as "Drawings, diagrams, schedules, and other data specially prepared for the work by the contractor and any subcontractor, manufacturer, supplier, or distributor to illustrate some portion of the work." The detailing, production, and supplying of shop drawings are the sole responsibility of the contractor or the contracted agent. However, the architect is responsible for verification that the supplied shop drawings correctly interpret the contract documents. Dimensions, quantities, and coordination with other trades are the responsibility of the contractor. Approved shop drawings become the critical working drawings of a project and are considered a part of the contract documents.

This Shipping Order Carbon, and retained by the Agent.

Shipper's No. ____________

____________________________ Carrier's No. ____________

(Name of Carrier)

RECEIVED, subject to the classifications and tariffs in effect on the date of the issue of this Bill of Lading.

at HALLANDALE, FLA. 19 From MEADOW STEEL PRODUCTS, INC.

the property described below, in apparent good order, except as noted (contents and condition of contents of packages unknown), marked, consigned, and destined as indicated below, which said carrier (the word carrier being understood throughout this contract as meaning any person or corporation in possession of the property under the contract) agrees to carry to its usual place of delivery of said destination, if on its own route, otherwise to deliver to another carrier on route to said destination. It is mutually agreed, as to each carrier of all or any of said property over all or any portion of said route to destination, and as to each party at any time interested in all or any of said property, that every service to be performed hereunder shall be subject to all the terms and conditions of the Uniform Domestic Straight Bill of Lading set forth (1) in Official, Southern, Western and Illinois Freight Classification in effect on the date thereof, if this is a rail or rail-water shipment, or (2) in the applicable motor carrier classification or tariff if this is a motor carrier shipment.

Shipper hereby certifies that he is familiar with all the terms and conditions of the said bill of lading, including those on the back thereof, set forth in the classification or tariff which governs the tranportation of this shipment, and the said terms and conditions are hereby agreed to by the shipper and accepted for himself and his assigns.

Consigned to ____________________________

(Mail or street address of consignee-For purposes of notification only.)

Destination ____________ State ________ Zip ________ County ________ Delivery Address * ____________

* To be filled in only when shipper desires and governing tariffs provide for delivery thereof.)

Route ____________________________

Delivering Carrier ____________ Car or Vehicle Initials ____________ No. ________

No. Packages	Kind of Package, Description of Articles, Special Marks, and Exceptions	*WEIGHT (Subject to Correction)	Class or Rate	Check Column
	REINFORCING STEEL ACCESSORIES		50	

*If the shipment moves between two ports by a carrier by water, the law requires that the bill of lading shall state whether it is carrier's or shipper's weight.

NOTE Where the rate is dependent on value, shippers are required to state specifically in writing the agreed or declared value of the property. Agreed or declared value of the property is hereby specifically stated by the shipper to be not exceeding.

per

† The fibre boxes used for this shipment conform to the specifications set forth in the box maker's certificate thereon, and all other requirements of the Consolidated Freight Classification.

Subject to Section 7 of Conditions of applicable bill of lading, if this shipment is to be delivered to the consignee without recourse on the consignor, the consignor shall sign the following statement:

The carrier shall not make delivery of this shipment without payment of freight and all other lawful charges.

(Signature of Consignor)

If charges are to be prepaid, write or stamp here: "To be Prepaid."

TO BE PREPAID

Received $ ______
to apply in prepayment of the charges on the property described hereon.

Agent or Cashier

Per ______
(The signature here acknowledges only the amount prepaid.)

Charges Advanced:

$ ______

† Shipper's imprint in lieu of stamp; not a part of Bill of Lading approved by the Interstate Commerce Commission.

MEADOW STEEL PRODUCTS, INC. Shipper, Per ______

Agent must detach and retain this Shipping Order and must sign the Original Bill of Lading.

Permanent post-office address of shipper **1804 SO. 31st AVE., HALLANDALE, FLA. 33009**

FIGURE 12.6 Typical bill of lading (Courtesy of Agusta MEADOW STEEL PRODUCTS, INC.).

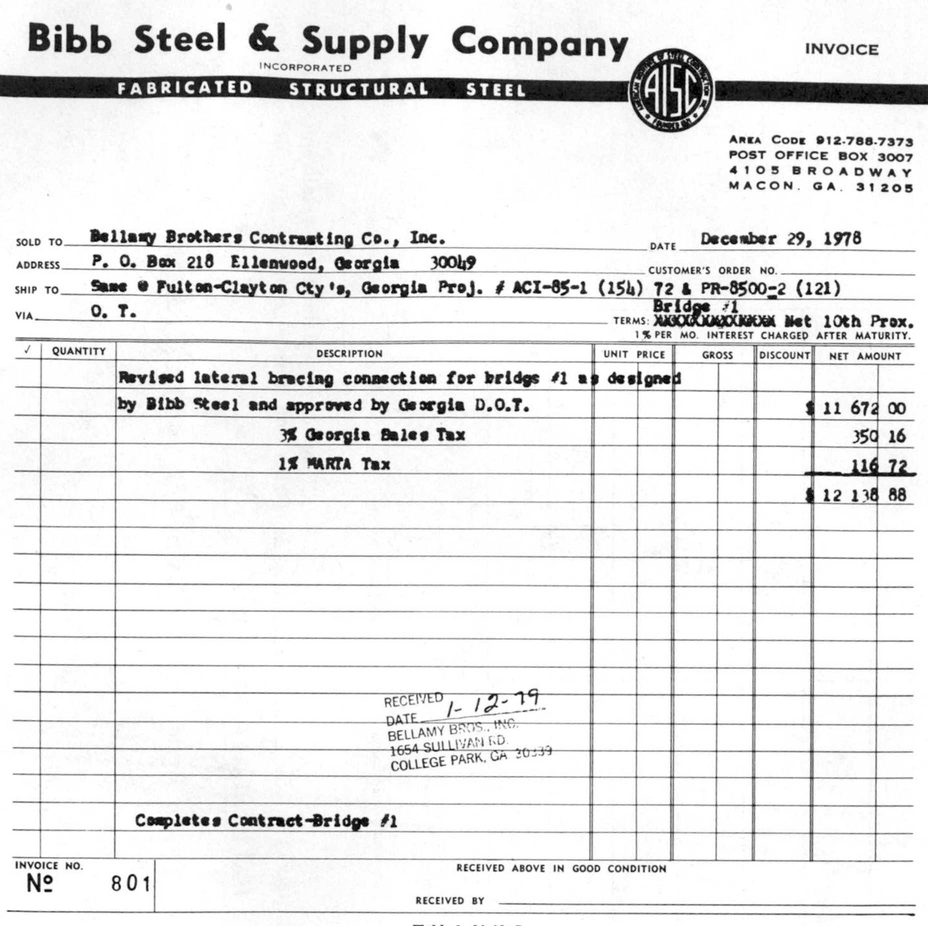

Bibb Steel & Supply Company INVOICE
INCORPORATED
FABRICATED STRUCTURAL STEEL
AISC

Area Code 912-788-7373
POST OFFICE BOX 3007
4105 BROADWAY
MACON, GA. 31205

SOLD TO Bellamy Brothers Contrasting Co., Inc. DATE December 29, 1978
ADDRESS P. O. Box 218 Ellenwood, Georgia 30049 CUSTOMER'S ORDER NO.
SHIP TO Same @ Fulton-Clayton Cty's, Georgia Proj. # ACI-85-1 (154) 72 & PR-8500-2 (121) Bridge #1
VIA O. T. TERMS: XXXXXXXXXX Net 10th Prox.
1% PER MO. INTEREST CHARGED AFTER MATURITY.

✓	QUANTITY	DESCRIPTION	UNIT PRICE	GROSS	DISCOUNT	NET AMOUNT
		Revised lateral bracing connection for bridge #1 as designed				
		by Bibb Steel and approved by Georgia D.O.T.				$ 11 672 00
		3% Georgia Sales Tax				350 16
		1% MARTA Tax				116 72
						$ 12 138 88
		Completes Contract-Bridge #1				

RECEIVED DATE 1-12-79
BELLAMY BROS., INC.
1654 SULLIVAN RD.
COLLEGE PARK, GA 30339

INVOICE NO. № 801 RECEIVED ABOVE IN GOOD CONDITION
RECEIVED BY

THANKS

FIGURE 12.7 Typical invoice (Courtesy of Bibb Steel & Supply Company).

Typically, shop drawings are submitted for materials such as reinforcing steel, formwork, precast concrete, structural steel, millwork, casework, metal doors, and frames.

Product data may be submitted to illustrate the performance characteristics of the material items described by the *shop drawings* or may be submitted as verification that a standard product meets the contract specifications. Paragraph 4.12.2 of the General Conditions defines Product Data as "illustrations, standard schedules, performance charts, instructions, brochures, diagrams, and other information furnished by the contractor to illustrate a material, product, or system for some portion of the work." Mill test reports, concrete mix designs, masonry fire rating tests, curtain wall wind test reports, and mechanical equipment performance tests are examples of product data.

Product data are particularly important when a subcontractor or supplier is submitting data on a product that is a variance from the contract specifications. The architect carefully analyzes the submitted data prior to rendering an approval of the substitution. Also, the product data are used extensively to coordinate the materials used by the mechanical and electrical subcontractors. The contractor must communicate the product data between these major subcontractors to ensure proper performance of their portion of the work.

Samples usually involve the finishes of a project and are physical examples of materials to be supplied. The architect may require samples of plastic laminate finishes for doors and counters, flooring, wall coverings, paint, stucco, precast concrete, ceilings, and other items. These are used by the architect in developing the overall building finish scheme.

The approval process involving shop drawings, product data, and samples has several substages that are critical to the material life cycle. These are: (1) submission by the subcontractor or supplier, (2) review of the submittal by the contractor, (3) review by the architect, and (4) return of submittal to the subcontractor or supplier.

At the time of awarding subcontracts and purchase orders, the contractor usually establishes the quantity, size, and other requirements for all submittals. In most cases, several blue line prints (usually six) are required when shop drawings are submitted for approval. The product data quantities required may range from three to six copies. The copies of a submittal may vary depending on the number of other subcontractors or vendors that must receive approved copies to coordinate their work. In all cases, careful planning of the quantity of submittals will expedite the other substages by eliminating the handling of unnecessary copies of submittals.

Timing of submittals is of utmost importance in the effective processing of material submittals. Subcontracts and purchase orders often contain language such as "all submittals must be made immediately" or "fifteen (15) days after execution of this agreement, all submittals must be made." In most cases, contractors do not preplan in detail the required submittal date from a subcontractor or supplier. The result is a landslide of submittals, most of which are not necessary, in the early stages of the project. Thus, field office personnel waste time sorting and determining the most critical submittals. A well-planned approach to scheduling submittals will assure timely processing and better control of required submittals.

Once a submittal is received by the contractor, the process of checking for conformance with the intent of the contract documents is performed. A submittal, whether it is a shop drawing, product data, or sample, is governed by the contract drawings and specifications. The contractor's field or main office personnel in charge of submittals may make notations and comments to the architect or his engineers to clarify portions of the submittal or to correct the submittal. The submittal represents specific details of the project and is of primary importance in coordination, as well as depicting exactly what a supplier or subcontractor is providing. The contractor is required by the general conditions to *clearly* note to the architect any variation from the contract documents.

The amount of time involved in the contractor's review of submittals may vary from one to five days, depending on the nature of the submittal and its correctness. Reinforcing steel and structural steel shop drawings typically require the greatest amount of time. Also, schedules such as doors, hardware, and door frames consume a great deal of time, because of the minute details that must be checked. However, the time expended in submittal processing by the contractor can most easily be controlled at this substage. It must be remembered that time spent in reviewing, checking, and coordinating submittals is one of the most effective methods of yielding a highly coordinated and smooth running project.

Once the contractor has completed the review of a submittal, the document is transmitted to the architect for approval. The contractor may indicate on the transmittal the date when approval is needed. Here again, the amount of time required for the architect to review a submittal depends on its complexity and whether or not other engineers (i.e., mechanical, electrical, or structural) must participate in the review. As a general rule, two to three weeks is a good estimate for the time required by the architect to complete the review and return the submittal.

The period when a submittal is in the hands of the architect is probably the most critical substage of the approval process for materials. During this critical substage, the contractor's submittal can be "lost in the shuffle" if the architect's activities are not monitored daily. The most common method of monitoring submittals is through the use of a submittal log, which indicates the date, description, and quantity of each submittal. From this log the contractor can develop a listing of critical submittals to monitor on a daily basis. Once the submittal leaves the contractor's control in the field office, its return must be followed constantly or valuable time will be wasted.

The final substage of the approval process for a material item is the return of the submittal to the supplier or subcontractor. The submittal may be in one of the following four states when returned to the architect:

1. Approved.
2. Approved with noted corrections, no return submittal needed.
3. Approved with noted corrections; however, a final submittal is required.
4. Not approved, resubmit.

The first through third designations would release the vendor or subcontractor to commence fabrication and delivery. The fourth stage would require that the approval process be repeated. In some cases the disapproval by the architect is due to a subcontractor or supplier not communicating clearly through the submittal the information needed. A meeting between all parties may then be arranged to seek a reasonable solution.

When the approval process is complete, the material has become an integrated part of the project. Its details have been carefully reviewed for conformance with the contract documents. Also, through this process, the item has been coordinated with all trades involved in its installation and incorporation into the project. The material is now ready for fabrication and delivery.

12.4 THE FABRICATION AND DELIVERY PROCESS

As a submittal is returned to the subcontractor or supplier, the needed delivery date to meet the construction schedule is communicated on the transmittal, verbally, or through other correspondence. In any event, delivery requirements are established and agreed on. The supplier or subcontractor may be required to return to the contractor corrected file and field-use drawings, product data, or samples. These are used to distribute to the contractor's field personnel (i.e., superintendent or foreman), and the other subcontractors and suppliers that must utilize these final submittals.

Of the four phases of a material's life cycle the fabrication and delivery process is the most critical. Generally, the largest amount of time is lost and gained in this phase. The duration of the fabrication and delivery process depends directly on the nature of the material and the amount of physical transformation involved. For these reasons, the contractor must employ every available method of monitoring materials throughout the fabrication and delivery process.

Contractors generally devote the largest amount of time and effort to controlling and monitoring the fabrication and delivery phase in the life cycle. The term *expediting* is most commonly used to describe monitoring methods in this phase of a material item's life. Methods used to ensure timely fabrication and delivery may range from using checklists developed from the job schedule to actually including this phase as a separate activity on a job schedule. Unfortunately, the fabrication and delivery only becomes an activity on the job schedule when the delivery becomes a problem. Extremely critical items requiring extended fabrication times often warrant visits by the contractor to the fabrication facility to ensure the material is actually in fabrication, and proceeding on schedule.

At the completion of fabrication, the delivery of the material is made and the final phase of the life cycle is begun. Materials delivered are checked for compliance with the approved submittal as regards quality, quantity, dimensions, and other requirements. Discrepancies are reported to the subcontractor or supplier. These discrepancies, whether they be shortages or fabrication errors, are subjected to the same monitoring and controlling process as the entire order. Occasionally they become extremely critical to the project and must be given a great deal of attention until delivery is made.

12.5 THE INSTALLATION PROCESS

The installation process involves the physical incorporation into the project of a material item. Depending on how effectively materials were scheduled and expedited, materials arriving at the jobsite may be installed immediately, partially installed and partially stored, or completely stored for later installation. When storage occurs, the installation process becomes directly dependent on the effective storage of materials.

One of the most important aspects of the effective storage of materials is the physical protection of material items. Careful attention must be given to protection from weather hazards such as prevention of water damage or even freezing. Another important aspect is protection against vandalism and theft. Finish hardware, for instance, is generally installed over a considerable time period. A secure hardware room is usually set aside where it is sorted, shelved, and organized to accomodate the finish hardware installation process.

Materials stored outside the physical building on the project site or within the building must be carefully planned and organized to facilitate effective installation. In high-rise-building construction material storage, each floor can be disastrous if careful planning is not used. For instance, materials stored concurrently on a floor may include plumbing and electrical rough-in materials, ductwork, window wall framing, glazing materials, drywall studs, and other items. The magnitude of the amount of materials involved warrants meticulous layout of materials. Equally important is the storage of materials to facilitate hoisting with a minimal amount of second handling. Reinforcing steel, for instance, may be organized in a "lay-down" area and then directly hoisted as needed. Adequate lay-down areas must be provided within reach of vertical hoisting equipment.

12.6 MATERIAL TYPES

Building construction materials can be logically grouped into three major categories. These categories are (1) bulk materials that require little or no fabrication, (2) manufacturer's standard items that require some fabrication, and (3) items that are fabricated or customized for a particular project. Grouping materials into categories such as the ones above can be of value in determining which materials warrant major contractor control efforts. Obviously, material items that require fabrication have longer life cycles because of submittal requirements and fabrication. These materials require a great deal of control by the contractor.

The bulk material category includes those materials that require very little vendor modification and can be delivered from vendor storage locations to the jobsite with very little fabrication delay. Table 12.1 lists examples of typical bulk materials in building construction projects. These materials usually require only a one- to five-day delivery time, following execution of purchase order or subcontract and approved submittals. Submittal requirements generally include only product and performance data.

Manufacturer's standard material items include materials that are usually stocked in limited quantities and are manufactured for the project after the order is executed and submittals are approved. Table 12.2 illustrates typical materials that are included in this category. Submittal requirements include detailed shop drawings, product and performance data, and samples. Finish materials such as paints, wallcoverings, floor coverings, and plastic laminates require a fully developed finish design for the project. Development of the finish design can have serious consequences on ordering and delivery of finish materials. Manufacturing and delivery time generally ranges from 3 to 12 weeks for these materials. This extended manufacturing and delivery time places considerable importance on planning and controlling these materials.

The fabricated category of construction materials must conform to a particular project's unique requirements. The fabricated item, however, is

Table 12.1 TYPICAL BULK MATERIALS

Paving materials
Fill materials—crushed stone, soil, sand, etc.
Dampproofing membrane
Lumber and related supplies
Form materials—plywood, post shores, etc.
Ready-mix concrete
Wire mesh
Stock reinforcing steel and accessories
Masonry
Stock miscellaneous metals
Soil and waste piping
Water piping
Electrical conduit
Electrical rough-in materials—outlet boxes, switch boxes, etc.
Caulking and sealants

Table 12.2 TYPICAL STANDARD MATERIAL ITEMS

General materials
Fencing materials
Formwork systems—metal and fiberglass pans, column forms, etc.
Brick paving
Brick or ceramic veneers
Standard structural steel members
Metal decking
Waterproofing products
Insulation products
Built-up roof materials
Caulking and sealants
Standard casework and millwork
Special doors
Metal-framed windows
Finish hardware and weather-stripping
Ceramic and quarry tile
Flooring materials
Acoustical ceilings
Paints and wallcoverings
Lath and plaster products
Miscellaneous specialties
Equipment-food service, bank, medical, incinerators, etc.
Building furnishings
Special construction items—radiation protection, vaults, swimming pools, integrated ceilings
Elevators, escalators, dumbwaiters, etc.

Mechanical and plumbing equipment and materials
Fire protection equipment
Water supply equipment
Valves
Drains
Clean-outs
Plumbing fixtures
Gas-piping accessories
Pumps
Boilers
Cooling Towers
Control systems
Air-handling equipment
Refrigeration units (chillers)

Electrical equipment and materials
Busduct
Special conduit
Switchboards and panels
Transformers
Wire
Trim devices
Lighting fixtures
Under-floor duct
Communications devices
Motors and starters
Motor control centers
Electric heaters
Fire alarm equipment
Lightning protection equipment

composed of or results from modification of standard components. Table 12.3 illustrates materials that fall into this category. Submittals required include highly detailed shop drawings, product data, and samples. Fabrication and delivery times range from 2 weeks for items such as reinforcing steel and precast concrete to 10 to 12 weeks for curtainwall systems, doors and frames, and similar items.

Table 12.3 TYPICAL FABRICATED MATERIAL ITEMS

Concrete reinforcement
Structural steel
Precast panels and decks
Stone veneers
Miscellaneous and special formed metals
Ornamental metals
Millwork
Custom casework and cabinetwork
Sheet metal work
Sheet metal veneers
Hollow metal doors and frames
Wood and plastic laminate doors
Glass and glazing
Storefront
Window walls and curtain walls

REVIEW QUESTIONS AND EXERCISES

12.1 Name four important items of information that should be on a typical purchase order.

12.2 What are four good sources of price information about construction materials?

12.3 What is meant by the following expressions?
(a) CIF
(b) 2/10 E.O.M.
(c) 2/10 net 30
(d) ROG
(e) Bill of lading

12.4 Visit a local architect's office and ascertain how product data are obtained and used.

12.5 Visit a local building contractor and determine how he handles control of submittals from subcontractors to architect/engineer. What system does he use to ensure the job will not be held up due to procurement and approval delays?

12.6 Visit a construction site and determine what procedures are used for verifying receipt arrival and ensuring proper storage of materials at the site.

12.7 Determine what procedures are used for removing waste material from a local construction (building) site. Is there any scrap value in these materials? Explain.

12.8 Determine the local prices for some bulk materials such as concrete, sand, cement, steel mesh, bricks, and lumber and compare them to the periodically published prices in the *Engineering News Record.*

12.9 Select a particular material item (e.g., concrete) and follow its material handling process from the local source through final installation in the building? What special equipment is needed (if any)?

12.10 What types of special materials handling equipment can be identified on local building sites? Do they take advantage of certain properties of the material being handled (e.g., the fluidity of concrete)?

CHAPTER THIRTEEN

Labor Relations

13.1 THE LABOR RESOURCE

The man-power component of the four M's of construction is by far the most variable and unpredictable. It is, therefore, the element that demands the largest commitment of time and effort from the management team. Manpower or labor has four major aspects that are of interest to management. To properly understand the management and control of labor as a resource, the manager must be aware of the interplay between:

1. Labor organization.
2. Labor law.
3. Labor cost.
4. Labor productivity.

The cost and productivity components were central to the discussion of equipment management in Chapters 9 and 11. Labor includes the added human factor. This element can only be understood in the context of the prevailing legal and organizational climate that is characteristic of the construction industry.

13.2 SHORT HISTORY OF LABOR ORGANIZATIONS

The history of labor organizations begins in the early nineteenth century, and their growth parallels the increasing industrialization of modern society. Initially tradesmen processing some skill or craft began organizing into groups variously called guilds, brotherhoods, or mechanics societies. Their objectives were to provide members, widows, and children with sickness and death benefits. In addition, these organizations were interested in the development of trade proficiency standards and the definition of skill levels such as apprentice and journeyman. They were often "secret" brotherhoods, since such organizations were considered unlawful and illegal conspiracies posing a danger to the society.

From the 1840s until the era of the New Deal* in the 1930s, the history of labor organizations is the saga of confrontation between management and workers, with the pendulum of power on the management side. With the coming of the New Deal and the need to rejuvenate the economy during the Depression period, labor organizations won striking gains which virtually reversed the power relationship between managers and workers. The American Federation of Labor (AFL) was organized by Samuel Gompers in 1886. This was the first successful effort to organize skilled and craft workers such as cabinetmakers, leather tanners, and blacksmiths. Since its inception, the AFL has been identified with skilled craft workers as opposed to industrial "assembly line" type of workers. The Building Trades Department of the AFL, which is the umbrella organization controlling all construction craft unions, was organized in 1908.

The semiskilled and unskilled factory workers in "sweat box" plants and mills were largely unorganized at the time Gompers started the AFL. Many organizations were founded and ultimately failed in an attempt to organize the industrial worker. These organizations, with euphonious sounding names such as Industrial Workers of the World and the Knights of Labor, had strong political overtones and sought sweeping social reforms for all workers. This was particularly attractive to immigrant workers arriving from the socially repressive and politically stagnant atmosphere in Europe. Such organizations attracted political firebrands and anarchists preaching social change and upheaval at any cost. Confrontation with the police was common and violent riots often led to maiming and killing. The most famous such riot occurred in the Haymarket in Chicago in 1886.

Gompers was seriously interested in protecting the rights of skilled workers and had little interest in the political and social oratory of the unskilled labor organizations. Therefore, separate labor movements representing skilled craft and semiskilled factory workers developed and did not combine until the 1930s. This led to different national and local organizational structures and bargaining procedures which are still utilized and strongly influence the labor picture even today.

In the 1930s, industrial (i.e., factory semiskilled) workers began to organize effectively with the support of legislation evolving during the postdepression period. The AFL, realizing such organizations might threaten their own

*The democratic administration of President Franklin D. Roosevelt.

dominance, recognized these organizations by bringing them into the AFL camp with the special designation of Federal Locals. Although nominally members, the industrial workers were generally treated as second-class citizens by the older and more established craft unions. This led to friction and rivalry that culminated in the formation of the Committee for Industrial Organizations (CIO). This committee was established in 1935 unilaterally by the industrial locals without permission from the governing body of the AFL. The act was labelled treasonous and the AFL board ordered the Committee to disband or be expelled. The AFL suspended the industrial unions in 1936. In response, these unions organized as the Congress of Industrial Organizations (CIO), with John L. Lewis of the United Mine Workers as the first CIO president. Following this rift between the industrial and craft union movements, the need to cooperate and work together was apparent. However, philosophical and personal differences prevented this until 1955, at which time the two organizations combined to form the AFL-CIO. This organization remains the major labor entity in the United States today.

13.3 EARLY LABOR LEGISLATION

The courts and legislative bodies of the land have alternately operated to retard or accelerate the progress of labor organizations. The chronology of major items of legislation and the significant events in the labor movement are shown in Table 13.1. At the outset, the law was generally interpreted in order to check organization of labor and, therefore, management was successful in controlling the situation. The most classic illustration of this is the application of the Sherman Antitrust Law to enjoin workers from organizing. The Sherman Antitrust Act had originally been enacted in 1890 to suppress the formation of large corporate trusts and cartels, which dominated the market and acted to fix prices and restrain free trade. The oil and steel interests formed separate cartels in the late nineteenth century to manipulate the market. More recently, International Business Machines (IBM) has been reviewed by the Justice Department for similar alleged activity in the computer market. To break up such market dominance, the Antitrust Law provides the government with the power to enjoin corporations from combining to control prices and restrict trade. In 1908 the Supreme Court ruled that the Antitrust Law could be applied to prevent labor from organizing. The argument ran roughly as follows: "If laborers are allowed to organize, they can act as a unit to fix wage prices and restrict free negotiations of wages. This is a restraint of trade and freedom within the labor market." Based on this interpretation, local courts were empowered to issue injunctions to stop labor from organizing. If a factory owner found his workers attempting to organize, he could simply go to the courts and request an injunction forbidding such activity.

Table 13.1 CHRONOLOGY OF LABOR LAW AND ORGANIZATION

LABOR LAW		LABOR MOVEMENT	
1890	Sherman Antitrust Act	1886	AFL—founded by Samuel Gompers — Knights of Labor organized factory workers
1908	Supreme Court supported application to Union Activity		
1914	Clayton Act Ineffective-individual basis —as court ruled	1905	Industrial Workers of the World
1931	Davis—Bacon Act On Federal contracts Wages and fringes pay at prevailing rate	1907	Building and Construction Trades Department of AFL founded
1932	Norris-LaGuardia (Anti-injunction Act)	1930s	Take in industrial workers as Federal Locals
1935	Wagner Act (National Labor Relations Act)	1935	Committee for Industrial Organization AFL ordered disbanding
1938	Fair Labor Standards Act Minimum wages, maximum hours defined	1936	Federal Locals (CIO) thrown out
1943	Smith-Connolly Act (War-Labor Disputes Act) Reaction to labor in wartime Ineffective	1938	Congress of Industrial Organizations
		1940s	Wartime strikes—accused of not supporting war effort Criminal activities alleged
1946	Hobbs Act "Anti-Racketeering law" Protect employer from paying kickbacks to labor	1955	AFL and CIO reconcile differences and recombine as AFL-CIO
1947	Taft-Hartley (Labor Management Relations Act)		
1959	Landrum Griffin Act (Labor Management Reporting and Disclosures)		
1964	Title IV Civil Rights Act		

In 1914, the Congress acted to offset the effect of the Sherman Antitrust Act by passing the Clayton Act. This act authorized employees to organize to negotiate with a particular employer. However, in most cases the employer could demonstrate that the organizing activity was directed by parties outside the employer's shop. This implied that the action was not a local one and, therefore, was subject to action under the Sherman Antitrust legislation. Therefore, the injunction remained a powerful management tool in resisting unionization.

13.4 THE NORRIS-LAGUARDIA ACT

The passing of the Norris-LaGuardia Act heralded the first major movement of the power pendulum away from management and toward labor. This act, sometimes referred to as the Anti-Injunction Act, accomplished what the Clayton Act had failed to do. It specifically stated that the courts could not intercede on the part of management so as to obstruct the formation of labor organizations. It effectively overrode the Supreme Court interpretation that the Sherman Antitrust Act could be applied to labor organizations. It curtailed the power of the courts to issue injunctions and protected the rights of workers to strike and picket peaceably. It also outlawed the use of so-called "yellow dog" contracts on the part of management. It was a common practice to have an employee sign a contract upon being hired, in which he agreed not to join or become active in any union organization. Such yellow dog contracts were declared illegal by the Norris-LaGuardia Act. This piece of legislation as interpreted by the Supreme Court during the period of the New Deal effectively freed labor from the constraints of the Sherman Antitrust Act.

13.5 THE DAVIS BACON ACT

In 1931, a very far-reaching piece of legislation was passed that even today has a significant impact on the cost of federally funded projects throughout the United States. The Davis Bacon Act provides that wages and fringe benefits on all federal and federally funded projects shall be paid at the "prevailing" rate in the area. The level of prevailing rates is established by the secretary of labor, and a listing of these rates is published with the contract documents so that all contractors will be aware of the standards. To ensure that these rates are paid, the government requires submittal by all contractors of a certified payroll each month to the federal agency providing the funding. These rates are reviewed to determine whether any violations of the Davis Bacon pay scale have occurred. This act is so far reaching in its effect because much of public construction at the state and local level may be funded in part by federal grants. A large municipal mass transit system or waste water treatment plant, for instance, may be funded in part by a federal agency. In such cases, the prevailing rates must be paid (see Figure 13.1). Since the Department of Labor generally accepts the most recently negotiated union contract rates as the prevailing ones, this allows union contractors to bid without fear of being underbid by nonunion contractors paying lower wage rates.

13.6 THE NATIONAL LABOR RELATIONS ACT

The National Labor Relations Act, also referred to as the Wagner Act, is a landmark piece of legislation that established a total framework within which labor-management relations were to be conducted. Its central purposes were to protect union-organizing activity and encourage collective bargaining. Employers are required to bargain in good faith with the properly chosen representatives of the employees. Among other things, it establishes the procedures by which labor can organize and elect representatives. Discrimination against an employee for labor-organizing activities or participation in a union is forbidden by this act.

Employer unfair labor practices defining precisely what actions are not acceptable in management dealings with labor are specified. These practices are summarized in Table 13.2. Comparable unfair practices on the part of labor in dealing with management were not defined. It was assumed that labor was the abused party and would act equitably in its dealings with management. This trust had to be specifically spelled out later in the Taft-Hartley Act.

The act also established a "watch dog" organization to ensure its provisions were properly administered. This organization is the National Labor Relations Board (NLRB). The NLRB acts as the clearinghouse for all grievances and issues leading to complaints by labor against management and vice-versa. It is the highest tribunal below the Supreme Court for settling labor disputes and rules on most issues affecting labor-management relationships.

The act also established the concept of a closed shop. For years, labor organizations had fought for the right to force all members of a particular work activity (shop) to be members of a union. If the majority voted for union membership, then in order to work in the shop, a new employee had to belong to the union. This is in contrast to the "open" shop, in which employees are not organized and do not belong to a union. The Wagner Act endorsed the concept of the "closed" shop and made it legal. This concept was later revoked by the Taft-Hartley Act and replaced by the "union" shop. The closed shop was attacked as illegal, since it infringed upon a person's "right to work" and freedom of choice regarding union membership. The union shop will be discussed later in the section on the Taft-Hartley Act.

2 CONTRACTS CLEARED

MARTA Wins a Round

By RALEIGH BRYANS

A new wage finding by the U. S. Department of Labor puts MARTA in position to let two additional construction contracts by mid-November, officials said Thursday.

And the finding—specifically, a "wage determination" establishing minimum wages that must be paid on MARTA construction jobs—was substantially in MARTA's favor.

The two MARTA (Metropolitan Atlanta Rapid Transit Authority) projects that evidently are coming unstuck now are among a half dozen that have been stalled for most of the past seven months while the transit agency hassled with the Labor Department over the levels the wage minimums would take.

Federal minimums established for the two projects are in most instances lower than those issued in April. The April determination had fixed wage minimums that MARTA officials felt would dangerously elevate the overall cost of the $2.1 billion rail rapid transit system it is about to begin to build.

And, because of that, MARTA chose to challenge the April wage determination, even if it meant a substantial delay in getting on with the construction of the MARTA system.

MARTA to date has begun only one construction job, a smaller one to develop an underpass at Arizona Avenue in the Kirkwood section of Atlanta.

The second project to go to contract will be one of those covered by the Wednesday Labor Department ruling. It is a $2-3 million project to lay the railbed of a small section of the proposed MARTA West Line.

Officials said Thursday that bids should be opened about Nov. 12 on this particular project, called Construction Contract Unit (CCU) 540.

The project already had been advertised but now bidders must be given an addendum specifying the federal minimum wages required.

A bid opening on a second project, identified by MARTA as CCU 170, probably will be scheduled a week after that for CCU 540, officials said. CCU 170 is a project

Turn to Page 12A, Column 7

MARTA WINS ROUND

Continued from Page 1A

to lay railbed for a portion of the East Line.

While MARTA officials like staff counsel Jeffrey Trattner expressed hope these two projects are now in the clear, this was not absolutely the case, since labor unions here conceivably might contest this new wage determination.

Officials of the North Georgia Building Trades Council have suggested they might challenge a recent ruling by the Wage Appeals Board which preceded the Wednesday decision.

T. D. Archer, president of the council, said he doesn't like the wage minimums established by the Labor Department. But he declined to say whether a new challenge is in the offing.

The Labor Department is empowered by the Davis-Bacon Act to establish wage minimums on projects, like MARTA, that are financed in whole or in part by the federal government.

Here is a chart showing how the latest wage minimums for certain skills compare with those for last November and last April:

	Nov. '74	Apr. '75	Oct. '75
Bulldozer optrs	$4.48	$7.65	$4.37
Carpenters	5.23	7.95	5.19
Cement masons	4.65	6.10	4.50
Crane optrs	7.85	8.00	5.50
Electricians	9.55	9.55	9.55
Iron workers	6.40	7.90	7.90
Loaders	4.15	4.24	5.02
Laborers	3.32	3.98	3.22

FIGURE 13.1 Illustrative application of Davis-Bacon Act (Reproduced with the permission of the Atlanta Journal).

Table 13.2 EMPLOYER UNFAIR LABOR PRACTICES

Under the National Labor Relations Act, as amended, an employer commits an unfair labor practice if he:

1. Interferes with, restrains, or coerces employees in the exercise of rights protected by the act, such as their right of self-organization for the purposes of collective bargaining or other mutual assistance [Section 8(a) (1)].
2. Dominates or interferes with any labor organization in either its formation or its administration or contributes financial or other support to it [Section 8(a) (2)]. Thus "company" unions that are dominated by the employer are prohibited, and employers may not unlawfully assist any union financially or otherwise.
3. Discriminates against an employee in order to encourage or discourage union membership[Section 8(a) (3)]. It is illegal for an employer to discharge or demote an employee or to single him out in any other discriminatory manner simply because he is or is not a member of a union. In this regard, however, it is not unlawful for employers and unions to enter into compulsory union-membership agreements permitted by the National Labor Relations Act. This is subject to applicable state laws prohibiting compulsory unionism.
4. Discharges or otherwise discriminates against an employee because he has filed charges or given testimony under the act [Section 8(a) (4)]. This provision protects the employee from retaliation if he seeks help in enforcing his rights under the act.
5. Refuses to bargain in good faith about wages, hours, and other conditions of employment with the properly chosen representatives of his employees [Section 8(a) (5)]. Matters concerning rates of pay, wages, hours, and other conditions of employment are called mandatory subjects, about which the employer and the union must bargain in good faith, although the law does not require either party to agree to a proposal or to make concessions.
6. Enters into a hot-cargo agreement with a union [Section 8(e)]. Under a hot-cargo agreement, the employer promises not to do business with or not to handle, use, transport, sell, or otherwise deal in the products of another person or employer. Only in the garment industry and the construction industry (to a limited extent) are such agreements now lawful. This unfair labor practice can be committed only by an employer and a labor organization acting together.

Source. From Clough, *Construction Contracting*, 3rd Ed., John Wiley & Sons, Inc., New York, 1975.

13.7 FAIR LABOR STANDARDS ACT

The Fair Labor Standards Act is commonly referred to as the minimum wage law. It was originally passed in 1938 and establishes the minimum wages and maximum hours for all workers. The minimum wage level is periodically changed to be consistent with the value of money and was recently increased to $3.10 per hour. The law defines the 40-hour work week with time over this considered to be overtime. It is, generally, an outgrowth of the "child" labor abuses that occurred in the nineteenth century. It also forbids discrimination by establishing the concept of "equal pay for equal work." Recent arguments against increasing the minimum wage have hinged on the ideas that certain menial and domestic tasks that could provide certain unskilled workers with employment have become so expensive that it no longer is reasonable to perform them. The clearing of refuse and cutting of grass along roadways was done in former times by hand labor at low wages. Increasing minimum wages makes this too expensive.

13.8 UNION GROWTH

Under the provisions of the union legislation of the 1930s, the labor unions began to flourish. As is often the case during periods of transition, where inflexible barriers previously existed, a vacuum in favor of labor developed. The hard line of management was broken, and labor rushed in to organize and exploit the new situation. Along with the benefits accruing to the worker from these events, the inevitable abuses of the unstructured and unrestricted growth soon became apparent. In 1938, the unbridled actions of the unions and their leaders started to swing public opinion against them. Some unions flaunted their newfound power by introducing *restrictive labor practices* and *wartime strikes,* which shut down plants producing critical military supplies. Criminal activities within the unions were widespread and virtually unchecked. In 1943, the Congress responded to this changing public perception of unions by passing the War Labor Disputes Act (Smith Connolly Act). This reflected public displeasure with the high-handed tactics and unpatriotic stance of the labor unions. It was designed to limit strikes in critical wartime industries and expedite settlement of disputes. It was largely ineffective, but did reflect increasing public support of legislation that would control the prerogatives of labor unions. By 1947, thirty-seven states had enacted some form of labor control bill.

The inroads made by criminal elements active in union activities were recognized by the Hobbs (Anti-Racketeering) Act of 1946. This legislation was enacted to protect employers from threats, force, or violence by union officials extorting payments for "services rendered." Payments requested included commissions for various types of aid and assistance, gifts for controlling labor trouble, and equipment rentals forced on the employers at exorbitant costs. These laws and the continuing difficulties developing from abuse of power on the part of unions set the stage for the enactment of the Taft-Hartley Act of 1947.

13.9 LABOR MANAGEMENT RELATIONS ACT

The Labor Management Relations (Taft-Hartley) Act, together with the Wagner Act, form the two cornerstones of American labor relations legislation. It amended the Wagner Act and reversed the swing of the power pendulum once more, still leaving labor in a very strong position, but pushing the pendulum more toward center. It is the first postdepression law to place effective constraints on the activities of labor. It restructured the makeup and operation of the National Labor Relations Board (NLRB), attempting to give management a stronger voice and to balance representation of labor and management. Section 7 of the bill defines the rights of workers to participate in or refrain from union activities. Section 8 provides the counterpoint to the Employer Unfair Practices Section of the Wagner Act. It defines *Union Unfair Labor Practices,* which specify tactics on the part of labor that are illegal (see Table 13.3). The law also established the Federal Mediation and Conciliation Services, which acts as a third party in trying to expedite a meeting of the minds between unions and management involved in a dispute. This service has been very visible in meeting with representatives of players unions and sports team owners in order to work out the terms of player contracts.

Under the Taft-Hartley legislation, the president is empowered to enjoin workers on strike (or preparing to strike) to work for a 90-day "cooling off" period during which time negotiators attempt to reach agreement on contractual or other disputes. This strike moratorium may be invoked in industries where a strike endangers the health of the national economy. Most recently this power was utilized by President Carter in ordering mine workers to return to work. It proved ineffective, however, since the workers refused to follow the court order and the government took no further action.

Section 14(b) is significant in that it redefines the legality of closed shop operations and defines the union shop. A totally closed shop is one in which the worker must be a union member before he is considered for employment. This is declared illegal by the Taft-Hartley Act. The union shop is legal. A union shop is one in which a nonmember can be hired. He is given a grace period (usually 30 days in manufacturing shops and a shorter period in the construction industry), during which time he must become a union member. If he does not become a member, the union can request he be released. Under the closed shop concept, it was much easier for the unions to block a worker from gaining employment. This could be used to discriminate against a potential employee. The union shop gives the worker a chance to join the

Table 13.3 UNION UNFAIR LABOR PRACTICES

Under the National Labor Relations Act, as amended, it is an unfair labor practice for a labor organization or its agents:

1. a. To restrain or coerce employees in the exercise of their rights guaranteed in Section 7 of the Taft-Hartley Act [Section 8(b) (1) (A)]. In essence Section 7 gives an employee the right to join a union or to assist in the promotion of a labor organization or to refrain from such activities. This section further provides that it is not intended to impair the right of a union to prescribe its own rules concerning membership.
   b. To restrain or coerce an employer in his selection of a representative for collective bargaining purposes [Section 8(b) (1) (B)].
2. To cause an employer to discriminate against an employee in regard to wages, hours, or other conditions of employment for the purpose of encouraging or discouraging membership in a labor organization [Section 8(b) (2)]. This section includes employer discrimination against an employee whose membership in the union has been denied or terminated for cause other than failure to pay customary dues or initiation fees. Contracts or informal arrangements with a union under which an employer gives preferential treatment to union members are violations of this section. It is not unlawful, however, for an employer and a union to enter an agreement whereby the employer agrees to hire new employees exclusively through a union hiring hall so long as there is no discrimination against nonunion members. Union-security agreements that require employees to become members of the union after they are hired are also permitted by this section.
3. To refuse to bargain in good faith with an employer about wages, hours, and other conditions of employment if the union is the representative of his employees [Section 8(b) (3)]. This section imposes on labor organizations the same duty to bargain in good faith that is imposed on employers.
4. To engage in, or to induce or encourage others to engage in, strike or boycott activities, or to threaten or coerce any person, if in either case an object thereof is:
   a. To force or require any employer or self-employed person to join any labor or employer organization, or to enter into a hot-cargo agreement

Table 13.3 (Continued) UNION UNFAIR LABOR PRACTICES

---

that is prohibited by Section 8(e) [Section 8(b) (4) (A)].

b. To force or require any person to cease using or dealing in the products of any other producer or to cease doing business with any other person [Section 8(b) (4) (B)]. This is a prohibition against secondary boycotts, a subject discussed further in Section 13.18. This section of the National Labor Relations Act further provides that, when not otherwise unlawful, a primary strike or primary picketing is a permissible union activity.

c. To force or require any employer to recognize or bargain with a particular labor organization as the representative of his employees that has not been certified as the representative of such employees [Section 8(b) (4) (C)].

d. To force or require any employer to assign certain work to the employees of a particular labor organization or craft rather than to employees in another labor organization or craft, unless the employer is failing to conform with an order or certification of the NLRB [Section 8(b) (4) (D)]. This provision is directed against jurisdictional disputes, a topic discussed in Section 13.12.

5. To require of employees covered by a valid union shop membership fees that the NLRB finds to be excessive or discriminatory [Section 8(b) (5)].
6. To cause or attempt to cause an employer to pay or agree to pay for services that are not performed or not to be performed [Section 8(b) (6)]. This section forbids practices commonly known as featherbedding.
7. To picket or threaten to picket any employer to force him to recognize or bargain with a union:
   a. When the employees of the employer are already lawfully represented by another union [Section 8(b) (7) (A)].
   b. When a valid election has been held within the past 12 months [Section 8(b) (7) (B)].
   c. When no petition for a NLRB election has been filed within a reasonable period of time, not to exceed 30 days from the commencement of such picketing [Section 8(b) (7) (C)].

Source. From Clough, *Construction Contracting*, 3rd Ed., John Wiley & Sons, Inc., New York, 1975.

union (see Figure 13.2). If he requests membership and the union refuses after 30 days,* management can ask the union to show cause why the employee has not been admitted to membership.

The law also recognizes the concept of "agency" shop. In such facilities a worker can refuse to join the union. He, therefore, has no vote in unior affairs. He must, however, pay union dues since he theoretically benefits from the actions of the union and the union acts as his "agent." If the union, for instance, negotiates a favorable pay increase, all employees benefit and all are required to financially support the labor representation (i.e., the union negotiators).

Because of the way in which the law regarding closed ship under the Wagner Act was implemented, many workers felt that their constitutional "right to work" was being abrogated. That is, unless they were already union members, they were not free to work in certain firms. They had no choice. They were forced to either join the union or go elsewhere. The Taft-Hartley allows the individual states to enact right to work laws that essentially forbid the establishment of totally union shops. States in the south and the southwest where unions are relatively weak have implemented this feature at the state level. Clough explains this as follows:

> Section 14(b) of the Taft-Hartley Act provides that the individual states have the right to forbid negotiated labor agreements that require union membership as a condition of employment. In other words, any state or territory of the United States may, if it chooses, pass a law making a union-shop labor agreement illegal. This is called the "right-to-work" section of the act, and such state laws are termed right-to-work statutes. At the present writing, 19 states have such laws in force.** It is interesting to note that most of these state right-to-work laws go beyond the mere issue of compulsory unionism inherent in the union shop. Most of them outlaw the agency shop, under which nonunion workers must pay as a condition of continued employment the same initiation fees, dues, and assessments as union employees, but are not required to join the union. Some of the laws explicitly forbid unions to strike over the issue of employment of nonunion workers.†

The fact that a right-to-work provision has been implemented can be detected by reading the language of the labor agreements within a given state. In states in which no right-to-work law is in effect, a clause is included indicating that a worker must join the union within a specified period. Generally in construction contracts, this is only seven days since construction labor is very transient in nature. Such a clause taken from an Illinois labor contract is shown in Figure 13.2. This clause would be illegal in Georgia.

*The period shown in Figure 13.2 is seven days. This is typical in the construction industry and recognizes the more transient nature of construction work.

**Alabama, Arizona, Arkansas, Florida, Georgia, Iowa, Kansas, Mississippi, Nebraska, Nevada, North Carolina, North Dakota, South Carolina, South Dakota, Tennessee, Texas, Utah, Virginia and Wyoming now have right-to-work legislation in effect.

†R. H. Clough, *Construction Contracting*, 3rd Ed., John Wiley & Sons, Inc., New York, 1975, p. 301.

All present employees who are members of the Union on the effective date of this agreement shall be required to remain members in good standing of the Union as a condition of their employment.

All present employees who are not members of the Union shall, from and after the 7th day following the date of execution of this agreement, be required to become and remain members in good standing of the Union as a condition of their employment.

All employees who are hired hereafter shall be required to become and remain members in good standing of the Union as a condition of their employment from and after the 7th day of of their employment or the effective date of this Agreement, whichever is later, as long as Union membership is offered on the same terms as other members.

Any employee who fails to become a member of the Union or fails to maintain his membership therein in accordance with provisions of the paragraphs of this Section, shall forfeit his rights of employment and the employer shall within two (2) working days of being notified by the Union in writing as to the failure of an employee to join the Union or maintain his membership therein, discharge such employee. For this purpose, the requirements of membership and maintaining membership shall be consistent with State and Federal Laws. The Employer shall not be deemed in default unless he fails to act within the required period after receipt of registered written notice.

(Excerpted from *Agreement Between Central Illinois Builders and The United Brotherhood of Carpenters and Joiners of America Local Union No. 44, Champaign-Urbana, Illinois June 1970*)

FIGURE 13.2 Contract typical member clause.

13.10 OTHER LABOR LEGISLATION

The Labor Management Reporting and Disclosure (Landrum-Griffin) Act was passed in 1959 to correct some of the deficiencies of previous legislation. Among its major objectives were (1) the protection of the individual union member, (2) improved control and oversight of union elections, and (3) an increased government role in auditing the records of unions. Misappropriation of union funds by unscrupulous officials and apparent election fraud were the central impetus in enacting this law. Under this law, all unions must periodically file reports with the Department of Labor regarding their organization finances and other activities. The act provides that employers cannot make payments directly to union officials. They can, however, pay dues and fringe benefits to qualified funds of the union for things such as health and welfare, vacation, apprenticeship programs, and the like. Records regarding these funds are subject to review by government auditors.

Title IV of the Civil Rights Act (enacted in 1964) establishes the concept of equal employment opportunity. It forbids discrimination on the basis of race, color, religion, sex, or national origin. It is administered by the Equal Employment Opportunity Commission (EEOC) and applies to discrimination in hiring, discharge, conditions of employment, and classification. Its application in the construction industry has led to considerable controversy. Individual workers can file an *unfair labor practice charge* against a union because of alleged racial discrimination. Unions found guilty face *cease and desist* orders as well as possible recision of their mandate to act as the authorized employee representative.

Executive Order 11246 issued by President Johnson in 1965 further amplified the government position on equal opportunity. It establishes affirmative action requirements on all federal government or federally funded construction work. It is administered by the Office of Federal Contract Compliance (OFCC). This office is instrumental in establishing the level of minority participation in government work. It has spawned a number of plans for including minority contractors in federally funded projects. The best known of these attempts to ensure minority participation in the construction market is the so-called "Philadelphia Plan." Executive Order 11375 (1968) extends Order 11246 to include sex discrimination. Contractors working on federally funded work are required to submit affirmative action reports to the OFCC. If the plan is found to be deficient, the OFCC can suspend or terminate the contract for noncompliance.

13.11 VERTICAL VERSUS HORIZONTAL LABOR ORGANIZATION STRUCTURE

The traditional craft unions are normally referred to as horizontally structured unions. This is because of the strong power base that is located in the union local. Contract negotiations are conducted at the local level and all major decisions are concentrated at the local level. Construction unions are craft unions with a strong local organization. The local normally is run on a day-to-day basis by the *business agent*. His representatives at the individual job sites are called job stewards. The local elects officers and a board of directors on a periodic basis. The local president and business agent may be the same individuals. The bylaws of the local define the organizational structure and particulars of union structure. At the time of contract negotiations, representatives from the local meet with representatives of the local union contractors to begin discussions. The Associated General Contractors (AGC) in the local area often act as the contractor's bargaining unit. This horizontal structure leads to a proliferation of contracts and a complex bargaining calendar for the contractors' association. If a contractors' group generally deals with 12 craft unions in the local area and renegotiates contracts on an annual or biennial basis, it is obvious that the process of meeting and bargaining can become complicated. Contracts are signed for each union operating in a given area. A résume of the contracts for unions operating in the area of the North Georgia Building Trades Council is shown in Figure 13.3. The national headquarters organizations for construction craft unions normally coordinate areas of national interest to the union, such as congressional lobbying, communication of information regarding recently negotiated contracts, national conventions, printing of newsletters and magazines, seminars, workshops, and other general activities. The real power in most issues, however, is concentrated at the local level. The horizontal organization then is similar to a confederation, with strength at the bottom and coordination at the top.

Vertically structured unions tend to concentrate more of the power at the national level. Significantly, labor contracts are negotiated at the national level. This means a contract is signed at the national level covering work throughout the country. This is considerably more efficient than the hundreds of locally negotiated contracts that are typical of horizontally structured unions. The industrial unions of the CIO have traditional organization in a vertical structure, while the construction unions of the AFL maintain the strong local horizontal structure. The construction elements within industrial unions usually follow the example of the parent union. The construction workers of the United Mine Workers (UMW) are an example of this. The workers doing construction in the mines are organized as a separate entity called District 50 of the United Mine Workers. They sign a national contract with the mine owners covering all of the crafts from operating engineers to electricians. A single list of scales covering all specialties (i.e., craft disciplines) is contained in the national contract. Since the members of the union are mine construction workers first and carpenters, operators, or electricians second, the jealousy regarding so-called craft "lines" and jurisdiction is less pronounced. It is not uncommon to see an equipment operator in District 50 get down from a tractor and do some small carpentry. This would be impossible in a horizontally organized craft union situation because the carpenters would immediately start a jurisdictional dispute.

13.12 JURISDICTIONAL DISPUTES

In addition to the fragmentation of contracts by craft and local area, one of the major difficulties inherent in the horizontal craft structured union is the problem of craft jurisdiction. Job jurisdiction disputes arise when more than one union claims jurisdiction over a given item of work. This is true primarily because many unions regard a certain type of work as a proprietary right and jealously guard against any encroachment of their traditional sphere by other unions. As technology advances and new products are introduced, the

CRAFT	WAGE RATE	FOREMAN	OVERTIME	W—WELFARE P—PENSION A—APPRENTICE V—VACATION IA—INDUSTRY ADVANCEMENT
1) Bricklayers Masons Union No. 8	7.80	+ .50	1½	W—0.30 P—0.30 V—0.30
2) Carpenters No. 225	7.40	+ .75	1st 2 hr @ 1½ then double	W—0.35 P—0.35 A—0.15
3) Operating Engineers No. 926	7.65	—	1½	W—0.13 A—0.07 P—0.15
4) Ironworkers No. 387	7.20	+ .75	1½	W—0.40 P—0.37 A—0.05

5) Millrights No. 1263	8.20	+1.00	2	
6) Laborer No. 438	4.85	+1.06	1½	W—0.15 P—0.20
7) Glaziers No. 1940	7.00		1½	W—0.35 P—0.25 V—0.25
8) Plasterers No. 148	7.32	+ .50	2	P—0.45 W—0.25 IA—0.04
9) Cement Masons No. 148	7.05	+ .50	1½	P—0.55 W—0.25 IA—0.10
10) Painters No. 38	7.45	+ .25	1½	A—0.03 P—0.40 W—0.40
11) Teamsters No. 528	5.15		1½	

FIGURE 13.3. Résumé of contracts for unions operating in the area of the North Georgia Building Trades Council, 1973.

question of which craft most appropriately should perform the work involved inevitably arises. A classical example in building construction is provided by the introduction of metal window and door frames. Traditionally, the installation of windows and doors had been considered a carpentry activity. However, the introduction of metal frames led to disputes between the carpenters and the metal workers as to which union had jurisdiction in the installation of these items. Such disputes can become very heated and lead to a walkout by one craft or the other. This may shut down the job. The contractor is sometimes simply an innocent bystander in such instances. If these disputes are not settled quickly, the repercussions for client and contractor can be very serious, as indicated by the following excerpt from the *Engineering News Record*.

The nozzle-dispute on the $1-billion Albany, N.Y. mall project has caused hundreds of stoppages on that job, which employs over 2,000 persons. The argument, which is not settled, revolves around whether the teamster driving a fuel truck or the operating engineer running a machine shall hold the nozzle during the fueling operation. Both unions claim the job. Because holding the nozzle involves a certain amount of work, the question is why either union should want it, since regardless of which man does the job, the other still gets paid. The answer undoubtedly is that the union that gets jurisdiction will eventually be able to claim the need for a helper. This particular dispute has been reported as plaguing contractors in many states, including West Virginia, Oklahoma, Missouri, California and Washington.*

Although this is a rather extreme example, it is indicative of the jealousies that can arise between crafts.

Concern on the part of unions for jurisdiction is understandable, since rulings that erode their area of work ultimately can lead to the craft slowly dwindling into a state of reduced work responsibilities and, eventually, into extinction. Therefore, the craft unions jealously protect their craft integrity. The following clause indicates how comprehensive the definition of craft responsibility can become.

Scope of Work

This Agreement shall cover all employees employed by the Employer engaged in work coming under all classifications listed under the trade autonomy of the United Brotherhood of Carpenters and Joiners of America.

The trade autonomy of the United Brotherhood of Carpenters and Joiners of America consists of the milling, fashioning, joining, assembling, erection, fastening or dismantling of all material of wood, plastic, metal, fiber. cork and composition, and all other substitute materials and the handling, cleaning, erecting, installing and dismantling of machinery, equipment and all materials used by members of the United Brotherhood.

Our claim of jurisdiction, therefore, extends over the following divisions and sub-divisions of the trade: Carpenters and Joiners; Millwrights; Pile Drivers: Bridge, Dock and Wharf Carpenters; Divers; Underpinners; Timbermen and Core Drillers;

*"Low Productivity: The Real Sin of High Wages," *Engineering News Record*, February 24, 1972.

Shipwrights, Boat Builders, Ship Carpenters, Joiners and Caulkers; Cabinet Makers, Bench Hands, Stair Builders, Millmen; Wood and Resilient Floor Layers, and Finishers; Carpet Layers; Shinglers; Siders; Insulators; Accoustic and Dry Wall Applicators; Shorers and House Movers; Loggers, Lumber and Sawmill Workers; Furniture Workers, Reed and Rattan Workers; Shingle Weavers; Casket and Coffin Makers; Box Makers, Railroad Carpenters and Car Builders, regardless of material used; and all those engaged in the operation of woodworking or other machinery required in the fashioning, milling or manufacturing of products used in the trade, or engaged as helpers to any of the above divisions or subdivisions, and the handling, erecting and installing material on any of the above divisions or subdivisions; burning, welding, rigging and the use of any instrument or tool for layout work incidental to the trade. When the term "carpenter and joiner" is used, it shall mean all the subdivisions of the trade.

The above occupational scope shall be subject to all agreements between International Representatives.*

Rulings regarding previous jurisdictional agreements are published by the Building Trades Department of the AFL-CIO in the "Green" Book. A similar publication issued by the Associated General Contractors is called the "Gray" Book. A typical ruling from the Green Book is shown in Figure 13.4.

Many of these disputes, however, are not covered by the Green Book, and settlement of such disputes by the NLRB can take years. The Taft-Hartley Act authorized unions and employers to establish boards to informally expedite agreements regarding jurisdictional disputes. The outgrowth of this for the construction industry was the establishment of the National Joint Board. The leading members of the National Joint Board, until recently, were representatives of the craft unions and the Associated General Contractors. The objective of NJB is to expedite definition of craft lines and minimize job shutdowns due to jurisdictional disputes. It provides what amounts to an "out of court" settlement in contrast to the time-consuming process of having the disputes decided by the NLRB. The National Joint Board has become less effective, however, with the withdrawal in 1969 of the AGC representation.

Jurisdictional disputes present less problem in vertically structured unions since craft integrity is not a matter that determines the strength of the union. All major automobiles are assembled by members of the United Automobile Workers (UAW). The UAW is a typical vertically structured union. Technological changes do not mean the work could be shifted to another union. Therefore, UAW workers can be installing windows today and can be moved to installation of electrical wiring next month. Craft integrity does not have to be jealously protected.

European construction workers are organized into vertically structured unions. National agreements in countries such as Germany cover all workers

*Excerpted from Agreement Between The United Brotherhood of Carpenters and Joiners of American Local No. 44 Champaign-Urbana, Illinois, and the Central Illinois Builders Chapter of Associated General Contractors of America, June 1970.

INSTALLATION OF CEILING SYSTEMS
JANUARY 15, 1968

This dispute, which was referred by the National Joint Board on November 23, 1965, was the subject of hearings before the Hearings Panel in Washington, D.C., March 4, 1966, and May 16 and 17, 1966. The decision of the Hearings Panel was issued August 24, 1966, but was stayed because of court litigation until January 15, 1968. The decision of the Hearings Panel, which was upheld by decisions of the U.S. District Court for the District of Columbia, the U.S. Court of Appeals for the District of Columbia Circuit and the Supreme Court of the United States, is as follows:

1. The decision of this Hearings Panel is limited to the jurisdictional disputes or work assignments in controversy between lathers and carpenters involved in the installation of ceiling systems. Nothing in this decision shall affect the jurisdiction or work assignment of any other trade, such as, but not limited to, the sheet metal workers, electricians, iron workers, etc., with a claim or interest in the installation of ceiling systems. Contractors shall use this decision as a basis for making work assignments only with regard to work in the installation of ceiling systems involving carpenters and lathers.

2. The installation of gypsum wallboard and other types of panels fastened directly to ceiling joists shall be performed by carpenters. In the event that a carrying channel is used with gypsum wallboard or other types of panels attached thereto, not withstanding paragraph 4(a) below, the contractor may use carpenters to install the carrying channel in areas not exceeding 300 square feet where there are no lathers on the job.

3. The installation of light iron work in ceiling systems with gypsum, Portland cement, acoustical or other plasters sprayed-on or trowel-applied over lath or directly to structural members shall be performed by lathers.

4. The following types of ceiling systems are included in this paragraph: Direct Hung Suspension System; Attached Concealed System without Backing Board; Furring Bar Attached System; Furring Bar Suspension System; Indirect Hung Suspension System or similar systems.

(a) The installation of the 1½″channel or similar carrying channel and hangers in any of the above types of systems shall be performed by lathers.

(b) The installation of all other work, including the installation of a ceiling system in its entirety if no 1½″ channel or other carrying channel is used, shall be performed by carpenters.

5. This decision shall be effective on all work assignments made on construction contracts let after January 15, 1968.

HEARINGS PANEL

PETER T. SCHOEMANN (Signed)
HUNTER P. WHARTON
WM. E. NAUMANN
ED S. TORRENCE
JOHN T. DUNLOP
(Impartial Umpire)

FIGURE 13.4 Typical Green Book ruling.

and are signed periodically defining wage scales and general labor management procedures. Each worker has a primary specialty and is paid at the rate established in the national agreement. Since craft jurisdiction is not a major issue, it is not unusual to see a worker who is operating a backhoe get down and work as part of a crew installing shoring. Similar mobility back and forth across craft lines is common in District 50 of the United Mine Workers since it is also vertically structured.

13.13 UNION STRUCTURE

The largest labor organization in the United States is the AFL-CIO. The building and construction trade unions are craft unions and as such are affiliates of the Building and Construction Trades Department of the AFL-CIO. The structure of affiliates from local to national level is shown schematically in Figure 13.5. A list of the construction unions that are within the AFL-CIO is given in Table 13.4. Most labor unions are presently affiliated with the AFL-CIO; the most notable exception is the Teamsters Union.

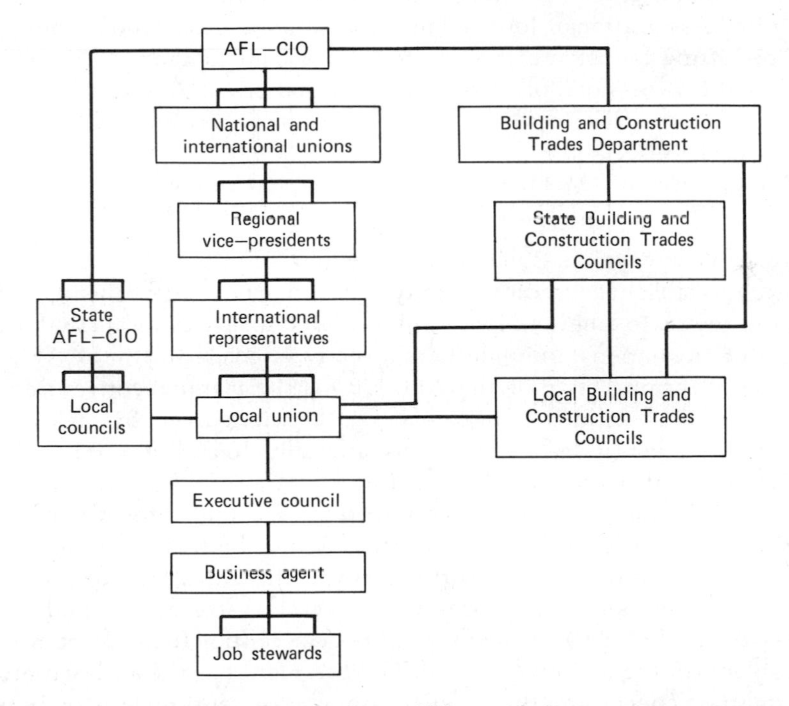

FIGURE 13.5 Structure typical of an affiliate of the Building and Construction Trades Department AFL-CIO.

Table 13.4 AFL-CIO CONSTRUCTION UNIONS

1. Bricklayers, Masons and Plasterers' International Union
2. Brotherhood of Painters, Decorators, and Paperhangers of America
3. Granite Cutters' International Association of America
4. International Association of Bridge, Structural and Ornamental Iron Workers
5. International Association of Heat and Frost Insulators and Asbestos Workers
6. International Association of Marble, Slate and Stone Polishers, Rubbers and Sawyers, Tile and Marble Setters Helpers and Terrazzo Helpers
7. International Brotherhood of Boiler Makers, Iron Ship Builders and Helpers of America
8. International Brotherhood of Electrical Workers
9. International Union of Elevator Constructors
10. International Union of Operating Engineers
11. Laborers International Union of North America
12. Operative Plasterers and Cement Masons' International Association
13. Sheet Metal Workers' International Association
14. United Association of Journeymen and Apprentices of the Plumbing and Pipe Fitting Industry of the United States and Canada
15. United Brotherhood of Carpenters and Joiners of America
16. United Slate, Tile, and Composition Roofers, Damp and Waterproof Workers' Association
17. Wood, Wire, and Metal Lathers International Union

There are two ways a national union may join the AFL-CIO. The first is for an already established union to apply for a charter. The other is for the federation to create a new union from a related group of locals that are not members of any national union but are directly associated with the AFL-CIO.

The top governing body of the AFL-CIO is the biennial convention. The decisions and instructions it makes are to be carried out by an executive council, which meets at least three times annually. In addition to the council, there is a general board, which meets at least once annually, and various presidentially appointed standing committees. The full-time administrative employees of the AFL-CIO are the president and the secretary-treasurer.

Between conventions, the executive council runs the affairs of the federation. The members are the president, secretary-treasurer, and several vice-presidents elected by the majority at the convention (usually from among the presidents of the national unions). The president has the authority to rule on any matters concerning the constitution or a convention decision between meetings of the council.

The AFL-CIO maintains trade departments at the level directly below the executive council. There are seven major sections whose main objective is to further unionization in the appropriate industry or trade. They also aid in the settlement of jurisdictional disputes between the members in their department (see Figure 13.4). Disputes with a union in another department are appealed to the executive council. Departments also represent their members before Congress and other government agencies. The Building and Construction Trades Department is responsible for all construction craft unions.

13.14 NATIONAL UNIONS

National unions are defined as those unions having collective bargaining agreements with different employers in more than one state, and federal employee unions with exclusive bargaining rights.* Because of their assumed role of collective bargaining in many areas, the national unions have become increasingly powerful. In construction unions, however, the locals still play the most important role in collective bargaining and, therefore, power still resides at the local level.

Each union has exclusive jurisdiction to function as the workers' representative in its trade or branch of industry. The jurisdiction of most unions is at least partially set forth in their charter and constitution. As the unions' outlook and purposes have changed or as their members' jobs have altered, many unions have changed their jurisdiction as well.

The daily conduct of union business is in the hands of the national president, whose influence is a big factor in deciding what issues the union executive board will discuss and vote on. What the president decides will have an effect on the general public as well as on the union. The president's more important powers are to decide on constitutional matters, issue or revoke local charters, hire or fire union employees, and sanction strikes. Most actions involving the powers of the president can be appealed to the board or to the convention.

The organizer or representative of the union provides contact between the locals and the national headquarters and attempts to gain new members for the union and to set up new locals. The organizer is the union advisor to all of the locals within his area, and must explain national policies to them. At the same time, he informs the national level of local problems.

13.15 STATE FEDERATIONS AND CITY CENTRALS

State federations are concerned mostly with lobbying for needed legislation and public relations on the state level. They are composed of locals whose national union is a member of the AFL-CIO. Conventions are held annually where programs of interest to all of the state's workers are concerned.

*Martin Estey, *The Unions* (New York: Harcourt, Brace and World, Inc., 1967), p. 37.

City centrals are concerned more with economics, serving as a clearinghouse for locals and aiding in dealings with employers. They have become increasingly involved in general community affairs and activities that may indirectly benefit their members.

Joint boards and trade councils are composed of locals involved in similar trades or industries. Their principal duty is to ensure that workers present a unified front in collective bargaining and obtain uniform working conditions in their area. A joint board or council is usually required for unions with more than three locals in the same region. The joint board is made up of all locals of the same national union, while the trades council is composed of locals of different national unions in related trades in the same industry.

The prototype for local trades councils is the Building and Construction Trades Council, which has its higher-level counterpart in the Building and Construction Trades Department of the AFL-CIO. Its problems are not limited to labor-management relations; it is often involved in settling ticklish jurisdictional disputes. The Building Trades Council provides craft unions with an important advantage characteristic to the industrial unions: the ability to present a united front in dealings with management. Some councils negotiate citywide agreements with employers or see that the agreements of their member locals all expire on the same date. They have a great deal of influence with the locals but may not make them act against national union policy.

13.16 UNION LOCALS

The locals are the smallest division of the national union. They provide a mechanism through which the national union can communicate with its members at the local level. Locals provide for contact with other workers in the same trade and are a means by which better working conditions are obtained, grievances are settled, and educational and political programs are implemented. They may be organized on an occupational or craft basis or on a plant or multiplant basis. In the building industry, it is common to have locals for each craft in large cities. The local officials who preside over the committees and the general meeting are the president, vice-president, treasurer, and various secretaries. They are usually unpaid or paid only a small amount and continue to work at their trade. They perform their union duties in their spare time. In small locals, a financial secretary will take care of the local books and records; but in large locals a trained bookkeeper is employed for this purpose.

The most important local official is the business agent, a full-time employee of the local. He exercises a great deal of leadership over the local and its affairs through the advice he provides to the membership and elected officials. He is usually trained and experienced in labor relations and possesses a large amount of knowledge of conditions on which other members are poorly informed.

The business agent's duties cover the entire range of the local's activities. He helps settle grievances with employers, negotiates agreements, points out violations of trade agreements, and operates the union hiring hall. He is also an organizer, trying to get unorganized workers into the union. Only locals with a large membership can afford a full-time agent, and over one-half of the locals employing agents are in the building trades where there is a greater need due to the transient nature of the work. For the locals who do not have enough money to employ their own business agent, an agent is usually maintained by the city central or state federation.

The shop steward is not a union official, but is the representative who comes in closest contact with the members. He must see that union conditions are maintained on the job, and handle grievances against the employer. The steward is a worker on the jobsite elected by his peers.

13.17 UNION HIRING HALLS

One of the salient features of construction labor is its transient nature. Construction workers are constantly moving from job site to job site and company to company. It is not uncommon for a construction worker to be employed by five or six different contractors in the same year. The union hiring hall provides a referral service that links available labor with contractor's requests. Following each job, a worker registers with the union hall and is referred to a new job site as positions become available. The procedures governing operation of the union hiring hall constitute an important part of the agreement between the union and the contractor. Articles of the labor contract specify precisely how the hiring hall is to operate. Although there are small variations from craft to craft and region to region, similar procedures are commonly used for referring workers through the union hall.

13.18 SECONDARY BOYCOTTS

The legality of boycotts to influence labor disputes has been an issue of primary importance throughout the history of labor-management relations. A boycott is an action by one party to exert some economic or social pressure on a second party with the intent of influencing the second party regarding some issue. A secondary boycott is one in which a party *A* who has a dispute with *B* attempts to bring pressure on *B* by boycotting a party *C* who deals with *B* and who can bring strong indirect pressure on *B* to agree to some issue. This is shown schematically in Figure 13.6. If the electrical workers in a plant fabricating small appliances go to the factory and form a picket line to get an

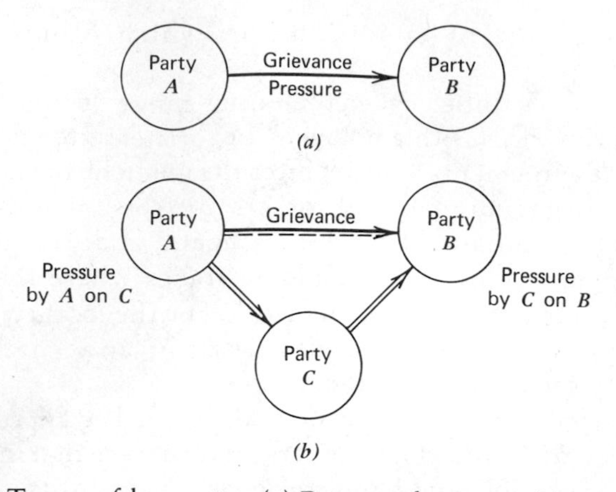

FIGURE 13.6 Types of boycotts. (*a*) Primary boycott. (*b*) Secondary boycott.

agreement, there is a primary boycott in progress. If, however, the workers send some of their members into the town and put pickets up at stores selling appliances from the plant, a secondary boycott is established. The store owners are a third party (C) being pressured to influence the factory to settle with the workers. The Taft-Hartley Act declared the use of a secondary boycott to be illegal.

In the construction industry, such secondary boycotts occur on sites with both union and nonunion workers when a union attempts to force a nonunion subcontractor to sign a union contract. In such cases, the union will put up a picket line at the entrance to the work site, in effect, to picket or boycott the nonunion subcontractor. Tradition among labor unions, however, demands that no union worker can cross another union's picket line. Therefore, the actual effect of the union picket line will be to prevent all union workers from entering the site. This may cause the shutdown of the entire site pending resolution of the nonunion subcontractor's presence on the site. In this situation, the general or prime contractor is a third party being pressured by the union to influence the nonunion subcontractor. This is called *common situs* picketing. In 1951, the U.S. Supreme Court ruled this practice to be a secondary boycott and, therefore, illegal under the Taft-Hartley Act. The high court made this ruling in the case of the Denver Building and Construction Trades Council.

Following this decision, the doctrine of "separate gates" was developed to deal with secondary boycott problems. Under this policy, the prime contractor establishes a separate or alternate gate for the nonunion subcontractor with whom the union has a dispute. The union is then directed to place its picket line at this gate rather than the main project gate. If it fails to comply, it can be enjoined from boycotting. Other union personnel entering the site can enter at the main gate without crossing the picket line of another union.

Certain interpretations of the secondary boycott have essentially provided exceptions in the construction industry. Unions normally have attempted to refrain from handling goods or products from nonunion shops. Such materials are called "hot cargo" and unions have bargained for hot cargo contract clauses that, in effect, prevent a contractor from handling such materials from nonunion fabricators. This is a secondary boycott in the sense that the contractor becomes an innocent third party in the dispute between the union and the fabricator or product supplier. The Landrum-Griffin Act provides that such hot cargo or subcontractor clauses that ban use of these materials or contact with these open shop units are illegal. An exemption is made, however, for the construction industry. As noted by Clough:

> Subcontractor clauses are widely used in the construction industry and typically require that work be awarded only to subcontractors who will comply with all the terms and conditions of the prime contractor's labor agreement . . .*

The Supreme Court also ruled in 1967 that prefabrication clauses that ban the use of certain prefabricated materials are exempted from the secondary boycott legislation if such prefabricated items threaten the craft integrity and eliminate work that would normally be done on site. Union carpenters, for instance, might refuse to install prefabricated door units, since the doors and the frames are preassembled in a factory offsite. This eliminates assembly work that could be done on-site and endangers the union scope of work. Use of such prefabricated items could lead to the decay of the craft's jurisdiction and integrity. Therefore, the use of such clauses in labor agreements is not considered to be an unlawful practice in these instances.

13.19 OPEN SHOP AND DOUBLE-BREASTED OPERATIONS

In recent years escalating union wage settlements have led to an upsurge in the number of open shop contractors successfully bidding on large contracts. Restrictive work rules and high wages have made it difficult for union contractors to be competitive in some market areas. In an open shop firm, there is no union agreement and workers are paid and advanced on a merit basis.

*R. H. Clough, *Construction Contracting*, 3rd edition, Wiley-Interscience, p. 304.

The largest group of open or merit shop contractors is represented by the Associated Builders and Constructors (ABC). Traditionally, open shop contractors have bid successfully in the housing and small building market where the required skill level is not high. Union contractors have dominated the more sophisticated building and heavy construction markets based on their ability to attract skilled labor with higher wages and benefits. In the early seventies, a trend on the part of construction unions to "price themselves out of the market" has given impetus to the open shop movement.

Damaging developments have occured in the area of restrictive work rules. One of the more controversial examples is the requirement in some operating engineer labor agreements that a compressor operator at full scale must be assigned to a compressor with no other duties. The operator starts the compressor in the morning and proceeds to attend the compressor for eight hours. At the end of the shift, he shuts the machine down. This has been criticized as "featherbedding" and is typical of work rule provisions that make it difficult for union contractors to remain competitive.

Large open shop contractors have been willing to meet or exceed the union wage rates in order to avoid the costly work delays associated with jurisdictional disputes and restrictive work rules. In some cases, the unions have responded by signing project agreements that relax certain work rules for the duration of a given job. Robert A. Georgine, President of the Building and Construction Trades Department of the AFL-CIO, says the unions want to make union contractors as competitive as possible. He states that:

> We can't abolish craft lines, but we are entering into project agreements whenever asked. There's a limit to what you can do.*

In order to be able to bid in both open shop and union formats, some firms have organized as *double-breasted* contractors. Large firms will have one subsidiary that operates with no union contracts. A separately managed company will be signed to all union contracts. In this way, the parent firm can bid both in union shop markets† and in markets in which the lower-priced open shop encourages more effective bidding. The following excerpt from the *Engineering News Record* gives a short summary of the development of the double-breasted concept:

*Quoted from "Open Shop, A Growing Force and a Catalyst for Change," *Engineering News Record*, November 2, 1972, p. 19.

†The owner may specify that union labor is to be used, or Davis-Bacon rulings may dictate that union rates will prevail.

NLRB defines groundrules for going double breasted

The National Labor Relations Board (NLRB), in the Gerace-Helger case, didn't map a fail-safe path for union contractors that want to go open shop. It did, however, identify those factors it considers to carry the most weight in determining whether both companies are actually the same employer.

The issue was whether two companies with common stockholders and directors, one union and one open shop, were separate employers and therefore did not violate Section 8(a)(5) of the Taft-Hartley Act when it refused to recognize unions representing employees of the first company as the bargaining representatives of the second company's employees.

Gerace Construction Co., a Midland, Mich., union general contractor, had trouble in competing for small projects since early in 1969. Gerace decided to form an open shop company when it learned of a remodeling job where the owner insisted that the contract was to be performed without interruption. Concerned that expiring labor contracts would result in strikes, Gerace formed Helger Construction Co. and successfully bid the work.

Several unions filed an unfair labor practice charge, claiming that the two companies were actually a single employer and, therefore, their collective bargaining agreements with Gerace also should cover Helger's employees.

The NLRB decision came on Oct. 8, 1971. The board noted that Helger had entered a new market, and that there was no interchange of employees between the companies. It held: "A critical factor in determining whether separate legal entities operate as a single employing enterprise is the degree of common control of labor relations policies. Thus, the board has found common ownership not determinative where requisite common control was not shown and . . . common control must be actual or active, as distinguished from potential control."

One month later, the board applied the decision in the Smith-Keuka case involving a New York state union contractor that had formed an open shop firm to compete for small residential and commercial jobs. The board rules that the issues posed were similar to those resolved in the Gerace case and dismissed the unfair labor practice charge.*

Recently, the unions have been successful in establishing that the management personnel in both union and open shop entities of a double-breasted firm were linked and, therefore, the separate companies were not truly separate. This has placed the future of double-breasted operation in some doubt and has forced some large companies to make the difficult decision of relinquishing the double-breasted approach and going either solely union or open shop.

*Ibid.

REVIEW QUESTIONS AND EXERCISES

13.1 What is meant by the following terms:
(a) Yellow dog contract.
(b) Agency Shop.
(c) Subcontractor clause.

13.2 What is a secondary boycott? Name two types of secondary boycotts. Does the legislation forbidding secondary boycotts apply to construction unions? Explain.

13.3 What is a jurisdictional dispute? Why does this kind of dispute present no problem in District 50 locals?

13.4 How do the National Labor Relations Board and the National Joint Board differ (at least two ways)?

13.5 What are the basic differences between the AFL as a labor union and the CIO type of union?

13.6 What will be the impact on double-breasted operations and the right-to-work provision of the Taft-Hartley legislation if labor is able to revoke existing practices regarding common situs picketing?

13.7 Answer the following questions true (T) or false (F):
(a) ___ Some state laws authorize use of closed shop.
(b) ___ A union can legally strike a job site in order to *enforce* the provisions of a subcontractor clause.
(c) ___ The teamsters union is the largest member of the AFL-CIO.
(d) ___ Open shop operations have caused construction labor unions to rethink their position vis a vis union contractors.
(e) ___ The right-to-work clause of the Taft-Hartley law allows the individual states to determine whether union shops are legal.
(f) ___ The unit price contract is an incentive-type negotiated contract.
(g) ___ After walking out in 1969, the Building Trades Department of the AFL-CIO refused to rejoin the National Joint Board thus reducing its effectiveness.
(h) ___ The local AFL craft unions have very little authority and are directed mainly by the national headquarters of AFL-CIO.
(i) ___ The Sherman Antitrust Law was originally designed to prevent the formation of large corporations or cartels which could dominate the market.
(j) ___ The business agent is the representative of the union charged with enforcing the work rules of the labor agreement.

13.8 Identify the local labor unions that operate in your region. List the relevant business agents and the locations of the hiring halls.

13.9 List the labor unions that you consider would be involved in a project similar to the gas station project of Appendix L.

13.10 Visit a local contractor and a local hiring hall and determine the procedure to be followed in the hiring of labor.

CHAPTER FOURTEEN

Labor Cost and Productivity

14.1 LABOR AGREEMENTS

Just as the contractor enters into a contract with the client, with vendors supplying materials (i.e., purchase orders), and with subcontractors working under his direction, if union labor is utilized, he also enters into contracts or labor agreements with each of the craft unions with whom he deals. These contracts usually cover a one- or two-year period and include clauses governing the reconciliation of disputes, work rules, wage scales, and fringe benefits. The wages are normally defined in step increases throughout the period of the contract. These step increases are normally contained in the addendum to the labor contract.

The opening sections of the agreement normally provide methods for reconciling disputes that can arise between the contractors and the union during the life of the contract. It has been noted that:

> The drawing up and signing of collective bargaining agreements does not mean that for the duration of the agreement the workers and management are going to live in an industrial atmosphere of sweetness and light.*

To handle disputes, articles in the contract typically set up a Joint Conference Committee to reconcile disputes and provide for arbitration procedures for

*Clyde E. Dankert, *An Introduction to Labor* (New York: Prentice-Hall, 1954)

disputes which cannot be settled by the Committee. Typical contracts also include provisions governing:

1. Maintenance of Membership
2. Fringe Benefits
3. Work Rules
4. Apprentice Program Operation
5. Wages (Addendum)
6. Hours
7. Worker Control and Union Representation
8. Operation of the Union Hiring Hall
9. Union Area
10. Subcontractor clauses (See Section 13.18)
11. Special Provisions

Fringe benefits are economic concessions gained by unions covering vacation pay, health and welfare, differentials in pay due to shift, contributions by the contractor to apprenticeship programs, and so-called industrial advancement funds. These are paid by the contractor in addition to the base wage and garnish the salary of the worker. As will be shown in Section 14.2 on cost, they have a considerable impact on the contractor's cost of labor. The building and construction trades councils for each union area normally print summaries of contract wage and fringe benefit provisions that assist in the preparation of payroll. Such a summary is shown in Figure 14.1.

Work rules are an important item of negotiation and have a significant effect on the productivity of workers and the cost of installed construction. A typical work rule might require that all electrical materials on-site will be handled by union electricians. Another might require that all trucks moving electrical materials on-site be driven by union electricians. Such provisions can lead to expensive tradesmen doing work that could be done by less expensive crafts or laborers. Therefore, work rules become major topics of discussion during the period of contract negotiation.

14.2 LABOR COSTS

The large number of contributions and burdens associated with the wage of a worker makes the determination of a worker's cost to the contractor a complex calculation. The contractor must know how much cost to put in the bid to cover the salary and associated contributions for all of the workers. Assuming that the number of carpenters, ironworkers, operating engineers, and other craft workers required is known and the hours for each can be estimated, the average hourly cost of each craft can be multiplied by the required craft hours

LABOR ORGANIZATIONS AND WAGE RATES

CRAFT AND BUSINESS REPRESENTATIVE	WAGE RATE PER HOUR	FOREMAN	OVER TIME RATE	W—WELFARE P—PENSION A—APPRENTICE V—VACATION	TRAVEL PAY SUBSIS-TENCE	AUTOMATIC WAGE INCREASES	AUTOMATIC FRINGE INCREASES	EXPIRA-TION DATE
Asbestos Workers Local No. 18 Robert J. Scott, BR 946 North Highland Indianapolis, Indiana 46202 317-638-4234	$7.20		Double	W—20c P—20c A—6c V—60c Deduct	**$11** **per day**			5-31-71
Boilermakers Local No. 60 George Williams, BR 400 North Jefferson Peoria, Illinois 61603 309-673-9131	$7.15	50c—F $1.00—GF	Double	W—40c P—65c A—01c	30-M—$6-D 60-M—$8-D	$1.00— 9-1-71		8-31-72
Boilermakers Local No. 363 Anthony Moceri B.M. 19 S. 97th St. Belleville, Illinois 62223 618-397-7779	$7.35	50c—F $1.00—GF	Double	W—40c P—65c A—01c	30-M—$5-D 60-M—$7-D			9-2-71
Carpenters Local No. 44 Gene Stirewalt, BR **212 W. Hill St.** Champaign, Illinois 61820 217-356-5463	$6.29	12%	Double	W—17½c P—30c A—05c IAF—02c ISC—½c		35c—10-15-70 40c—4-15-71 35c—10-15-71	P—10c 4-15-71	4-15-72
Carpenters Local No. 347 **Lee V. Foreman, BR** **P.O. Box 774** Mattoon, Ill. 61938	$5.93	**50c**	Double	W—25c P—15c A—02c		80c—6-1-71	P—05c 6-1-71	6-1-72
Cement Finishers Local No. 143 Francis E. Ducey, BR **212½ South First St.** Champaign, Illinois 61820 Office 217-356-9313 Home 217-485-3515	$6.62½	50c 15% GF	Double	W—17½c		30c—1-24-71		7-24-71
Electricians Local No. 601 **Jack Hensler, BR** **212 South First St.** Champaign, Illinois 61820 217-352-1741	$6.35	**10%** **20%-GF**	Double	W—20c A—²⁄₁₀%		45c—11-1-70 70c—5-1-71 30c—11-1-71		4-30-72
Electricians Local No. 146 **Larry Lawler, BM** **2955 N. Woodford** Decatur, Illinois 62526 217-877-4604	$6.60	10%—F 15%—GF	Double	W—20c V—15c Deduct A—¹⁄₁₀% P—1%		60c—2-21-71		2-20-71
Electricians Local No. 489 **William Dittamore, BM** 106 S. 19th St. Mattoon, Illinois 61938	$6.30	10%—F 20%—GF	Double	W—20c P—1% A—$25		45c—1-1-71 35c—7-1-71 45c—1-1-72		8-31-72

FIGURE 14.1 Labor organizations and wage rates.

to arrive at the total labor cost. The hourly average cost of a worker *to the contractor* consists of the following components:

1. Direct wages.
2. Fringe benefits.
3. Social security contributions (FICA).
4. Unemployment insurance.
5. Workmen's compensation insurance.
6. Public liability and property damage insurance.
7. Subsistence pay.
8. Shift pay differentials.

The direct wages and fringe benefits can be determined by referring to a summary of wage rates such as the one shown in Figure 14.1.

All workers must pay social security on a portion of their salary. For every dollar the worker pays, the employer must pay a matching dollar. The worker pays a fixed percent on every dollar earned up to a cutoff level. After the annual income has exceeded the cutoff level the worker (and the worker's employer) need pay no more. An indication of the increasing level of contribution since the Federal Insurance Contributions Act (FICA) was enacted in 1937 is shown in Figure 14.2. The FICA contribution in 1976 was required on the first $15,300 of annual income at the rate of 5.85 %. Therefore, a person making $15,300 or more in annual income would contribute $895.05 and the person's employer or employers would contribute a like amount. The amount of contribution was increased in 1980 to 6.13% on the first $25,900 of income and will continue to increase through 1984. The contractor must match the amount of FICA paid by the worker.

Unemployment insurance contributions are required of all employers. Each state sets a percent rate that must be paid by the employer. The premiums are escrowed on a monthly or quarterly basis and sent periodically to the state unemployment agency. The amount to be paid is based on certified payrolls submitted by the employer at the time of paying this contribution. The fund established by these contributions is used to pay benefits to workers who are temporarily out of work through no fault of their own.

The states also require employers to maintain Workmen's Compensation Insurance for all workers in their employ. This insurance reimburses the worker for injuries incurred in the course of employment. Labor agreements also specifically state this requirement. This recognizes the employer's responsibility to provide a safe working environment and the employer's obligation to provide support to disabled workers. Without this insurance, workers injured in the course of their work activity could become financially dependent on the state. The rates paid for workmen's compensation are a function of the risk associated with the work activity. The

Dear Social Security:

Is It True? Yes, Taxes for Many to Rise

WASHINGTON (AP) — If you earn more than $14,100 next year, you'll be paying up to $70.20 more in Social Security taxes.

The government announced Wednesday that it will levy Social Security taxes on the first $15,300 of your earnings beginning next Jan. 1, up from the $14,000 taxable wage base this year.

The result will be that an estimated 18 million workers will be paying higher Social Security taxes next year, to a maximum of $895.05 or $70.20 more than this year. Their employers will pay a like amount.

Self-employed persons will pay a maximum of $1,208.70 next year, up $94.80.

The higher taxable wage base will yield about $2.1 billion to help pay part of the 8 per cent cost of living benefit increase that began flowing to 31.3 million social Security recipients last July. General revenues paid for the same 8 per cent increase for four million supplemental Security Income recipients.

The added revenue is not expected to ease the projected Social Security deficits of $3 billion this year and $6 billion next year, nor even offset the first-year cost of the benefit increase.

Social Security Commissioner James B. Cardwell said the heavier tax on workers "will mean higher benefits for them and their families in the event of retirement, disability or death than would have been possible without an increase in the base.

"In return for the increase in taxes, these affected workers will have greater protection because a larger amount of their earnings will be credited toward benefits than before."

The base increase, mandated by law and based on a formula gauging the average wage increase, also raises the limit on outside income that retirees can earn without losing some Social Security benefits.

The government estimates that about 1.3 million retirees will benefit by the 1976 provision allowing them to earn $2,760 in outside income, a $240 increase over this year. Every $2 earned over that limit will result in a $1 reduction in Social Security payments.

Beneficiaries will be permitted to earn $230 a month next year without losing benefits, $20 more than this year.

The tax rates of 5.85 per cent each on employes and employers and 7.9 per cent on the self-employed will not change next year.

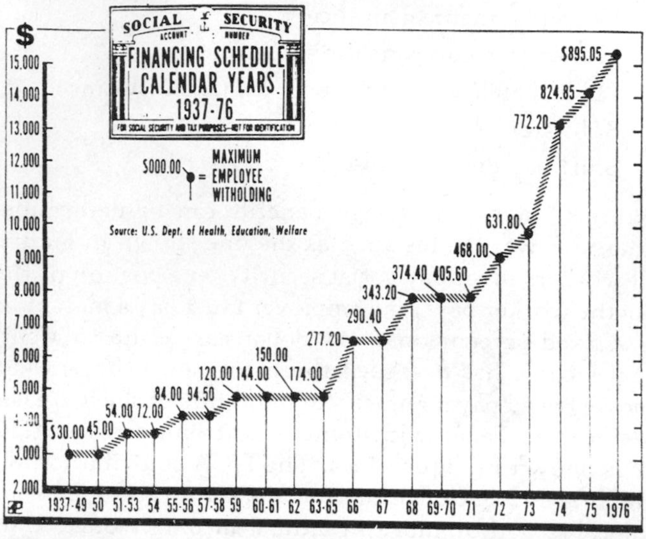

FIGURE 14.2 Levels of social security contributions (Associated Press, 1975).

contribution for a pressman in a printing plant is different from that of a worker erecting steel on a high-rise building. A typical listing of construction specialties and the corresponding rates is given in Table 14.1. Similar summaries are printed in the Quarterly Cost Roundup issues of the *Engineering News Record*. The rates are quoted in dollars of premium per $100 of payroll. The rate for iron or steel erection, for example, is $12.82 per hundred dollars of payroll paid to ironworkers and structural steel erectors.

The premium pay for public liability and property damage (PL and PD) insurance is also tied to the craft risk level and given in Table 14.1. When a

Table 14.1 MANUAL RATES FOR WORKMEN'S COMPENSATION, PUBLIC LIABILITY, AND PROPERTY DAMAGE INSURANCE FOR CERTAIN COMMON CONSTRUCTION CLASSIFICATIONS (California rates on October 1, 1968)[a]

Classification	Workmen's Compensation		PL and PD[b]			Total
	Code Number	Rate	Code Number	Bodily Injury	Property Damage	
Bridge building — metal	5040	$16.35	3452	$1.50	0.94	18.79
Bridge and trestle construction — wood	6209	17.70	6209	0.40	0.26	18.36
Caisson work — not pneumatic	6252	10.55	3438	0.39	*xcu*[d]	
Canal construction	6361	4.24	6229	0.29	0.55*xu*	
Carpentry — not otherwise covered	5403	8.35	3457	0.40	0.26	9.01
Carpentry — shop only	2883	3.81	2464	0.10	0.05	3.96
Carpentry — new, private residences	5645	4.70	5645	0.29	0.18	5.17
Churches, clergy, professional assistants	8840	0.52		—	—	
Churches, all other employees	9015	3.23		—	—	
Concrete construction — bridge and culverts with clearance over 10 ft	5222	9.62	5213	0.38	0.27	10.27
Concrete construction — including forms, etc.	5213	7.09	5213	0.38	0.27	7.74
Concrete building construction — tilt-up method	5214	2.69		—	—	
Concrete or cement work — sidewalks, floors, etc.	5200	2.69		—	—	
Contractors — executive supervisors[c]	5606	1.54	3759	0.21	0.11	1.86
Dam construction — concrete	5207	7.07	5213	0.38	0.27	7.74
Dam construction — not otherwise covered	6011	5.21	6019	*e*	*e*	
Electric wiring — within buildings	5190	2.74	5190	0.17	0.21	3.12
Engineers, consulting	8601	0.78	3759	0.21	0.17	1.16
Excavation — rock, no tunneling[e]	1605	8.03	3470	0.93	1.00*xcu*	
Iron or steel erection	5059	12.82	3452	1.50	0.94	18.79
Iron or steel erection — not otherwise covered	5057	11.88	5057	0.99	0.80	13.16
Iron, steel, brass, aluminum erection — nonstructural, within building	5102	4.19	3442	0.37	0.36	4.92

Table 14.1 MANUAL RATES FOR WORKMEN'S COMPENSATION, PUBLIC LIABILITY, AND PROPERTY DAMAGE INSURANCE FOR CERTAIN COMMON CONSTRUCTION CLASSIFICATIONS (California rates on October 1, 1968)[a]

Classification	Workmen's Compensation		PL and PD[b]			Total
	Code Number	Rate	Code Number	Bodily Injury	Property Damage	
Iron or steel works — steel shop	3030	8.36	3431	0.21	0.20	8.77
Logging	2702	13.57	2702	0.18	0.18	13.93
Painting	5474	5.11	3429	0.13	0.44	5.68
Painting — steel structures or bridges	5040	16.35	3452	1.50	0.94	18.79
Pile driving — building foundation only	6003	13.81	3470	0.93	1.00 *xcu*	
Pile driving — including timber wharf building	6003	13.81	3430	0.24	0.83 *cu*	
Pile driving — sonic method	6003	13.81	3764	—	—	
Plumbing	5183	2.97	3434	0.24	0.59 *u*	
Railroad construction	6701	9.75	3444	0.36	0.28 *x*	
Railroad construction — laying track and contractor maintenance	7855	6.28	3444	0.36	0.28 *x*	
Reinforcing steel installation — placing for concrete construction	5225	7.09		—	—	
Sewer construction	6306	7.58	3449	0.89	1.00 *xcu*	
Street and road construction — paving, etc.	5506	5.33	5506	1.00	0.55 *xcu*	
Street and road construction — grading	5507	5.20	3450	1.40	0.91 *xcu*	
Tunneling[f] — all work to completion, including lining	6251	17.47	3438	0.39	*(5) *xcu*	
Tunneling[f] — pneumatic, all work to completion	6260	29.89	3438	0.39	*(5) *xcu*	
Water mains or connections — construction	6319	4.71	3449	0.89	1.00 *xcu*	
Excavation — general, not otherwise covered[e]	6217	$ 3.48	3470	$0.93	1.00 *xcu*	

[a] Premium rates are to be applied to the base of $100 payroll. In California premium can be computed on the straight-time portion (i.e., no overtime premium pay) of the employee's pay, if such information is a part of the payroll records; otherwise gross payroll must be used. If the total annual charge at the rate shown is less than a given minimum, the contractor must pay this minimum. In some cases where the contractor's accident record is good, sizeable refunds or dividends may be paid back to him after the insurance period is over.

[b] Limits of Coverage:

Public liability. Maximum coverage under these rates—$5000 per person or $10,000 per accident. Cost of higher coverage $10,000/$20,000, 1.26 x basic rate; $25,000/$50,000, 1.47 x basic rate, $50,000/$100,000, 1.59 x basic rate; $300,000/$300,000, 1.78 x basic rate.

Property damage. Maximum coverage under these rates—$5,000 per accident and $25,000 per policy. Cost of higher coverage $25,000/$100,000, 1.23 x basic rate; $50,000/$100,000, 1.30 x basic rate.

Workmen's compensation. The employer may be required to pay to an injured employee an additional 50% of the compensation award ($7,500 maximum + costs up to $250) where accident was caused by employer's serious and willful misconduct.

Territory covered by quoted rates. Territory No. 01 (Alameda, Contra Costa, San Francisco, Santa Clara) as of August 21, 1968.

[c] Foremen and superintendents in charge of erection or construction work, watchmen, timekeepers, or cleaners shall be assigned to the governing classification.

[d] Symbol meaning:

x — explosion hazard c — collapse hazard u — underground hazard

The presence of "xcu" hazards results in increases in property damage rates commensurate with the hazardous condition.

[e] Schedule rating — type of merit rating by which basic manual rates are modified to fit the physical conditions of the individual plant in accordance with the industrial compensation rating schedule.

[f] Tunneling subject to basic pneumonoconiosis surcharge.

construction project is underway, accidents occuring as a result of the work can injure persons in the area or cause damage to property in the vicinity. If a bag of cement falls from an upper story of a project and injures persons on the sidewalk below, these persons will normally seek a settlement to cover their injuries. The public liability arising out of this situation is the responsibility of the owner of the project. Owners, however, normally pass the requirement to insure against such liability to the contractor in the form of a clause in the general conditions of the construction contract. The general conditions direct the contractor to have sufficient insurance to cover such public liability claims. Similarly, if the bag of cement falls and breaks the windshield on a car parked near the construction site, the owner of the car will seek to be reimbursed for the damage. This is a property damage situation that the owner of the construction project becomes liable to pay. Property damage insurance carried by the contractor (for the owner) covers this kind of liability. Insurance carriers normally quote rates for PL and PD insurance on the same basis as for workmen's compensation insurance. Therefore, to provide PL and PD insurance, the contractor must pay $1.50 for PL and $0.94 for each one hundred dollars of steel erector salary paid on the job. These rates vary over time and geographical area and can be reduced by maintaining a safe record of operation. The total amount of premium is based on a certified payroll submitted to the insurance carrier.

Subsistence is paid to workers who must work outside of the normal area of the local. As a result, they incur additional cost because of their remoteness from home and the need to commute long distances or perhaps live away from home. If an elevator constructor in Atlanta, Georgia must work in Macon for two weeks, he will be outside of the normal area of his local and will receive subsistence pay to defray his additional expenses.

Shift differentials are paid to workers in recognition that it may be less convenient to work during one part of the day than during another. Typical provisions in a sheet metal worker's contract* are given in Figure 14.3. In this example, the differential results in an add-on to the basic wage rate. Shift differential can also be specified by indicating that a worker will be paid for more hours than he works. A typical provision from a California ironworkers contract provides as follows (See Figure 14.4):

For shift work the following standards apply:

(a) If two shifts are in effect: Each shift works 7½ hours for 8 hours of pay.

(b) If three shifts are in effect: Each shift works 7 hours for 8 hours of pay.

This means that if a three-shift project is being worked the ironworker will receive overtime for all time worked over seven hours. In addition, he will be

*Agreement between Atlantic Building Systems, Inc. and Local Union 93, Sheet Metal Worker's International Association.

A shift differential premium of ten (10) cents per hour will be paid for all time worked on the afternoon or second shift, and a shift differential of fifteen (15) cents per hour will be paid for all time worked on the night or third shift as follows:

(1) *First Shift*. The day or first shift will include all Employees who commence work between 6 a.m. and 2 p.m. and who quit work at or before 6 p.m. of the same calendar day. No shift differential shall be paid for time worked on the day or first shift.

(2) *Second Shift*. The afternoon or second shift shall include all Employees who commence work at or after 2 p.m. and who quit work at or before 12 midnight of the same calendar day. A shift differential premium of ten (10) cents per hour shall be paid for all time worked on the afternoon or second shift.

(3) *Third Shift*. The night or third shift shall include all Employees who commence work at or after 10 p.m. and who quit work at or before 8 a.m. of the next following calendar day. A shift differential premium of fifteen (15) cents per hour shall be paid for all time worked on the night or third shift.

(4) *Cross Shift*. Where an Employee starts work during one shift, as above defined, and quits work during another shift, as above defined, said Employee shall not be paid any shift differential premium for time worked, if any, between the hours of 7 a.m. and 3 p.m.; but shall be paid a shift differential of ten (10) cents per hour for all time worked, if any, between the hours of 3 p.m. and 11 p.m., and a shift differential premium of fifteen (15) cents per hour for all time worked, if any, between the hours of 11 p.m. and 7 a.m.

FIGURE 14.3 Shift work provision.

paid eight hours pay for seven hours work. Calculation of shift pay will be demonstrated in the next section.

14.3 AVERAGE HOURLY COST CALCULATION

A typical summary of data regarding a statewide ironworkers contract negotiated in California is shown in Figure 14.4.* A work sheet showing the calculation of an ironworker's hourly cost to a contractor is shown in Figure 14.5. It is assumed that the ironworker is working in a subsistence area on the second shift of a three-shift job during June 1970.

The ironworker works 10-hour shifts each day for six days or 60 hours for the week. It is important to differentiate between those hours that are straight-time hours and those that are premium hours. Insurance premiums and fringe benefit contributions are based on straight-time hours. Social security and unemployment insurance contributions are calculated using the total income figure. The column labeled *Hours Work* breaks the weekday and

*The wage data are not current but are used here only to show the procedure needed in calculating the average hourly cost to the contractor.

IRON WORKERS

(California Statewide & Portions of Nevada)
(Agreement Expires August 15, 1970)

WAGE RATES:	8/16/68	8/16/69
Reinforcing	$6.10	$6.37
Structural and Ornamental	6.23	6.48
Fence Erector	6.03	6.28

Foremen: Shall be paid $0.45 per hour more than the regular hourly rate over which they have supervision. When two or more Iron Workers are employed, one shall be selected by the Individual Employer to act as Foreman and shall receive a Foreman's wages.

APPRENTICES: 1) Structural and Ornamental apprentices with three years apprenticeships and Reinforcing apprentices with a three year term of apprenticeship:

1-6 mo	77%	12-18 mo	85%	24-30 mo	93%
6-12 mo	81%	18-21 mo	89%	30-36 mo	97%

2) Reinforcing apprentices who have a two year term of apprenticeship:

1-6 mo	77%	6-12 mo	83%	12-18 mo	89%	18-24 mo	95%

HEALTH AND WELFARE: 8/16/68—$0.305 8/16/69—$0.33

PENSION: 8/16/68—$0.30 8/16/69—$0.325

VACATION: 8/16/68—$0.25

APPRENTICESHIP PLAN: 8/16/68—$0.02

OVERTIME: Double time for all overtime worked, including before 8:00 a.m. and after 5:00 p.m. weekdays.

HOURS: 8:00 a.m.-5:00 p.m.; 40 hours work Monday through Friday.

SHIFT WORK: Two Shifts—Each shift works 7½ hours for 8 hours pay. Three Shifts—Each shift works 7 hours for 8 hours pay.

LUNCH PERIOD: Not more than 4½ hours of continuous work without a meal period, or else overtime pay. (See Agreement for details)

SHOW-UP TIME: 2 hours pay

MINIMUM PAY PROVISIONS: Four (4) hours pay if employee works; hours beyond four based on actual hours worked.

HOLIDAYS: Northern California and Northern Nevada: New Year's Day, Washington's Birthday, Decoration Day, Independence Day, Labor Day, Admission Day, Thanksgiving Day, and Christmas Day.
Southern California and Southern Nevada: New Year's Day, Decoration Day, Independence Day, Labor Day, Veterans Day, Thanksgiving Day and Christmas Day.

PARKING: San Francisco, Oakland, Sacramento, Fresno, San Diego and Los Angeles: The employer shall provide or pay for parking facilities when free parking is not available within three blocks of the job.

TRAVEL AND SUBISTENCE: Where a job is located 35 miles from the City Hall of San Francisco, Oakland, San Jose, Sacramento, Stockton, Fresno, Bakersfield, Eureka, Redding, Napa, Los Angeles, San Diego, San Bernardino, Ventura, El Centro, Reno and Las Vegas (Nevada), based on the city of which the workman is a bona fide resident, workmen will be compensated at the rate of $9.00 per scheduled work day. No local resident of the area in which a job is located shall receive subsistence unless he is actually required to travel more than 35 miles. In addition, workmen shall receive $0.11 per mile for transportation and time paid for traveling at the straight time rate at the beginning and completion of the job. Employer shall pay bridge, ferry, and toll road fares.

FIGURE 14.4 Ironworkers agreement. Summary.

Compute the average hourly cost to a contractor of an ironworker involved in structural steel erection in a subsistence area. The iron worker works on the second shift of a three-shift job and works six ten-hour days per week. Additional PL and PD insurance for $50,000/$100,000 coverage is desired. Use 5.85% FICA and 5.0% for unemployment insurance. Make calculations for June 1970.

	HOURS WORK	STRAIGHT TIME HOURS (ST)	PREMIUM TIME (PT)
Monday-Fri	5 × 7 = 35	5 × 8 = 40	
	5 × 3 = 15	5 × 3 = 15	1 × 5 × 3 = 15
Saturday	1 × 7 = 7	1 × 8 = 8	1 × 1 × 8 = 8
	1 × 3 = 3	1 × 3 = 3	1 × 1 × 3 = 3
	60	66	26

Base Rate = $6.48
ST 66 hours @ $6.48 = $427.68
PT 26 hours @ $6.48 = $168.48

Gross Pay $596.16

Fringes:		
	Health and Welfare	0.33 x 66 = 21.78
	Pension	0.325 x 66 = 21.45
	Vacation	0.25 x 66 = 16.50 (Deferred Wage)
	Apprenticeship Training	0.02 x 66 = 1.32
		$0.925 x 66 = 61.05

WC = 12.82
PL 1.59 × 1.50 = 2.385
PD 1.30 × 0.94 = 1.222
Total $16.43 per $100 Payroll

WC, PL, and PD = $16.43 \times \frac{427.68}{100}$ = $70.27

FICA = 0.0585 × ($596.16 + $16.50) = $35.84
Unemployment = 0.05 × ($596.15 + $16.50) = $30.63
Subsistence = 6 days × $9.00/day = $54.00

TOTAL COST = BASE + FRINGES + WC, PL, PD + UNEMPL + SUBS = $847.95
AVERAGE HOURLY COST (TO CONTRACTOR) = $\frac{\$847.95}{60}$ = $14.13

FIGURE 14.5 Sample wage calculation.

Saturday hours into straight-time and premium-time components. Since the worker receives a shift differential, the first seven hours are considered straight time and the other three are overtime at premium rate. The corresponding straight-time hours corresponding to the hours worked are shown in the second column. Eight hours are paid for the first seven hours worked. The overtime rate is given as double time. The single-time portion or first half of this rate is credited to straight time. The second half of the double time is credited to the premium time column. Based on the column totals the worker works 60 hours and will be paid 66 straight-time hours and 26 premium hours.

By consulting Figure 14.4, it can be determined that the base wage rate for structural and ornamental ironworkers was increased to $6.48 per hour on 16 August 1969. Since the period of this calculation is June 1970, this is the rate to be used. This yields straight-time wages of $427.68 (66 hours) and premium pay of $168.48 (26 hours). Total gross pay is $596.16.

Fringes are based on straight-time hours, and the rates are given in the contract wage summary. The fringes paid by the contractor to union funds amount to $0.925 per hour. The vacation portion of the fringe is considered to be a deferred income item and, therefore, is subject to FICA. It is also used in the calculation of unemployment insurance contribution.

The amounts to be paid to the insurance carrier for workmen's compensation, PL, and PD can be taken from Table 14.1 under code 5059—iron or steel erection. The contract calls for increased PL and PD rates. The bodily injury (PL) portion and the property damage coverage are to be increased to cover $50,000 per person/$100,000 per occurence. This introduces a multiplier of 1.59 for the PL rate and 1.30 for the PD rate (see footnotes at the bottom of Table 14.1). The total rate per $100 of payroll for WC, PL, and PD is $16.43. This is applied to the straight-time pay of $427.68 and gives a premium to be escrowed of $70.27.

Both FICA and unemployment insurance are based on the total gross pay plus the deferred vacation fringe. Subsistence is $9.00 per day not including travel pay and time to travel to the site (not included in this calculation). By summing all of these cost components, the contractor's total cost becomes

Gross pay	$596.16
Fringes	61.05
WC, PL, PD	70.27
FICA	35.84
Unemployment	30.63
Subsistence	54.00
Total	$847.95

Hourly rate = $847.95/60 = $14.13

This is considerably different from the base wage rate of $6.48 per hour. A contractor relying on the wage figure only to come up with an estimated price will grossly underbid the project and "lose his shirt."

It is particularly important to verify that the WC,PL,PD rate being used for a worker is the correct one. Particularly hazardous situations result in rates as high as $44 per $100 of payroll (e.g., tunneling). However, if a worker is simply installing miscellaneous metals he should not be carried as a structural steel erector. The difference in the rates between the two specialties can be significant. It should also be noted that the rates given in Table 14.1 are for a particular geographical area and are the so-called "manual" rates. The manual rate is the one used for a firm for which no safety or experience records are available. These rates can be substantially reduced for firms that evidence over years of operation that they have an extremely safe record. This provides a powerful incentive for contractors to be safe. If the WC, PL, PD rate can be reduced by 30%, the contractor gains a significant edge in bidding against the competition.

The calculation of the hourly average wage indicates the complexity of payroll preparation. A contractor may deal with anywhere from 5 to 15 different crafts, and each craft union has its own wage rate and fringe benefit structure. Union contracts normally require that the payroll must be prepared on a weekly basis further complicating the situation. In addition, all federal, state, and insurance agencies to which contributions or premiums must be paid, require certified payrolls for verification purposes. Because of this, most contractors with a work force of any size use the computer for payroll preparation. Data are collected by field personnel using time cards such as the one shown in Figure 14.6. These time records are submitted to clerical personnel who prepare them for submittal to the computer. Large firms have in-house computers for this purpose. Medium to small firms may utilize service bureaus to provide this payroll preparation function. Charges for this service run in the vicinity of ½ to 1% of the total payroll amount.

14.4 STANDARD LABOR COST ESTIMATES

In order to arrive at a total cost figure for the job, the number of hours worked by each craft must be determined. In most firms, a system of cost accounting on previous jobs is utilized to arrive at labor cost per unit of production. Implicit in these labor cost figures is a certain level of productivity on the part of the crafts involved. If data from previous jobs are not available or appropriate, standard references are available that allow the determination of labor cost directly from the estimated quantity to be installed or constructed. The standard references normally give a nationally averaged price per unit. A multiplier is used to adjust the national price to a particular area. These

JOB #

WEEK ENDING   CODE   HOURS

EMP #	Name	CLASS CODE	CLASSIFICATION	RATE CODE	RATE	S	M	T	W	T	F	S	S	M	T	W	T	F	S	Total Hrs.

EMP #	Name	CLASS CODE	CLASSIFICATION	RATE CODE	RATE	S	M	T	W	T	F	S								Total Hrs.

EMP #	Name	CLASS CODE	CLASSIFICATION	RATE CODE	RATE	S	M	T	W	T	F	S								Total Hrs.

EMP #	Name	CLASS CODE	CLASSIFICATION	RATE CODE	RATE	S	M	T	W	T	F	S								Total Hrs.

FIGURE 14.6 Time card (Courtesy of Bellamy Brothers, Inc., Ellenwood, Ga.).

references are updated on an annual basis to keep them current. Among the largest and best-known of these services are:

1. *Dodge Construction Pricing and Scheduling Manual.*
2. R. S. Means *Building Construction Cost Data.*
3. F. R. Walker's *The Building Estimator's Reference Book.*
4. The *Richardson General Construction Estimating Standards.*

These references contain listings of cost *line items* similar to the cost account line items a contrator would maintain. Figure 14.7 shows two pages from the *Dodge Construction Pricing and Scheduling Manual** with line items pertaining to brick masonry installation and concrete block work. A problem showing the development of a cost estimate for masonry work on a school building in St. Louis indicates how the line items are used (Figure 14.8). Quantities for the appropriate line items have been developed from the plans for the school building and are shown in the QUANTITY column. Consulting the appropriate line item listings in Figure 14.7 it is determined that the estimated labor price per unit of EXTERIOR FACE BRICK (HEADER IN SIXTH) is $133 per thousand (M).† The associated material cost is $71 per thousand. The cost for labor for the estimated quantity of 29,500 bricks is $3925 (rounded to the nearest $5). Similarly, the materials cost for this line item is $2095. By consulting other line item listings, unit prices for the other items are determined, listed, and line item labor and material costs by item are developed. As shown the total labor cost for the EXTERIOR MASONRY package is $7195. The total material cost is $4060.

The next step in this method of cost development is to adjust these national average prices to the area in which the construction is to be done. Since the school is to be built in the St. Louis area, the adjustment will be based on a St. Louis multiplier. Adjustment factors for major work areas are given in Figure 14.9. The factors are arranged in a matrix format by city and craft. By checking the Masonry column for the St. Louis area it can be determined that masonry labor costs are approximately 98 % of the masonry rates nationwide. Material costs, however, run above the national average by 6 % (table reading of 106). Therefore, the labor cost can be reduced by 2%, but the material cost should be increased by 6%. This leads to a $140 labor deduction and an increase in material cost of $245. The adjusted labor and material costs are shown in Figure 14.8 as $7055 and $4305, respectively.

After adjusting the labor costs for geographical area, an additional factor is used to calculate the fringe benefits to be included in the bid. A listing of

*This material is not current and is presented here for illustration purposes only. Current editions of the references cited are available in most technical book stores or by direct order from the publisher.

†M is used as an abbreviation for 1000 units.

8.6 MEANS 8 MASONS SUPPORTED BY 6 LABORERS

OUTPUT PER DAY			ITEM	COST		
CREW	QUANTITY	UNIT		UNIT	LABOR	MATERIAL
1 MASON (8:6)	525	PCES	FACE BRICK: STANDARD @ 70./M—STRETCHER BOND	M	$128.00	$71.00
1 MASON (8:6)	480	PCES	STRETCHER—HEADER SIXTH	M	133.00	71.00
1 MASON (8:6)	460	PCES	ALL HEADERS IN SIXTH	M	144.00	71.00
1 MASON (8:6)	425	PCES	FLEMISH BOND	M	154.00	71.00
1 MASON (8:6)	400	PCES	ENGLISH BOND	M	170.00	71.00
1 MASON (8:6)	420	PCES	GLAZED @ 110./M—STRETCHER BOND	M	155.00	113.00
			EXTRA FOR QUOINS	PCE	0.03	0.09
			EXTRA FOR BULLNOSED	M	0.03	0.07
1 MASON (8:6)	385	PCES	ROMAN @ 95./M—STRETCHER BOND	M	175.00	97.00
1 MASON (8:6)	360	PCES	NORMAN @ 105./M—STRETCHER BOND	M	187.00	108.00
1 MASON (8:6)	330	PCES	SCR—STANDARD SIZE 6 in. THICK	M	195.00	160.00
			ADD TO FACE BRICK IF STACK BOND	M	12.00	5.00
			ADD TO FACE BRICK IF BACK PARGED 3/8 in.	SF	0.05	0.03
1 MASON (3:1)	830	SF	WASH DOWN FACE BRICK	SF	0.06	0.01
			LABOR ONLY, LAYING FACE BRICK PATTERNS	SF	0.25	
			FACE BRICK ARCHES: SPLAYED, FLAT	lin ft	1.95	1.15
			SEMICIRCULAR 1 RING	lin ft	5.30	0.95

FIGURE 14.7 Masonry line items from *Dodge Construction Pricing and Scheduling Manual* (Copyright © 1969, McGraw-Hill, Inc.).

FIGURE 14.7 (Continued)

1 MASON (8:6)	570	PCES	COMMON BRICK: BACK UP—CLAY BRICK—4 in.	M	117.00	51.00
1 MASON (8:6)	600	PCES	CLAY BRICK—8 in.	M	112.00	51.00
1 MASON (8:6)	430	PCES	CLAY JUMBOS 3½ x 11½ - 4 in.	M	156.00	72.00
1 MASON (8:6)	570	PCES	SAND LIME BRICK	M	117.00	36.00
1 MASON (8:6)	570	PCES	CONCRETE BRICK	M	117.00	47.00
1 MASON (8:6)	550	PCES	PARTITIONS—CLAY BRICK	M	123.00	50.00
1 MASON (8:6)	550	PCES	CONCRETE BRICK	M	123.00	47.00
1 MASON (8:6)	525	PCES	CHIMNEYS—CLAY BRICK	M	128.00	50.00
1 MASON (1:1)	560	PCES	FIRE BRICK FLUE (9 x 2 x 4½)	M	132.00	186.00
1 MASON (1:1)	125	LF	FIRE CLAY FLUE LINING: 8 x 8	LF	0.65	0.75
1 MASON (1:1)	105	PCES	8 x 12	LF	0.72	0.85
1 MASON (1:1)	90	PCES	12 x 12	LF	0.82	1.25
1 MASON (1:1)	70	PCES	18 in. diam.	LF	1.08	3.95
1 MASON (1:1)	60	PCES	24 in. diam.	LF	1.30	7.20
1 MASON (1:1)	400	PCES	FIREPLACE—FACE BRICK	M	185.00	85.00
			CHIMNEY CLEAN-OUT 12 x 12 C.I.	EA	4.25	9.00
			CHIMNEY DAMPER C.I. 36 in. 12 in. THROAT	EA	8.00	14.50
			CHIMNEY CAP 4 in. CONCRETE	SF	1.00	0.80
			CHIMNEY SPARK ARRESTOR (12 x 12 FLUE)	EA	3.50	4.75
			CHIMNEY BIRD SCREEN (12 x 12 FLUE)	EA	2.50	2.20

1 MASON (1:1)	550	PCES	REINFORCED MASONRY: LABOR ONLY, BRICKWORK	M	134.00	
			REINFORCING STEEL	TON	78.00	192.00
			CONCRETE FILL	CF	0.31	0.65
			GROUT FILL	GF	0.31	0.85
1 MASON (1:2)	800	PCES	ACID-RESISTING BRICK PAVING	M	130.00	100.00
1 MASON (8:6)	185	PCES	CONCRETE BLOCK: REGULAR 8 x 16—2 in.	PCE	0.37	0.20
1 MASON (8:6)	170	PCES	4 in.	PCE	0.40	0.19
1 MASON (8:6)	150	PCES	6 in.	PCE	0.46	0.23
1 MASON (8:6)	120	PCES	8 in.	PCE	0.57	0.25
1 MASON (8:6)	220	PCES	10 in.	PCE	0.62	0.37
1 MASON (8:6)	110	PCES	12 in.	PCE	0.72	0.40
			ADD FOR SOLID BLOCKS	PCE	$ 0.05	$0.12

EXTERIOR MASONRY WORK PLEASANT SCHOOL, ST. LOUIS

ITEM	QUANTITY	UNIT LABOR	UNIT MATERIAL	LABOR	MATERIAL
EXTERIOR FACE BRICK (HEADER IN SIXTH)	29.5 M	$133.	$71.	$3925.	$2095.
COMMON BRICK BACKUP	3.0 M	117.	51.	350.	155.
8-in. CONCRETE BLOCK BACKUP (8 × 16)	3800. PCS.	0.57	0.25	2165.	950.
6-in. CONCRETE BACKUP (8 × 16 in.)	600. PCS.	0.46	0.23	275.	140.
EXTERIOR SCAFFOLD	4100. SF	0.08	0.06	330.	245.
MORTAR 1-2-9	33. CY	$ 4.50	$14.35	150.	475.
				$7195.	$4060.

		Labor	Material
ST. LOUIS ADJUSTMENT—LABOR DEDUCT 2%		-140	
ST. LOUIS ADJUSTMENT—MATERIAL ADD 6%			+245.
		7055.	4305.
		→	7055.
ADD: FRINGE BENEFITS 17% × $7055			1200.
INSURANCES—W. COMP., P.L. & P.D.	?% × $7,055		______
PAYROLL TAXES	?% × $7,055		______
JOB OVERHEAD:			______
TOTAL COST			$______
+ FEE (PROFIT)			$______
MASONRY—GRAND TOTAL			$______

NOTES.

EXTENSIONS TO NEAREST $5 ONLY.

In the case of a complete estimate, the labor and material adjustments and the fringe benefits items would go at the end of each trade estimate sheet; insurances and payroll tax items would be embodied in the job overhead detail sheet.

FIGURE 14.8 Sample estimate using Dodge pricing manual (Copyright © 1969, McGraw-Hill, Inc.).

THE NUMBER 100 EQUALS UNITS USED IN THE MANUAL; THUS THE INDICES APPEARING BELOW REFLECT EACH CITY'S PERCENTAGE OF MANUAL PRICES.

TRADE	EXCAVATION		FORMWORK		CONCRETE		REIN STEEL		STR STEEL		MASONRY		CARPENTRY	
CITY	LAB.	MAT.	LAB.	MAT.	LAB.	MAT.	LAB.	MAT.	LAB.	MAT.	LAB.	MAT.	LAB.	MAT.
ATLANTA	85	95	90	98	85	90	90	98	94	108	94	88	92	94
BALTIMORE	98	98	96	95	90	96	98	101	98	100	96	96	96	96
BIRMINGHAM	85	96	85	102	80	95	88	96	90	94	90	90	90	105
BOSTON	98	104	100	105	98	98	105	102	102	103	104	100	100	106
BUFFALO	102	106	95	102	103	100	104	98	102	98	102	102	102	104
CHICAGO	104	100	106	105	104	96	106	94	106	95	104	98	105	106
CINCINNATI	102	98	96	98	102	102	97	102	96	100	98	96	96	95

CLEVELAND	115	95	108	100	116	102	106	98	104	103	106	102	108	105
DALLAS	90	102	90	102	88	90	90	106	94	108	94	100	96	100
DENVER	92	107	95	104	94	86	95	110	95	125	98	104	98	98
DETROIT	106	98	102	110	108	96	98	100	98	98	104	110	105	106
KANSAS CITY, MO.	96	100	95	112	94	95	95	92	95	104	96	98	98	106
LOS ANGELES	99	104	104	90	100	90	105	98	105	108	95	96	105	85
MINNEAPOLIS	102	100	96	100	105	98	96	98	94	102	95	102	95	96
NEW ORLEANS	85	96	90	96	85	102	90	106	92	103	94	106	94	98
NEW YORK	122	108	120	100	120	110	130	106	122	102	110	106	115	105
PHILADELPHIA	102	102	98	95	96	102	102	102	104	103	102	100	104	96
PITTSBURGH	100	95	110	98	103	104	105	98	106	102	108	100	110	102
ST. LOUIS	107	104	108	102	108	95	104	98	102	98	98	106	106	104
SAN FRANCISCO	106	105	104	90	102	104	110	90	102	115	108	124	105	90
SEATTLE	103	102	96	85	101	98	98	108	96	110	100	120	96	90

FIGURE 14.9 Adjustment index for major trades (Copyright © 1969, McGraw-Hill, Inc.).

FIGURE 14.9 (Continued)

THE NUMBER 100 EQUALS UNITS USED IN THE MANUAL; THUS THE INDICES APPEARING BELOW REFLECT EACH CITY'S PERCENTAGE OF MANUAL PRICES.

TRADE	MISC IRON		PLASTER		PAINTING		UTILITIES		PLUMBING		HVAC		ELECTRICITY	
CITY	LAB.	MAT.	LAB.	MAT.	LAB.	MAT.	LAB.	MAT.	LAB.	MAT.	LAB.	MAT.	LAB.	MAT.
ATLANTA	95	105	96	104	96	95	88	85	96	95	96	95	92	102
BALTIMORE	100	102	98	96	98	100	92	100	98	100	98	98	95	98
BIRMINGHAM	92	96	90	102	92	102	85	85	94	96	95	98	94	102
BOSTON	102	104	102	98	104	100	98	104	106	102	106	102	100	98
BUFFALO	102	100	104	102	100	98	102	103	102	104	104	100	100	102
CHICAGO	105	96	106	100	104	96	106	102	106	100	104	102	102	96
CINCINNATI	96	102	98	94	96	98	100	115	98	96	100	98	98	98

CLEVELAND	105	104	112	98	108	95	110	98	106	102	108	102	100	100
DALLAS	95	110	96	102	95	102	92	90	94	98	95	96	92	102
DENVER	98	120	98	105	96	105	94	92	98	104	98	102	98	105
DETROIT	102	100	104	96	100	100	104	100	102	102	104	100	100	100
KANSAS CITY, MO.	96	102	102	98	104	95	96	102	100	98	98	102	98	104
LOS ANGELES	107	105	98	100	102	100	98	105	108	105	106	104	104	102
MINNEAPOLIS	98	104	97	110	98	98	102	112	98	104	98	102	98	102
NEW ORLEANS	94	104	90	105	88	102	85	96	95	96	94	100	96	102
NEW YORK	112	108	115	98	112	100	125	104	110	100	108	102	110	100
PHILADELPHIA	102	104	104	94	98	100	105	96	102	98	104	100	108	100
PITTSBURGH	105	100	107	100	106	102	100	108	106	102	105	100	104	98
ST. LOUIS	104	104	104	105	100	104	108	95	106	96	106	98	100	102
SAN FRANCISCO	104	110	100	102	106	105	106	110	120	106	125	104	125	106
SEATTLE	96	112	98	115	102	102	102	104	98	106	100	105	100	106

composite fringe-benefit, add-on factors is given in Figure 14.10. Since fringe benefits in the masonry trade for the St. Louis area are running at an average of 17% ($0.17 on the dollar) for all masonry crafts, this amount is added to the labor cost (only). Examination of the levels of fringe benefit factors for various cities gives a fairly good indication of the strength of labor and unions around the country. The 17% level in St. Louis indicates that the masonry unions in the St. Louis area are strong and have been successful in negotiating relatively strong fringe benefit packages. By comparison, the masonry fringe benefit in Birmingham is 0.5% indicating little or no masonry union strength in bargaining for fringe benefits in the Alabama area. Looking along the row of fringe benefit figures for Birmingham, it is clear that the construction crafts have not negotiated fringes. This indicates a relatively weak union area, which is to be expected, since Alabama is a "right to work state" (see Section 13.9) and a strong open shop area. The masonry trades in New York, by contrast, must be estimated with a 26% increase for fringes. Obviously, the unions have been active in this area and have negotiated comprehensive benefit packages. It should be noted that open shop contractors do not sign contracts and, therefore, do not pay fringe benefits of this variety.

To complete the calculation of exterior masonry work on the Pleasant School (Figure 14.8), the sample estimate shows a space for addition of WC, PL, and PD expense as well as FICA and unemployment taxes (payroll taxes). As discussed in the previous sections, these are a function of the individual contractor's safety record and the state in which he is operating. Job overhead is added to arrive at TOTAL COST, and a fee (profit) item is added to complete the estimated cost of masonry work.

14.5 RESOURCE ENUMERATION

In some instances, a contractor is confronted with a situation for which he has no in-house labor cost data available. Standard references with appropriate cost data are either not available or the construction is so nonstandard that it is not listed in estimating data handbooks. In such a case, the resource enumeration approach of arriving at labor cost can be used. This approach is used in describing cost development in Walker's *Building Estimator's Reference Book* and is discussed in detail in traditional estimating texts.*

In this method, a crew or set of crews required for the work is defined, and an hourly cost for each crew is developed based on hourly direct wage rates. This rate does not include fringes, insurances, and other burdens. A production rate for each crew is established based on an assessment by the estimating section. Quantities of work for each line item of unique work are developed in the conventional way from plans and specifications. An evaluation of job

*See R. L. Peurifoy, *Estimating Construction Costs*, 2nd Ed., McGraw-Hill.

conditions may dictate the use of an efficiency factor to adjust the ideal production assumed to the work site conditions. Based on the revised productivity rate, a unit price is developed and multiplied by the required quantity. These steps are summarized as follows:

1. Assume a crew composition.
2. Calculate hourly crew rate.
3. Calculate required quantities.†
4. Make an assumption regarding efficiency factor.
5. Calculate effective unit price.
6. Calculate line item total price.

An example of a resource enumeration calculation of labor cost on a concreting operation is given in Figure 14.11. In this example a concrete placement crew consisting of a carpenter foreman, two cement masons, a pumping engineer (for operation of the concrete pump), and seven laborers for placing, screeding, and vibrating the concrete has been selected. A concrete pump (i.e., an equipment resource) has also been included in the crew. Its hourly cost has been determined using methods described above. The total hourly rate for the crew is found to be $80.60. The average assumed rate of production for the crew is 12 cu yd/hr. This results in an average rate per cubic yard of concrete of $6.72. The line items requiring concrete are listed with the quantities developed from the plans and specifications. Consider the first item that pertains to foundation concrete. The basic quantity is adjusted for material waste.* The cost per unit is adjusted to $7.47 based on an efficiency factor for placement of foundation concrete estimated as 90% (0.90).

The development of the efficiency is based on "engineering intuition." That is, it is an estimate of the reduction in productivity based on the work face environment for the work item being considered. Some methods for formalizing this process have been developed. The Dallavia method described in Figure 14.12 is typical of such methods. Major job environment and economic factors are scored and weighted. Based on this evaluation, a composite efficiency factor is calculated.

The total activity labor cost is calculated by multiplying the required quantity by the effective unit price and rounding to the next higher dollar value. For the 14 concrete line items shown, the total direct labor cost is $2620. This total would be increased by appropriate factors for fringes, insurance, and other distributed costs.

As an aid in determining which work items should be estimated for a particular job, most companies have a set of standard estimating/cost accounts

†Adjust Quantities for waste factor as required.

*The wastage will affect the total material cost but does not impact labor cost since it is not installed.

TRADE / CITY	EXC.	FORMS	CONC.	MASONRY	STRUCT STEEL	CARP'Y.	MISC. IRON	PLAST.	MECH.	ELEC.
ATLANTA	4.5	2.0	4.0	6.0	6.0	0.5	5.5	1.5	6.5	7.0
BALTIMORE	7.0	4.0	8.0	6.0	8.0	4.0	8.0	5.0	7.0	3.0
BIRMINGHAM	NONE	NONE	NONE	0.5	4.5	NONE	4.5	NONE	9.0	6.0
BOSTON	8.0	7.0	8.0	7.0	8.0	7.0	8.0	14.0	8.0	6.0
BUFFALO	14.0	12.0	12.0	13.0	11.0	13.0	7.0	2.0	8.0	6.0
CHICAGO	9.0	8.5	9.0	8.0	10.0	8.0	8.0	8.0	7.0	12.0
CINCINNATI	4.5	6.0	3.5	3.5	8.5	6.0	7.0	4.5	11.0	14.0
CLEVELAND	7.5	16.0	8.0	9.0	7.0	18.0	8.0	NONE	12.5	14.0

DALLAS	5.0	2.0	4.0	5.0	6.0	1.0	6.0	12.0	6.5	7.0
DENVER	NONE	4.0	1.0	2.5	3.0	5.0	3.0	1.5	10.0	3.0
DETROIT	12.0	16.0	14.0	15.0	20.0	18.0	17.0	14.0	14.0	22.0
KANSAS CITY, MO.	8.0	5.0	7.5	4.5	3.5	5.0	3.0	6.0	10.0	8.0
LOS ANGELES	18.0	16.0	18.0	14.0	13.0	14.0	13.0	17.0	20.0	11.0
MINNEAPOLIS	4.0	11.0	3.0	12.0	3.5	12.0	3.5	14.0	10.0	15.0
NEW ORLEANS	5.0	5.0	5.0	6.0	5.0	6.5	5.0	5.0	8.0	3.0
NEW YORK	15.0	15.0	18.0	26.0	20.0	15.0	18.0	19.0	22.0	25.0
PHILADELPHIA	7.0	13.0	7.0	10.0	12.0	15.0	11.0	9.0	8.0	5.0
PITTSBURGH	6.0	7.0	8.0	7.0	7.5	6.0	7.0	7.0	12.0	14.0
ST. LOUIS	8.0	4.0	8.0	17.0	8.0	3.0	9.0	8.0	24.0	19.0
SAN FRANCISCO	18.0	22.0	18.0	19.0	15.0	21.0	14.0	23.0	6.0	6.0
SEATTLE	9.0	9.0	8.0	7.0	7.0	9.0	6.0	10.0	13.0	8.0

NOTE: CHANGES THAT MAY OCCUR IN FRINGE BENEFITS DURING 1968 ARE NOT ANTICIPATED IN THIS TABLE.

FIGURE 14.10 Percentage additions to labor for fringe benefits (Copyright © 1969, McGraw-Hill, Inc.).

CONCRETE PLACING CREW

QUANTITY	MEMBER	RATE	TOTAL/HOUR
1	Carpenter Foreman	$9.42	$9.42
2	Cement Masons	8.36	16.72
1	Pumping Engineer	8.45	8.45
7	Laborers	5.43	38.01
1	Concrete Pump	$8.00	$8.00
		CREW HOURLY RATE	$80.60

Production rate of crew under normal circumstances (Efficiency factor 1) = 12 cu yd / hr)

Average Labor Cost/cubic yard = $80.60 / 12 = $6.72

Area	Quantity	Percent Waste	Labor Efficiency Factor	Activity Labor Cost/Cubic Yard	Total Activity Cost
(1) Foundation	53.2	15	0.9	$7.47	$398
(2) Wall to Elevation 244.67	52.9	12	0.8	8.40	445
(3) Slab 10 in.	1.3	30	0.3	22.40	30
(4) Beams Elevated 244.67	10.5	15	0.7	9.60	101
(5) Beams Elevated 245.17	9.1	15	0.7	9.60	88
(6) Slab Elevation 244.67	8.7	10	0.7	9.60	84
(7) Interior Wall to 244.67	5.5	15	0.4	16.80	93
(8) Slab Elevation 254.17	6.3	10	0.75	8.96	57
(9) Walls 244.67 -254.17	57.2	10	0.8	8.40	481
(10) Walls 254.17 -267	42.0	10	0.8	8.40	353
(11) Floors elevated 267	8.9	10	0.9	7.47	67
(12) Manhole Walls	27.3	10	0.85	7.91	216
(13) Roof	14.0	15	0.7	9.60	135
(14) Headwall	8.5	10	0.8	$8.40	72
				Total Direct Labor Cost for Concrete	$2620

FIGURE 14.11 Labor resource enumeration.

Production Range Index

	Production effiency, percent		
PRODUCTION	25 35 45 55	65 75 85	95 100
ELEMENTS	Low	Average	High
1 General economy	prosperous	normal	hard times
local business trend	stimulated	normal	depressed
construction volume	high	normal	low
unemployment	low	normal	high
2 Amount of work	limited	average	extensive
design areas	unfavorable	average	favorable
manual operations	limited	average	extensive
mechanized operations	limited	average	extensive
3 Labor	poor	average	good
training	poor	average	good
pay	low	average	good
supply	scarce	average	surplus
4 Supervision	poor	average	good
training	poor	average	good
pay	low	average	good
supply	scarce	average	surplus
5 Job conditions	poor	average	good
management	poor	average	good
site and materials	unfavorable	average	favorable
workmanship required	first rate	regular	passable
length of operations	short	average	long
6 Weather	bad	fair	good
precipitation	much	some	occasional
cold	bitter	moderate	occasional
heat	oppressive	moderate	occasional
7 Equipment	poor	normal	good
applicability	poor	normal	good
condition	poor	fair	good
maintenance, repairs	slow	average	quick
8 Delay	numerous	some	minimum
job flexibility	poor	average	good
delivery	slow	normal	prompt
expediting	poor	average	good

EXAMPLE: After studying a project on which he is bidding, a contractor makes the following evaluations of the production elements involved:

PRODUCTION ELEMENT	PERCENT EFFICIENCY
1. Present economy	75
2. Amount of work	90
3. Labor	70
4. Supervision	80
5. Job conditions	95
6. Weather	85
7. Methods and equipment	55
8. Delays	75
TOTAL	625

As the total of the eight elements is 625, the average value will be 625/8, or 78 percent.

Detailed estimates must be made for the individual operations required by that project. Each of these operations must be evaluated in a similar manner and adjusted against the overall rating before the production of the typical shift crew can be determined in actual units of scheduled work. In making such evaluations, those production elements which do not seem directly applicable to a particular operation may be ignored.

(Excerpted from *Estimating General Construction Costs* by Louis Dallavia, published by F. W. Dodge, New York, 1957. Used with permission of McGraw-Hill Book Company).

FIGURE 14.12 Dallavia method.

that are utilized as a checklist when estimates are prepared. Various standard lists have been developed and have become accepted as industry-wide standards. One of the better-known building construction standards is the Associated General Contractors — Construction Specifications Institute (AGC-CSI) list of cost accounts. This listing is reprinted in Appendix K. The major categories in this breakdown are typical of those used by building contractors.

REVIEW QUESTIONS AND EXERCISES

14.1 Name three types of insurance (applying to each person) that must be considered when making out the company payroll.

14.2 A partition wall 8 ft. high by 20 ft. in length is to be constructed of 8″ × 16″ × 6″ block. The job is located in Cincinnati, Ohio. Estimate the cost of the wall to include labor (with fringes) and material. (Use Dodge Information in text).

14.3 Compute the average hourly cost to a contractor for an operating engineer. The job involves excavating a spillway. The operator works 10 hours per day on a 20 cu yd shovel 5 days a week. The last two hours are considered overtime. The contractor carries $300K/$300K PL and $50K/$100K PD insurance. Use wage rates and other data as given in Table 14.2. Assume overtime paid at time and a half.

14.4 Compute the average hourly cost of a carpenter to a contractor. Assume the work is in a subsistence area and the daily subsistence rate is $9.50. The carpenter works the second shift on a two-shift project where a project labor contract establishes a "work 7 pay 8 hour" pay basis for straight time. He works six days, 10 hours a day. In addition to time and a half for overtime Monday through Friday, the contract calls for double time for all work on weekends. Use 6.05% FICA and 4.5% for unemployment insurance. Assume all data relating to the W.C., P.L., P.D., fringes and wage are as given in Table 14.2.

14.5 Assume the walls in the vault shown in Problem 9.10 are to be constructed using 8″ × 16″ × 6″ block. Each wall section will consist of a double thickness of block. Estimate the cost of block to include labor (with fringes) and material. Assume the vault is being constructed in Atlanta, Georgia (use Dodge Information in text).

14.6 Develop the face brick quantities for the Small Gas Station shown in Appendix L. Using the Dodge information in this chapter, calculate the cost of labor and materials required to place the face brick. Assume the station is being constructed in a suburb of Chicago.

Table 14.2 BUILDING CRAFT WAGE AND INSURANCE RATES

Locals	Wages	Pension	Health and Welfare	Vacation	Apprentice Training	Miscellaneous	Workmen's[a] Compensation	Public[b] Liability	Property[b] Damage
Asbestos workers	$10.15	$0.60	$0.55		$ 0.10		$ 6.09	$1.00	$0.55
Boilermakers	10.25	0.75	1.05		0.02		6.46	0.37	0.36
Bricklayers	9.35	0.50	0.55	$0.65		$0.08 prom.	3.55	0.38	0.27
Carpenters	9.45	0.45	0.50		0.02		5.67	0.40	0.26
Cement masons	8.90	0.55	0.40			0.20 bldg.	2.51	0.41	0.29
Electricians	10.45	1.1%	0.9%	0.8%	0.05%		2.19	0.17	0.21
Operating engineers	9.35	0.75	0.50		0.07	0.10 admin.	5.61	0.93	1.00
Ironworkers	9.60	0.57	0.65		0.07		14.59	1.50	0.94
Laborers	6.25	0.33	0.20			0.05 educ.	3.75	0.19	0.20
Painters	9.45	0.65	0.65		$250/yr		3.59	0.13	0.44
Plasterers	9.17	0.55	0.40			0.20 bldg. 0.10 prom.	3.48	0.39	0.27
Plumbers	10.75	0.50	0.65		0.11	0.06 prom. 0.02 natl.	2.80	0.29	0.59
Sheet metal	$10.20	$0.70	$0.50		$ 0.04	$0.09 ind.	$ 3.57	$0.21	$0.20
Unemployment 4.5%									
Social security 5.85%									

a Rates are applied per $100 of pay.

b *Public liability*. Maximum coverage under these rates — $5000/person, $10,000 per accident.
For higher coverage — $10,000/20,000: 1.26 x basic rate
$25,000/50,000: 1.47 x basic rate
$50,000/100,000: 1.59 x basic rate
$300,000/300,000: 1.78 x basic rate

Property damage. Maximum coverage under these rates — $5000/person, $25,000 per accident.
For higher coverage — $25,000/100,000: 1.23 x basic rate
$50,000/100,000: 1.30 x basic rate

14.7 Using the plans given in Appendix L, calculate the volume of floor slab concrete to be placed in the Small Gas Station. Assuming a 12% waste factor and a labor efficiency factor of 0.85, what will be the approximate labor cost for placing the floor slab concrete? Assume rate of placement is 15 cu yd/hr.

14.8 What labor agreements are operating in your area?

14.9 Determine typical crew sizes and mixes for the following crews:

Masonry work
Concreting
Formwork erection
Formwork stripping
Rebar placement

14.10 Obtain a copy of a local labor agreement, identify the parties to the agreement and the date of expiration, and prepare a list of clauses that refer to working conditions and pay rates.

CHAPTER FIFTEEN

Construction Planning

15.1 THE PROJECT PLANNING AND CONTROL PROCESS

Project planning and control has as its broad and overall objective the prescribing, and field attainment, of an orderly progression, within budget and time, toward the completion of the project facilities. These overall project planning and control efforts establish the sequence, time, and cost frameworks within which construction activity is constrained. Thus, the construction planning and control framework is influenced by decisions at the owner, contract administrator, and contractor levels.

From the owner's point of view, there is a need to establish the time and cost framework for the various project components so that the completion of each can be funded, scheduled, and integrated into the project's start-up and business development programs. Thus, there is management interest in the identification of significant project *milestones* that can serve as indicators of project progress. Accordingly, the monitoring of the progressive achievement of milestones, especially in relation to the interaction of anticipated and actual cash flows on budgets and available finance, is of vital interest. Generally, in this environment, the owner delegates to a competent contract administrator (such as a consulting architect/engineer) the responsibility for ensuring that for each dollar spent during construction, there is a return in value achieved of construction-in-place. From the construction point of view, decisions made in the prebid phase on the order of sequence and time frame for the completion of each project component significantly affect the definition of the construction plan and the rate at which construction will proceed.

From the contract administrator's point of view, there is a need to establish a project time framework that is compatible with the needs of the owner and the magnitude of construction effort involved. Within this time framework, the administrator must provide general direction and supervision to the contractor and must monitor and assess construction progress, quality of work achieved, and value of work in place. In order to perform these functions, the contract administrator may indicate to the contractor the various significant progressive stages in the construction of a facility that he will recognize and the basis of evaluation that he will adopt for verification of the value of work progress achieved in each stage. These factors and decisions significantly affect the manner in which the contractor will execute the construction plan and the scheduling of work effort to achieve milestones.

For the contractor, construction planning also involves the determination of a construction strategy for the project. This formulation involves the selection of construction methods and equipment, their sequencing in relation to technological requirements, and the desired order of construction of the major components of the project. Thus, the construction plan must also correlate the number and type of critical equipment items that are required for the performance of the various construction operations with the material quantities or work effort involved and the time available to meet project and contract time schedules. The contractor's construction plan emerges from a series of interacting decisions, since methods, equipment sizes and efficiencies, work rates, and work volumes are interrelated. Several passes or iterations through these interacting decisions are generally necessary before an acceptable construction plan emerges.

Construction planning at the contractor level can therefore be summarized as including:

1. The determination of what has to be done. This involves a careful study of the contract terms and documents and the assessment of work content for each work-task or project activity involved in the construction plan,

2. The prescribing of how each work-task is to be performed. This involves the selection of construction methods and related support and management procedures,

3. The enumeration of required resources as a result of decisions relating to the selection, level, and extent of resource availabilities to the project.

4. The order in which things are to be done. This involves an understanding of the technological processes involved in the construction methods,

5. Decisions as to when things are to be done. These decisions on desirable sequences are pertinent to the attainment of recognizable work progress and the satisfaction of the contractor's internal cash flow requirements. They establish the rate at which project work is to proceed.

Although planning is essential to management, the implementation and control of a plan is far more difficult and demanding on managerial skill because of the many complex factors and events that enter into and perturb the execution of a plan. Consequently, in addition, project control is concerned with ensuring that:

6. The proper resources (materials, equipment, and labor) are available at the right place at the right time,
7. The relevant field decision makers and agents are properly briefed as to what has to be done, when, and how,
8. Resources are used effectively,
9. The resulting work output meets all contractual and quality control requirements.

The initial section in this chapter introduces the concepts of physical development networks and milestone planning concepts as a means of portraying the impact of overall project planning on construction planning. On this broad base of project planning, later sections and the following chapter develop in detail the networking techniques and their use in terms of various construction planning and control problems that are commonly met in practice.

15.2 MILESTONE PLANNING

An initial step in the formulation of a project construction plan and the strategy for its accomplishment is to break the project down into a number of independent or readily recognizable components and to establish the order, or sequence, in which it is planned that they will be built. In large projects, involving a number of buildings and facilities, this breakdown is simple and pertinent, since in many cases these individual components will be the subject of individual construction contracts, and the orderly development of power and service systems for the project together with the installation, and run-in requirements, of process lines in industrial and power plant construction will help to establish project development sequences. In other projects involving only one facility (a building, for example) this dissection is not possible. In these cases a dissection rationale may be developed by focusing on the different physical stages in the construction of the building, and component identification could follow that of the normal construction process (i.e., site clearance, excavation, foundations, the structural framework, enclosure cladding, etc.) and relate to the work performed by speciality subcontractors or trades. In either case the project is dissected in such a way that a *physical sequence network* can be developed that portrays the sequence in which *physical objects* are planned to appear.

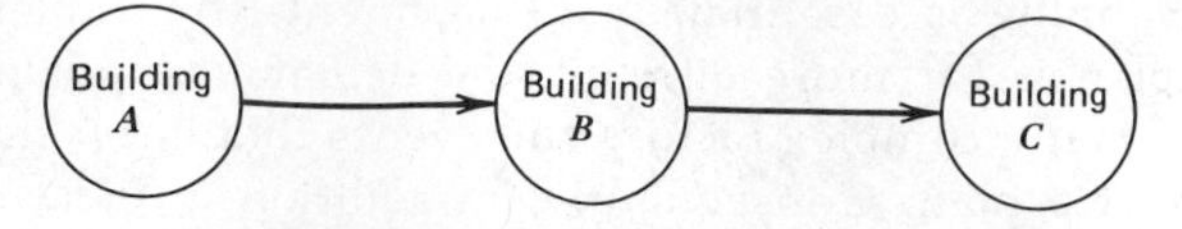

FIGURE 15.1 A physical sequence model (arrows indicate sequence of construction).

A simple illustration of a physical sequence network is shown in Figure 15.1 for a project involving the construction of three buildings (i.e., buildings *A, B,* and *C*), with a construction strategy of building Buildings *A, B,* and *C*, respectively. In this modeling of the construction plan the nodes portray (or model) physical objects and the directed arrows indicate the sequence in which the physical objects will appear on the construction site. Thus Figure 15.1 is also an example of a *precedence* diagram.

If, in addition to project component breakdown and sequencing, realistic time estimates are made for the construction periods thought adequate for each component, then a *master schedule* can be developed for the proposed construction effort. In this way the duration and scheduling of the construction effort for each component can be located on a project time scale in terms of *project milestones.* These milestones are usually associated with the initiation of, or time completion of, the construction effort for each project component or with critical phases in their construction that relate to subsequent and interfacing components. This initial fix on project construction planning is referred to as *milestone planning.*

A simple example of milestone planning and the portrayal of the master schedule for the project of Figure 15.1 is shown in Figure 15.2. Notice that an interface milestone is indicated for the completion of building *B* and the commencement of construction for building *C*. This interface milestone may refer to the requirement to complete building *B* before the commencement of

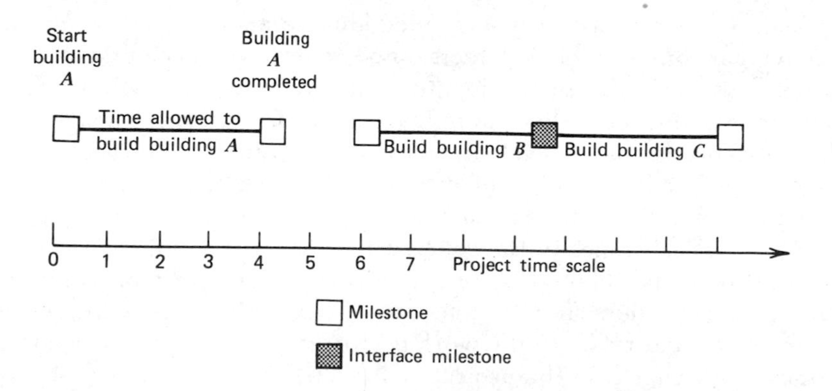

FIGURE 15.2 The Master scnedule (lines indicate time allowed to build the various components).

building C will be permitted. As a further, and more complex, example consider the development of a master schedule for the construction of a rock-fill dam. An initial and typical dissection of the proposed dam into physical objects together with construction period estimates (in months) for each item is given below:

1. Access roads and field camp (2),
2. The river diversion tunnel (2),
3. The quarry and its crushing plant (2),
4. The concrete batch plant (2),
5. The excavated hole for the dam (4½),
6. The rockfill for the dam (including banks) (12½),
7. The grout curtain ensuring water tightness of the substructure (3),
8. The inlet tower(2),
9. The valve house structure (3),
10. The mechanical fittings for the inlet tower and the valve house (3),
11. The concrete face slabs (4½),
12. The excavated hole for the spillway (5),
13. The spillway concrete (5),
14. The permanent roads associated with the dam maintenance (1),
15. The diversion tunnel plug (1).

A simple precedence diagram indicating the order in which these physical components might appear is given in Figure 15.3. For example, as indicated, the quarry and crushing plant must be erected on site before dam rockfill. Similar reasoning exists for the other component orderings.

A master schedule based on Figure 15.3 and the estimated construction periods is shown in Figure 15.4. It shows, for example, that site access and the field camp will be built in the first two months and that the concrete face slabs for the rockfill dam are planned to be built in the period covering the eleventh to the fifteenth project months. The construction period for each physical component is indicated by the time period between its two milestone dates. Many milestones have been shown as interface milestones indicating either the interfacing of related physical components and sequential construction effort or the contractual relationships between agents involved in the completion and follow-up contracts. Thus, for example, in Figure 15.4 the completion of the grout curtain signals commencement of concreting work on the spillway, and may relate to the needs of different contractors or the shift in equipment and formwork requirements of one contractor. The master schedule of Figure 15.4 thus portrays the major characteristics of the planned construction effort and the sequence in which the rockfill dam components are to be built. It thus provides the basic framework from which the contractor(s) can develop the details of the construction plans for their accomplishment.

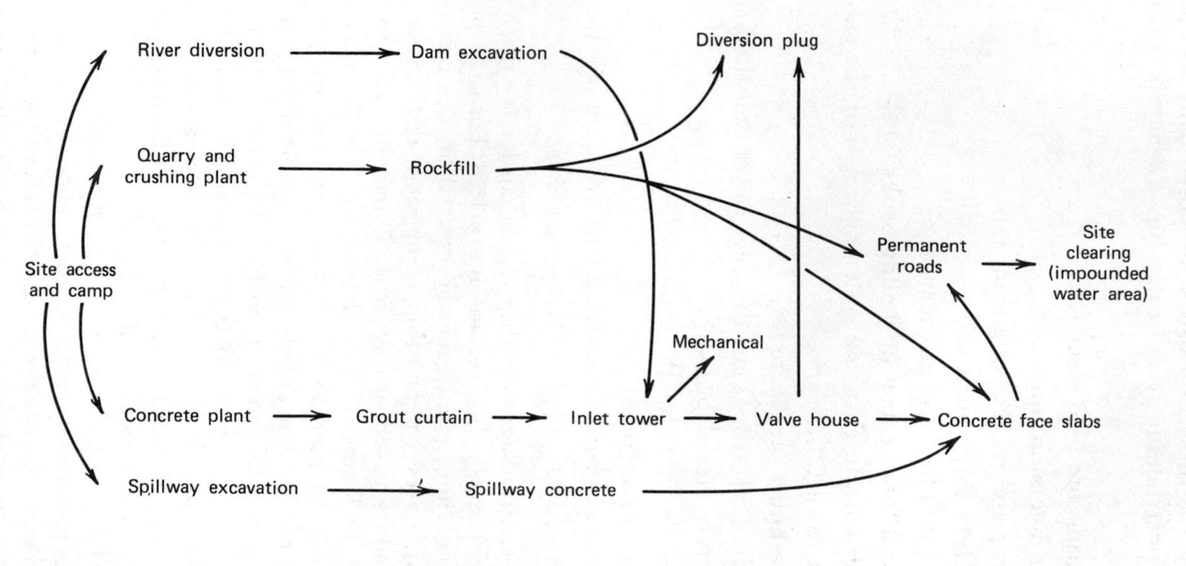

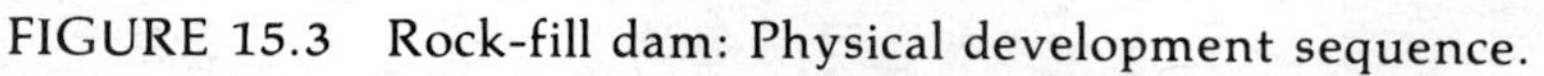

FIGURE 15.3 Rock-fill dam: Physical development sequence.

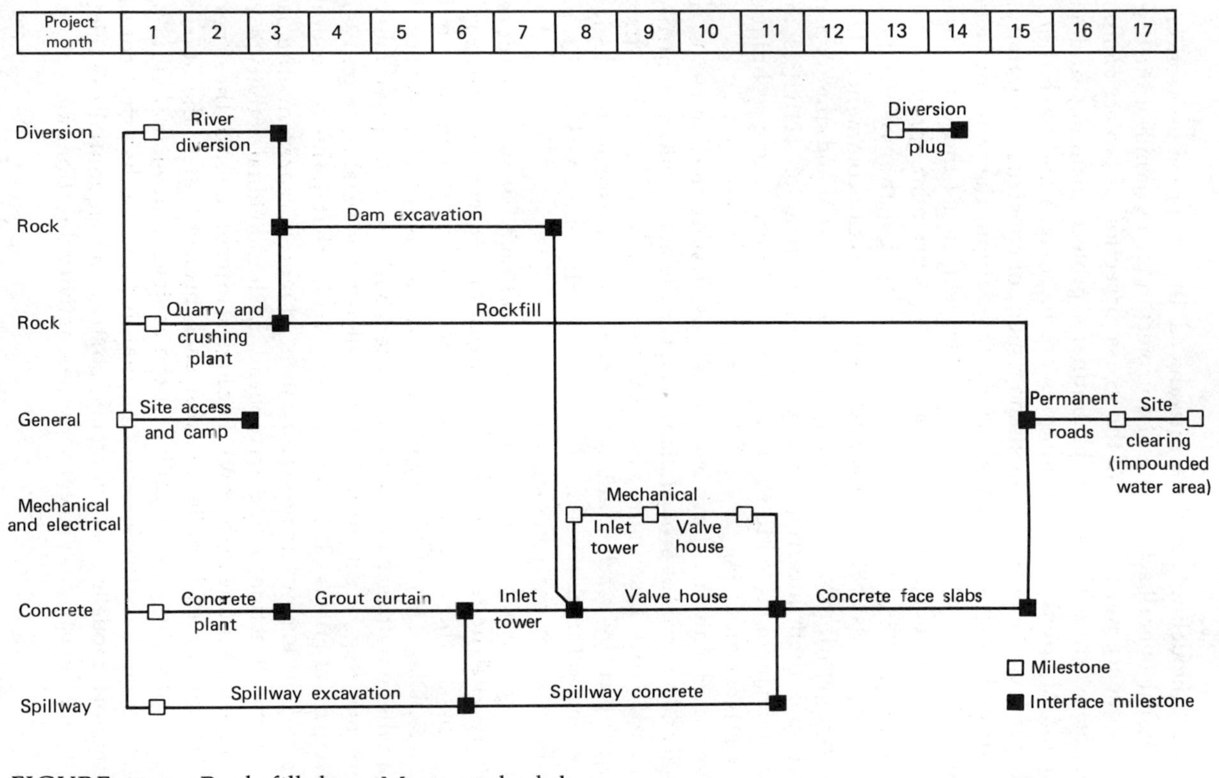

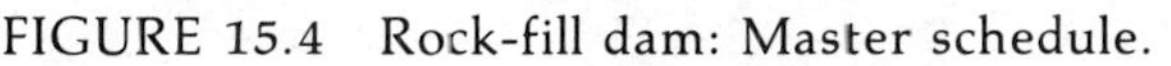
FIGURE 15.4 Rock-fill dam: Master schedule.

15.3 NETWORK LOGIC

The project modeling concepts introduced in the previous sections have focused on the overall scheduling of a project in terms of milestone planning, physical sequence networks, and the gross dissection of project components into major work packages as preliminary project planning and control techniques. Although these techniques are important and have their place, they do not focus in detail on the management of construction activity at the contractor and field management levels. In addition, they cannot be readily adapted to consider and integrate into a project model those activities that impact on construction effort, (such as procurement, preparation, and approval of drawings, inspections, etc.), nor to focus on the essential preconditions that must exist before construction activities can commence. The need for the consideration by management of these essential and time-consuming activities and their interrelationships with construction activities in construction management led to the development of networking techniques as a basis for project modeling.

Networking techniques have been widely used, since the late 1950s, for the portrayal and analysis of project plans, and for the management control of the project work effort. Essentially a network project model is a representation of the project plan by a schematic diagram in network form that depicts the sequence and interrelationship of those activities involved in the project considered relevant to its management.

In an *activity-oriented* network (or *arrow diagram*) each line or *arrow* represents one activity, and the relation between activities is represented by the relation of one arrow to the others; each circle (or node) represents an event (see Figure 15.5). Diagrams of this type are shown in Figure 15.6. Notice in this case that the activities have been identified both by descriptive labels and alphabetic tags [e.g., erect forms (*A*)] as well as by node labels (e.g., 12-13 for the activity Erect Forms). It is important to realize that in this representation the length of the arrow has no significance*; it merely represents the passage of time in the direction of the arrowhead. Each individual activity is represented by a separate line (or arrow). The logical interpretaion of these network models is that for each node (i.e., for each project state), the start of all activities leaving the node depends on the achievement of that state by the completion of all the activities entering that node. Since the prime consideration in network modeling is to correctly portray the sequential and relational logic between the various project activities, activity durations are subservient to this need.*

*Nevertheless, activity or arrow networks are often drawn to a time scale thereby taking on a bar chart format. In these cases, provided careful attention is paid to the consideration and portrayal of network logic, the necessity for float calculations as discussed in the following chapter is obviated.

The network diagrams in Figure 15.6 illustrate some of the logical procedures adopted by CPM. In diagram (1), it is obvious that *A* must precede *B*, and *B* must precede *C*. In (2), *D* must precede both *E* and *F*. In (3), *G* and *H* must precede *I*. In (4), *J* must precede *K*, and *L* must precede *M*. In (5), *N* must precede *O* and *P*, and *Q* must precede *P*; this necessitates using a connecting arrow (called a *dummy*) to maintain the logical sequence of events between *N* and *P*. *Dummy activities* have zero time; they are shown by broken arrows. Dummies may also be required to maintain specific activity identification between events, as shown in (6), where *R* must precede *S* and *T*, and *S* and *T* must precede *U*. Events and activities should be labeled, and they are usually numbered for computer identification of the network.

There is another type of network diagram in which the nodes represent activities and the lines or arrows represent relationships between activities. These networks, called *event-oriented* networks, *precedence diagrams*, and *circle diagrams* are constructed in a manner similar to activity-oriented networks. Like these, the length of the arrows has no significance, since the arrows merely portray the logical precedence of one activity to another and point in the direction of increasing time. Circle diagrams, however, eliminate the need for dummy activities, identify each activity with a single reference number, and can be readily adapted to changes in logical relationships between activities. Figure 15.7 shows the elements of a circle network, corresponding precisely to the situations presented in Figure 15.6 for an arrow network.

15.4 DEVELOPING THE NETWORK MODEL

The first step in developing a plan for a construction project is to break the project down into the separate operations or activities necessary for its completion. No specific order of precedence is necessary when developing this breakdown, but a systematic dissection and listing of project activities by trades, location, or plant requirements is often helpful. The degree of breakdown will vary for each project and will be influenced by the nature of work and class of labor involved, the location of the work on the site, and the broad general sequence of the project.

When a list of all the activities in a project has been prepared the next step is to determine the essential relationship between these activities. This involves a precise statement of the relationships between the activities as a

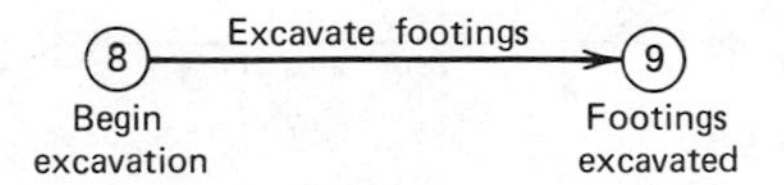

FIGURE 15.5 Activity on arrow model.

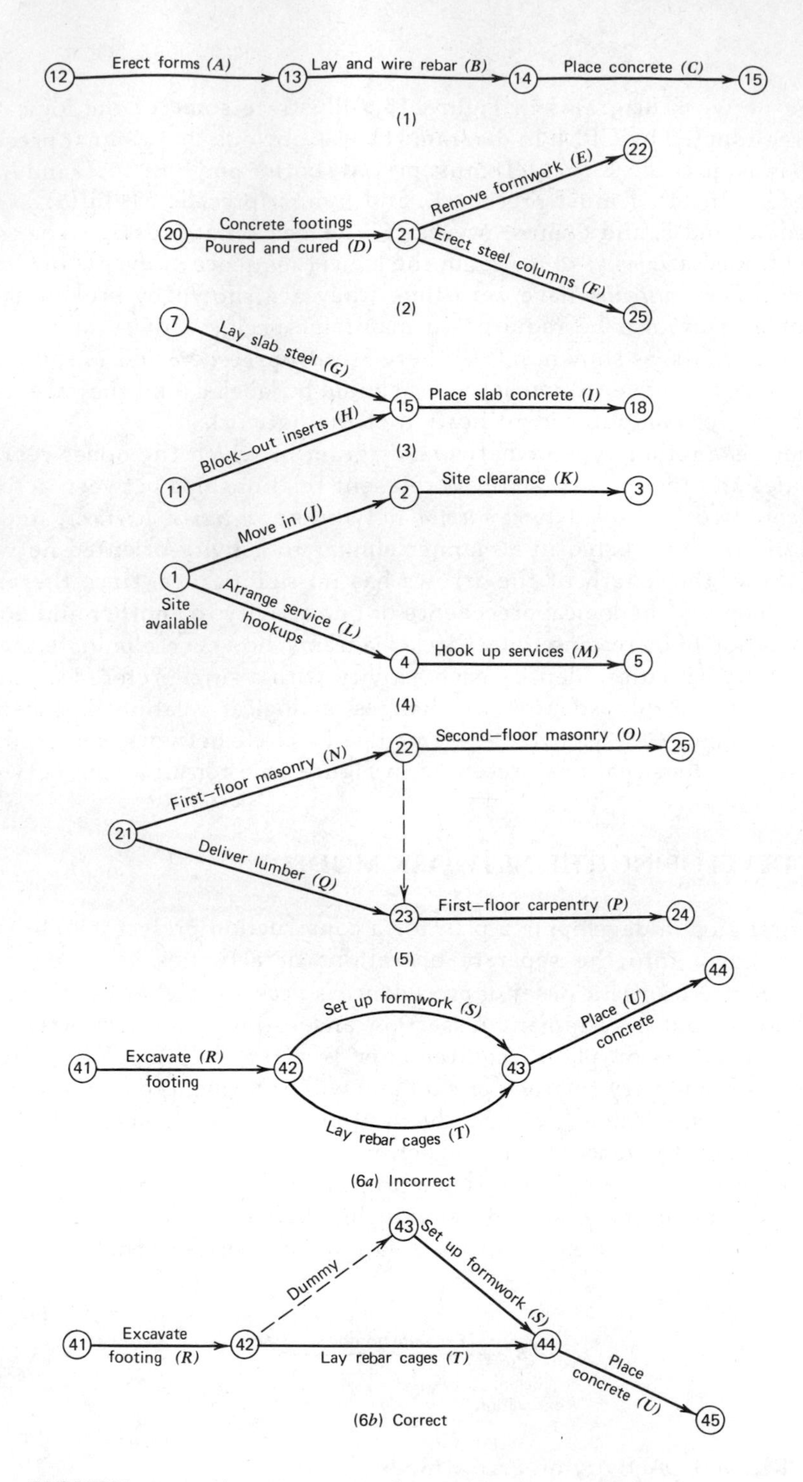

 FIGURE 15.6 Arrow network segments

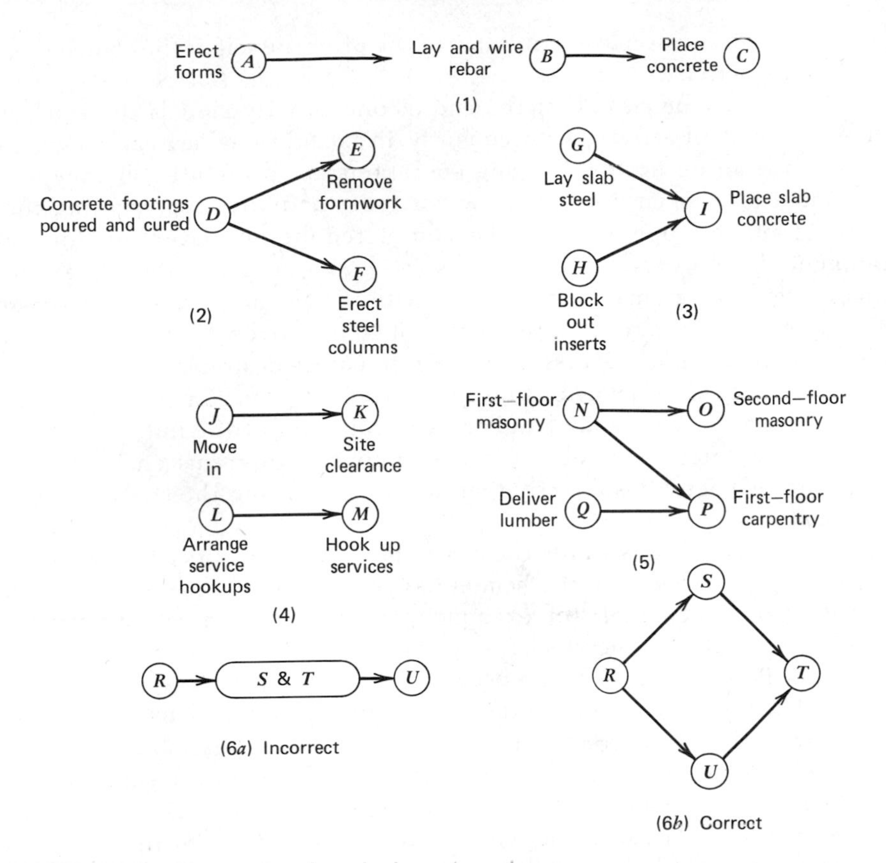

FIGURE 15.7 Elements of a circle network.

means of formulating the construction technology and prescribing management options. A general ordering of activities within the project is not difficult, since their description often implies a relative location within the job; the specific ordering, however, is more difficult and requires very careful consideration. Although many activities may proceed concurrently, certain ones must be constrained to a given sequence or chain; for example, casting of concrete presupposes formwork erection and reinforcement installation, and pipelaying presupposes pipe delivery. These are examples of *technological* and *physical constraints* applicable to activities, and are apparent as soon as each activity in the project is subjected to the following questions.

1. What activities must precede this activity?
2. What activities must follow this activity?
3. What activities may be done concurrently with this activity?

In this way each activity is examined and the necessary sequence of activities is determined. Every activity therefore has a definite event to mark its possible beginning; this event may be the start of the project itself or the completion of a preceding activity.

It should now be clear that the end of one activity signals the start of a related dependent activity. Consequently, in a diagram where each activity or operation is an entity, overlapping operations are prohibited. If they occur they must be broken down into two or more activities representing those portions of the operation to be completed before later portions are commenced. The overlapping of work as shown on conventional bar chart construction programs is impossible with CPM, and hence networking techniques offer a much greater degree of control over all the operations on the site. In many practical cases, however, it may be desirable to commence an *unrelated* activity after a portion only of another has been completed. Although this situation can be readily handled by subdivision of the initiating activity into parts, as mentioned above, some computer programs enable users to relate the two activities directly thereby circumventing the strictness of the network logic.

Besides physical constraints there may be other types of constraints. *Safety constraints* may necessitate the sequential separation of activities that could otherwise be concurrent: for example, ground floor concreting operations may be prohibited while steel framework is being assembled immediately overhead. *Resource constraints* can occur when it is essential to delay because resources for certain operations cannot be made available: for example, release of equipment required from another project may not be possible until a certain date, and consequently some activities must be delayed until the equipment will be available, whereas otherwise concurrent activities may proceed; or it may be essential for certain activities to be completed earlier than physically necessary in order to earn progress payments to finance otherwise concurrent activities. These decisions influence the sequence of operations and place constraints on the project. *Crew constraints* may also occur: for example, specialist welding crews may be hard to obtain, and all the welding activities may have to be done in sequence with a small crew, whereas they could otherwise have been done concurrently; another example is when a single operation may have to be divided into several activities to emphasize that different labor skills are required. Finally, there are *management constraints* when, for example, the sequence of otherwise independent activities is controlled by a management decision, or when normally concurrent activities are ordered to be done in a certain sequence, simply because management arbitrarily wants them done that way.

All these aspects must therefore be carefully studied by the planner when the project is broken down into its essential activities, and when considering the various chains of activities that must be maintained. To the extent that the project can be represented by a diagram, the diagram is representative of the project, and considerable skill is sometimes required in designing a diagram to satisfy all the requirements imposed by physical, safety, crew, equipment, finance, and management constraints.

A good approach to specific ordering is first to determine the obvious physical and safety constraints, then the crew and other resource constraints, and finally the management constraints. The physical constraints initially lead to chains of activities, simply determined and coupled. The consideration of other constraints and the detailed determination of physical requirements usually lead to the branching and intermingling of the chains into networks. It is often helpful to tabulate the activities systematically in order to note those that must precede each activity, those that must follow each activity, and those that may be carried out simultaneously. The network layout is then determined by trial and error, first satisfying some of the conditions and then refining the portions of the network that violate the remainder.

Designing a network that satisfies all the constraints requires a great deal of skill. In some cases, managerial decisions are extremely difficult to formulate in a diagram. However, it is reasonably simple to test a given network. Consequently, to determine improvements to the diagram, it is easier to start with a rough network (incorporating finer details successively) than to attempt a detailed diagram at the outset.

For example, consider the simple construction of concrete footings, which involves earth excavation, reinforcement, formwork, and concreting. A preliminary listing of activities might be:

A. Lay out foundations.
B. Dig foundations.
C. Place formwork.
D. Place concrete.
E. Obtain steel reinforcement.
F. Cut and bend steel reinforcement.
G. Place steel reinforcement.
H. Obtain concrete.

Examination of the list of activities shows that some grouping is obvious. Thus, considering physical constraints only, the following physical chains are developed.

1. From a consideration of the actual footings: *A, B, C, G, D.*
2. From a consideration of the steel reinforcement: *E, F, G, D.*
3. From a consideration of the concrete only: *H, D.*

When the project is seen from these different viewpoints, individual chains of activities emerge; but, on viewing the job as a whole, it is obvious that interrelationships exist. For example, it is useless to pour concrete before the steel reinforcement is placed and the formwork is installed. Therefore, all the chains must merge before pouring the concrete. And if steps are to be taken to obtain the steel and the concrete immediately when work begins (this would be a management decision or constraint), then the chains all start at the same point or event with the laying out of the foundations.

The development of a preliminary network for the project is possible at this stage because (1) a list of activities has been defined and (2) a rough construction logic has emerged.

The actual representation and appearance of the network depend on the modeling form adopted and on the spatial locations of the symbols as drawn. As mentioned previously, there are two basic ways in which activities can be modeled: (1) when the activities are represented by arrows in an activity-oriented network and (2) when the activities are represented by nodes.

In Figure 15.8 a preliminary network is developed, in both arrow and circle forms, from the above information.

15.5 EXAMPLE OF A ROCK-FILL DAM*

This rock-fill dam, approximately 1000 ft long and 175 ft high, involved the winning, placing, and sluicing of 500,000 cu yd of rock in the main bank, with 35,000 cu yd of packed rock on the upstream face, supporting a reinforced concrete face slab (11,000 cu yd). The spillway, located at one end of the dam, required 18,000 cu yd of concrete and an additional 1250 cu yd would be needed for outlet works, and so on. A good quarry site was available at an average haul of half a mile, and it was estimated that about 85% of quarry rock would be suitable for use. The river had a reasonably regular behavior, with a definite flood season. It was specified that bank construction begin at the end of a flood season and that the rock-fill be completed to RL 1400 before the next floods; placing of rock-fill could then continue to RL 1475, leaving a suitable flood gap, and thereafter the embankment could be completed to final level (RL 1500).

River diversion was to be handled in four stages: (1) through a temporary diversion culvert under the dam, which would ultimately be part of the outlet works (this could carry normal river flow and minor freshes); (2) after construction of the inlet tower and spillway and after the bank construction had reached RL 1475 (with flood gap), the stream could be taken into a temporary opening in the base of the inlet tower and then through the culvert; (3) on completion of the main bank and a reasonable proportion of the face slabs, the temporary opening in the inlet tower could be plugged and the reservoir permitted to fill; (4) the outlet pipeline from the inlet tower to the downstream valve house would be installed in the culvert.

*This material is taken with acknowledgements from Antill and Woodhead, Chapter 9, *Critical Path Methods in Construction Practice*, 2nd Edition, John Wiley & Sons, Inc., New York, 1970.

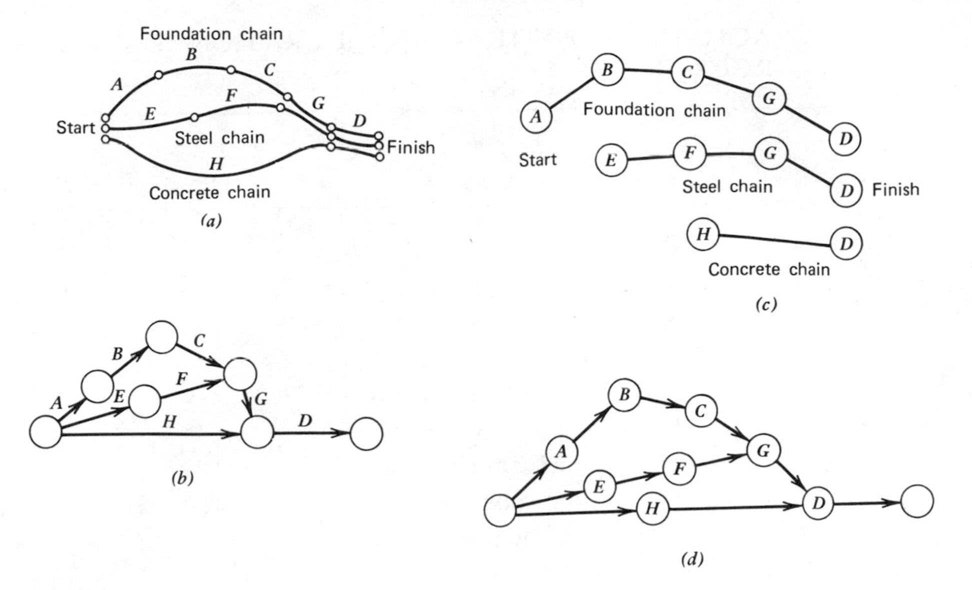

FIGURE 15.8 Preliminary network diagram. (*a*) Initial sketch, arrow network. (*b*) First draft. (*c*) Initial sketch, circle network. (*d*) First draft.

Because construction labor was in short supply, it was decided to attract workmen to the site by providing a long working week and hence higher wages. The construction time allowed was 20 months. A list of the major activities involved in the project and their durations is given in Table 15.1.

Many construction works require careful assessment of site hazards. The effects of such hazards, and the constraints that they produce on relevant activities, may be shown on network diagrams by means of artificial "hazard activities." This technique is particularly applicable to the constraints induced by river behavior in dam construction works, as this rock-fill dam project illustrates. Specified constraints, additional to those already stated, were: (1) activity *F* must be completed before *E*2 begins; (2) *G* must follow *E*3, and *H* must follow *B*3; and *J*2 cannot begin until *E*2 is finished.

The flood season lasts for 6 months with a month's uncertainty as to starting and finishing dates. It was decided to assume a flood period of 31 weeks. Adopting a 6-day week for the project (312 working days per annum), it was assumed for estimating purposes that the flood season would cover 190 working days and the clear season 120. By the use of "hazard activities," the influence of floods can be shown in the network diagram as constraints on the activities affected by flood conditions.

The specified starting date for the bank construction controlled the project starting time. With a 6-day working week, the total construction time permitted was 520 working days. In Figure 15.9 the network diagram developed

Table 15.1 ACTIVITY DEFINITION AND DURATION: DATA FOR ROCK-FILL DAM

		Activity	Time (Days)
A		Preliminary works	50
B1		River diversion, stage 1	60
B2		River diversion, stage 2	30
B3		River diversion, stage 3 (plug)	10
B4		River diversion, outlet pipeline	35
C		Excavation, dam site	130
D		Excavation, spillway	150
E		Excavation, quarry; and construction of bank:	(405)
	E1	Rock-fill to RL 1400	125
	E2	Rock-fill to RL 1475	125
	E3	Rock-fill to RL 1500	155
F		Drill and grout dam site	80
G		Permanent roadworks	20
H		Valve house embankment fill	15
J1		Concrete in spillway	150
J2		Concrete face-slabs to dam (*Note:* 25% may precede end of floods by working on sides of valley)	165
J3		Concrete in inlet tower	70
J4		Concrete in outlet valve house	60
J5		Concrete closure, inlet tower	24
K1		Metal work in inlet tower	20
K2		Metal work in valve house	20
L		Clean up and move out	25

Source. Antill and Woodhead, *Critical Path Methods in Construction Practice*, John Wiley & Sons, Inc., New York, 1970.

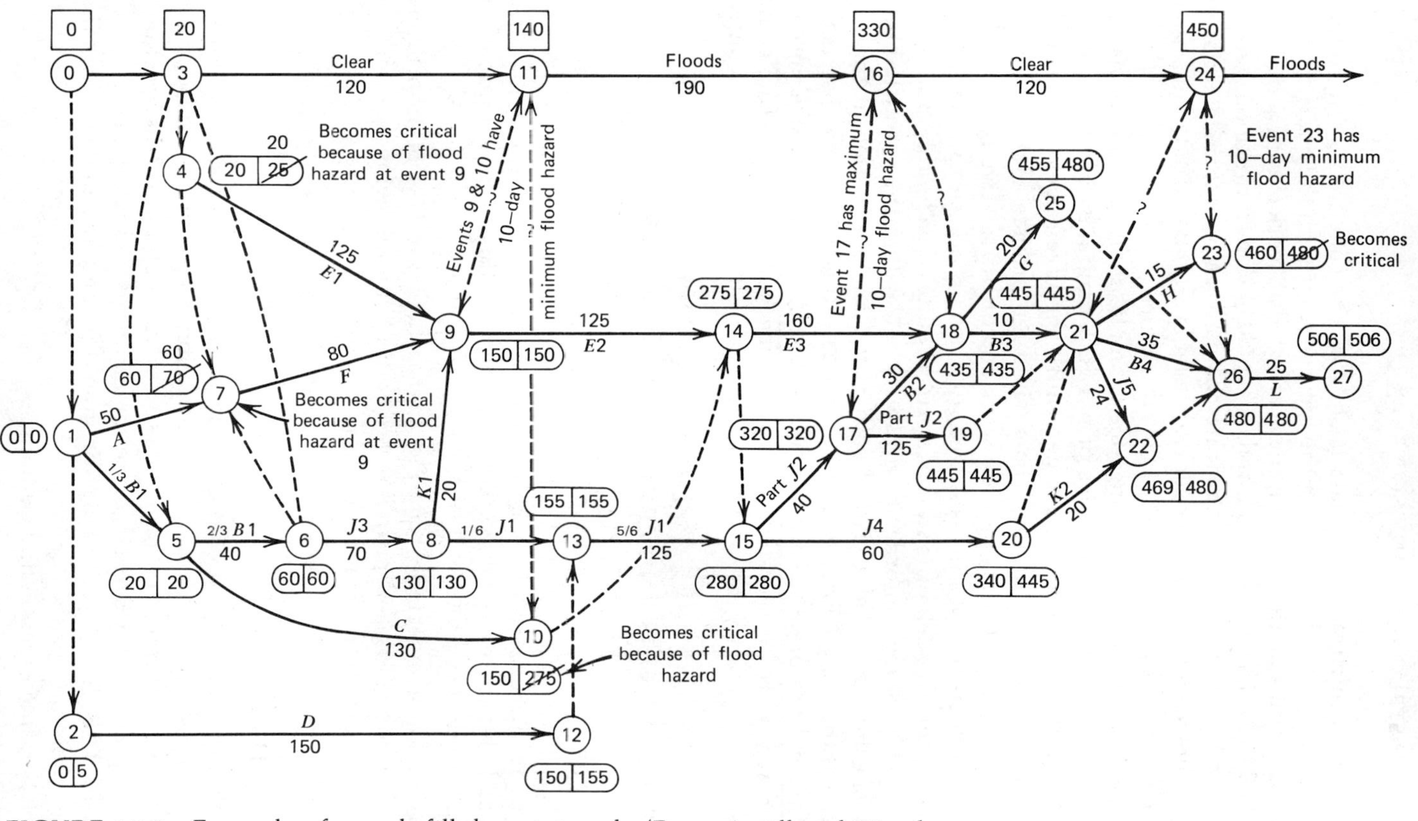

FIGURE 15.9 Example of a rock-fill dam network. (From Antill and Woodhead, *Critical Path Methods in Construction Practice*, 2nd Edition, John Wiley & Sons, Inc., New York, 1970.)

from the above constraints and data is shown. The constraints due to flood to flood hazards are indicated along the top, together with the hazard activity dummies and notes on the diagram; these affected all work except activities *A*, *B1*, *B4*, *D*, *E2*, *E3*, *G*, *J1*, *J1*, *J4*, *J5*, *K2*, and *L* (*H* could be done if the flood were not severe). For clarity in following separate chains of activities, the start of the project has been divided into three separate events, numbered 0, 1, and 2; these are not connected by dummies to retain the logic.

It should be added, before leaving this example, that when major hazards exist on a construction site, a series of network diagrams may be drawn up (during the preliminary planning) to determine the advantages of one starting date in comparison with another. This will often provide additional insight into the effects of the hazards on the work and of the relative risks involved in the various alternatives investigated. On certain occasions it may prove economical overall to delay the start of a project quite considerably, and then to adopt a crash program, in order to finish the job within the permitted time. Networking methods and a critical path analysis, as presented in the next chapter, are the only logical ways to investigate this aspect of construction planning. Furthermore, they can be employed to assess and compare the costs of risks that have to be tolerated. Networking methods have therefore gained considerable popularity as a method for project planning and control.

REVIEW QUESTIONS AND PROBLEMS

15.1 Consider the Gas Station Project of Appendix L. Prepare a list of activities for the project. For each activity (where applicable) list the relevant construction techniques and equipment that could be used and the trades involved. Do you consider this a good approach to planning the project? If not, why not?

15.2 Develop a milestone planning approach to the Gas Station Project that you think is appropriate for each of the following agents:
(a) The construction contractor.
(b) The owner.
(c) The gas distributor.
(d) The station manager's wife.

15.3 Prepare simple arrow notation network segments for the following:
(a) Site access, clearing and foundation preparation—add in a network segment that considers services, hookup, and the erection of a temporary security zone for materials and equipment.
(b) The delivery, installation hookup, and testing of gas tanks and pumps.
(c) Frame erection and cladding.
(d) Mechanical gear.

15.4 Visit a local construction or building site and observe closely a particular field operation. Then prepare a detailed network model (in either arrow or circle notation) as a means of fully describing the field activity to someone unfamiliar with the situation.

15.5 Develop a network model for the location and erection of column formwork, steel cage insertion, concreting, curing, and formwork stripping. Then expand the network model detail to focus more specifically on the activities of the formwork, steel, and concrete crews.

15.6 Visit a local contractor's office and determine if network methods are employed by the contractor or demanded in contracts. If network methods are used, determine the level of detail, number of activities, and modeling format used in the project network.

15.7 Different network models can be developed for the same project that reflect the different interests of the agents involved in the construction process. Develop simple network models for the Gas Station Model that reflect the following interests:

(a) Simple portrayal of physical progress.

(b) A progress payment control network based on physical progress and the various subcontracts that may be let based on trade classification.

(c) The portrayal of the contractor's construction plan as reflected by the contract.

(d) The contractor's portrayal of the construction plan to his own agents.

15.8 To what extent should a project network model be resource, equipment, and crew oriented? Should the model detail try to be exhaustive and present road strip map for field agents or representatives or be presented in gross detail only so as to guide field agents and give them flexibility of action? Give reasons for your preference. How would you react if you were the responsible field agent?

15.9 Redraw the rock-fill dam example network of Figure 15.9 in circle (precedence) network format.

15.10 A 148-unit apartment complex (similar to that discussed in Chapter 7) is to be built by a general prime contractor.

(a) Using the construction cost breakdown illustrated in Figure 7.2*b* as a guide, list the various trades involved in the building project and the major activities that each trade would perform.

(b) If the prime contractor is considering the use of specialty subcontractors on the project, identify the major subcontractors that could be involved.

(c) If the contractor wishes to perform those construction activities that enable him to control the quality finish of the project, which trade crews should he maintain on his work force strength?

15.11 In the building project of Problem 15.10 identify the major contractual, physical, and construction milestones involved in the project. How can these milestones be used:

(a) To plan the overall duration of the project?

(b) To schedule and manage the various subcontractors involved?

CHAPTER SIXTEEN

Project Scheduling

16.1 THE PROJECT SCHEDULING PROCESS

The overall scheduling requirements for a project are established when the various project milestones are identified, located in calendar time (or in terms of project weeks from some datum), and tied down into specific contract durations. Within this legal time framework, the contractor has to both formulate his construction plan and schedule his work effort in order to complete all contract works within the allowed construction period. In many cases, the contractor is required by contract conditions (see Appendix F) to supply a detailed network model and analysis of the construction plan as a visible indication of the manner and sequence in which he proposes to build the project. Whether this effort is required by contract conditions or not, most contractors (especially on complex industrial, power, and heavy engineering construction projects) find it helpful to define their construction plan in terms of a network model and to establish their construction schedule by network analysis.

Before the contractor can establish his schedule, he must determine the estimated duration of each project activity. This is often done by enumerating the resources (i.e., crew sizes, number of crews, number of equipment pieces, etc.) that are to be committed to each activity and by establishing the period (i.e., *activity duration*) for which the crews will be involved in the working of the activity by relating productivities to the work content of the activity.

Given the resource-duration data for each project activity and a network model of the contractor's construction plan, network analysis establishes the overall project duration and resource requirements. In this way, by suitable manipulation of the timing of *some* activities, and by a redefinition of the resources that are to be committed to certain activities, the contractor can establish a desirable project schedule that meets both contract durations and his resource availability profiles.

This chapter introduces network analysis concepts and techniques by initially focusing on the determination of the *critical path* activities and project duration. Float measures for noncritical activities are then introduced and used as the basis for the leveling and scheduling of project resources. In this way, the reader is led to an understanding of the problems and issues involved in establishing the construction schedules for a project.

16.2 NETWORK ANALYSIS CONCEPTS

The modeling of project networks produces directed paths in both arrow and circle notations. In arrow notation networks, the directed arrows indicate the direction of time progress through the activities, whereas in circle graphs the directed arrows point to following activities. Consequently, connected and directed paths exist in CPM networks, and all these paths commence at the project start node and terminate at the project finish node.

The number of different directed paths spanning the network model from start to finish nodes depends on the construction logic, since this establishes the graph structure. The construction logic has the effect of connecting, intermingling, and dividing the various unique paths that exist in the graph.

In normal CPM calculations, interest focuses on isolating and identifying only one of these paths, the *critical path*. Specifically, time attributes are associated with the activities (i.e., working durations) and nodes (i.e., event occurrence times) throughout the graph, and the critical path is identified as that chain of activities with the largest requirement for working time between the project start and finish nodes. All noncritical paths have summed durations less than that of the critical path, and accordingly they have available additional times (referred to as "float") for the completion of their activities.

Critical path calculations require the determination of the earliest times (i.e., T^E, or Earliest Finish Time, EFT) and latest times (i.e., T^L, or Latest Finish Time, LFT) for each network node (i.e., project event). These node times can be determined by simple numerical calculation. The simple calculations are equivalent to the application of two algorithmic procedures—a forward pass algorithm and a backward pass algorithm. In the forward pass algorithm, each node is labeled with the summed time durations along that specific path from the start node to the node being considered, which requires the largest work duration for all its chain activities. This requirement follows in network analysis, since a project event is satisfied only when all project activities terminating at that event have been completed. Network logic demands that progress past an event is impossible if any of the activities included in the specification of that event (i.e., terminating in its representative network node) is unfinished. Hence, the earliest possible occurrence time for an event (T^E) is the time taken to finish all the activities on the most time-consuming path from the start of the project to the event under consideration. Consequently, in the forward pass algorithm, a family of paths is determined, each commencing (or rooted) at the start node and uniquely terminating at each node in the network. This collection of paths identifies the *earliest start tree graph* (a portion or subgraph of the original network), so called because it is similar to a tree in appearance. Similarly, in the backward pass algorithm, a *latest start tree graph* can be identified rooted at the finish node of the network. This latest start graph is obtained in a similar manner to that of the forward pass but works backward from the last event in the project network model to each network node. In the backword pass algorithm a time origin equal to the project duration obtained in the forward pass calculations is used as a datum.

16.3 NETWORK ANALYSIS: AN EXAMPLE

Figure 16.1 illustrates the network calculations on an arrow network diagram for a simple hypothetical project involving 13 activities. The first draft of the network would appear as in part (*a*). Assignment of activity durations to the activity arrows is seen in part (*b*). Alongside each arrow is written the length of time (hours, shifts, or days, as desired) necessary to complete the work involved in the activity.

Next, by proceeding through the event nodes from start to finish, simple addition will give the earliest possible time at which the activities entering each event can be completed; this is then the earliest finish time (EFT) for the event. The EFT for each event (recorded in the left side of the oval time box adjacent to that event) has been shown. After proceeding right through the network, the EFT of the last event is derived; this is the earliest possible time in which the project can be completed, and is the sum of the durations of the longest time path through the network from start to finish. In Figure 16.1 this is 63 days. Figure 16.1 also shows the resulting earliest start tree generated by those activities determining event times and those other non-tree or link branches (e.g., 3-4, 3-5, 5-7, 6-7, 7-9 and 8-9). These latter activities have excess or float time available, since they do not enter into the determination of the earliest timing of events. Accepting at the moment 63 days as the project duration that must not be lengthened, the next step is to work backward from the final event time by subtracting activity durations to find the latest finish time (LFT) permissible for each event if the project is to be completed by the EFT of the final event. The latest finish time for an event is determined after considering the latest starting time of all the activities starting from the event concerned, and is the minimum figure so obtained. If the event is not completed by its LFT, the project will be delayed. The LFT value has been shown entered in the right

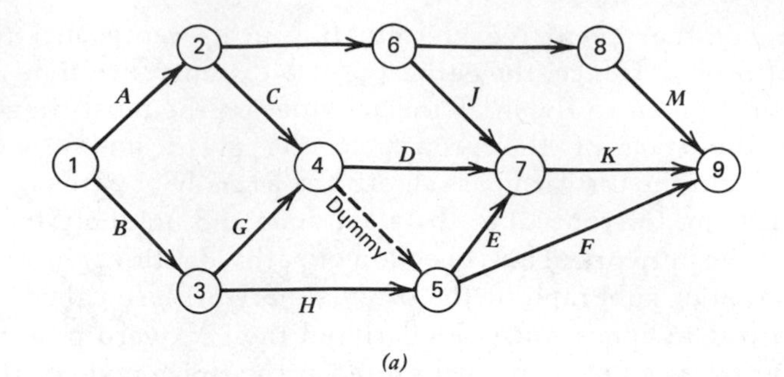

(a)

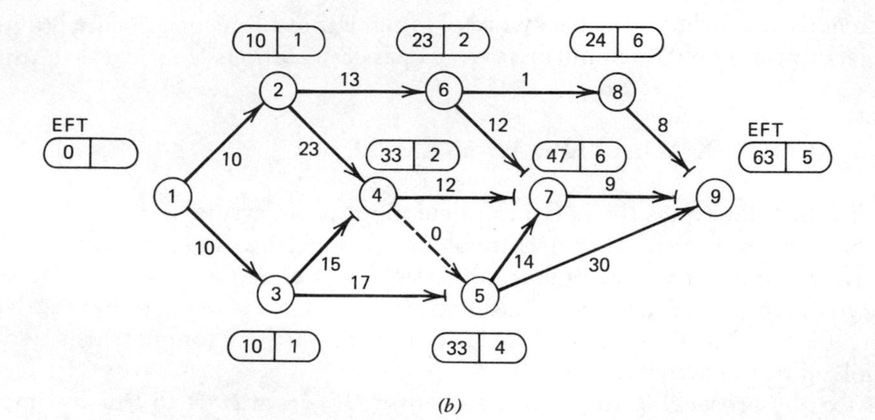

(b)

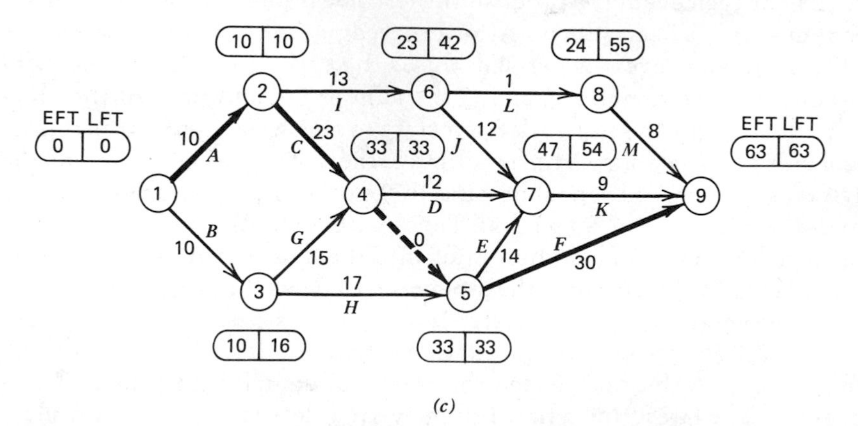

(c)

FIGURE 16.1 Steps in determination of critical path. (*a*) Network diagram, first draft. (*b*) Earliest event times and generation of the earliest start tree. (*c*) Critical path for all-normal duration.

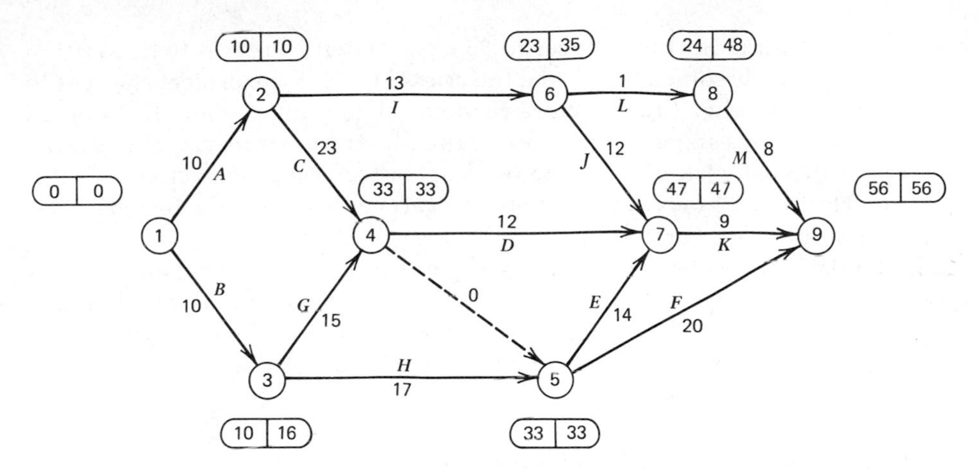

FIGURE 16.2 Alternative critical path.

side of the oval time box adjacent to each event in Figure 16.1*c*. There are now two figures in each time box of Figure 16.1*c*, giving, respectively, the EFT and LFT for each event; the difference between them is the leeway available for delays, etc., and is called *float*. For some events the same figure appears in both sides of the box, indicating the same time for latest and earliest finish; in these cases there is no float at all. These are the critical events that must be completed on schedule if the project is to be finished in the minimum total time. The path joining these critical events is therefore the critical path for this network, under the conditions for which it was drawn. It is shown in heavy lines in Figure 16.1*c*, and passes through events 1, 2, 4, 5, and 9. Activities along this critical path are called *critical activities*.* In general, however, activities lying on the critical path are not so easily determined. First, to be eligible for consideration, they must start and finish at critical events; second, the duration of the activity must equal the difference in event times between the head event and the tail event.

When the critical path through a network has been established, it may be desirable to check the effect of an alternative method of construction for one activity on the network as a whole. If so, another diagram is drawn, showing the new duration for this activity. For instance, in the case of the project shown in Figure 16.1, if activity *F* (shown with a duration of 30 days) could be carried out in 20 days by using a different method or equipment (irrespective of cost), then the EFT of event 9 would reduce to 53 days through activity *F*. It would, however, be 56 days through activity *K*, as seen in Figure 16.2. The

*It is axiomatic that critical activities have no float; hence the direct path between any two critical events is not necessarily a critical path. Observe activities 4-7 and 6-7 in Figure 16.2; these have float and are not critical, the critical path in this case being 1-2-4-5-7-9.

total project duration thus reduces to 56 days, activity *F* ceases to be a critical activity, *E* and *K* become critical, and the critical path of the project changes to 1-2-4-5-7-9. Whether this is a more economical method of doing the work is not relevant at this stage; the purpose of this illustration is merely to show the effect on the network of a change in the plan for doing one activity of this project. Notice also how the float of noncritical events has been affected by this alternative proposal for activity *F*.

It should now be clear that, within any given network, the critical path is dependent solely on the durations of its activities, so that, by suitable planning, it can be made to follow specific chains of activities. The constructional advantages of this manipulation of critical paths are obvious.

16.4 NONCRITICAL ACTIVITIES AND FLOATS

Although critical activities must be completed as fast as possible to prevent prolonging the project duration, this is not true with the noncritical activities, since they have more time available for their completion than is strictly necessary. Thus their starting and finishing times, limited, of course, by the timing of their starting and finishing events, can be altered without affecting the project duration; these noncritical activities and noncritical chains are therefore able to float about within the total time available for their completion.

In Figure 16.1*c* activity 6-8 has a duration of only one day. It is also apparent that since event 6 could occur as early as the 23rd day, whereas event 8 could be as late as the 55th day, there might be 32 days available in which this one-day operation could be done. This activity therefore has a total leeway, or float, of 32 days; it is not a critical activity. On the other hand, activity 2-4 is critical, having no float at all.

In noncritical chains, the float may either be used entirely before beginning any activity in the chain, kept in hand until all activities in the chain have been finished, or interspersed between the various activities in the chain, as suited to the planning. Obviously, some interdependence exists in the float times available for noncritical chains passing through an event, as will be shown later.

Since floats can be used as safety margins to delay the application of resources, to average out manpower requirements, etc., it is not surprising that a variety of float measurements exists including *total float* (TF), *free float* (FF), *interfering float* (IF), and *independent float* (IndF).

The full amount of time by which the start of an activity may be delayed without causing the project to last longer is called *total float* (TF). This delay may cause delays in some of the activities that follow it, but will not retard the project. It follows that critical activities have zero total float; in fact, critical paths may be defined as those chains of activities having zero total float.

Noncritical chains will always contain total float; however, the larger the number of activities in the chain and the smaller the total float, the closer the chain comes to being critical. Noncritical chains with small total float must be carefully watched during construction and are spoken of as near-critical paths.

For a specific activity (I-J starting at node i with event times T_i^E and T_i^L and finishing at node j with event times T_j^E and T_i^L) whose duration is d_{ij} the total float becomes

$$\begin{aligned} TF_{ij} &= T_j^L - (T_i^E + d_{ij}) \\ &= LFT_{ij} - EFT_{ij} \\ &= LST_{ij} - EST_{ij} \end{aligned} \tag{16.1}$$

Free float is the additional time available to complete an activity, assuming that all other activities commence and finish as early as possible. Full use may be made of available free float without disturbing the following activities, which may still begin at their earliest starting times. Thus free floats are usually shown concentrated at the end of noncritical activities or chains, where they become a safety margin to offset any unavoidable delays. Free float can be shared only by previous activities in the chain and is that quantity of the total float that may be consumed without affecting subsequent activities. Since critical activities have no total float, they automatically have zero free float.

Therefore, free float becomes

$$\begin{aligned} FF_{ij} &= T_J^E - (T_i^E + d_{ij}) \\ &= EST\ (\text{its following activity}) - EFT_{ij} \end{aligned} \tag{16.2}$$

Interfering float is the difference between total and free floats for any activity. If an activity is delayed by an amount greater than its free float, but less than or equal to its total float, the lateness of this activity will not, of course, delay the project; it will, however, interfere with the start of some subsequent activity. Thus, if any part of the interfering float is consumed, it will be necessary to reschedule all the activities following in that chain. If the interfering float is fully used, subsequent activities in the chain will become critical; if it is exceeded, the project duration will be increased.

Therefore, interfering float becomes

$$IF_{ij} = TF_{ij} - FF_{ij} \tag{16.3}$$

Independent float is the amount of time that an activity may be delayed or displaced, regardless of the state of the preceding or following activities within the project, without affecting the project duration. The independent float of an activity cannot be shared with any other activity.

Therefore, independent float becomes

$$\text{Ind}F_{ij} = T_j^E - (T_i^L + d_{ij}) \tag{16.4}$$

To determine the floats available to any activity it is first necessary to compute its earliest and latest start times (EST and LST), as well as its earliest and latest finish times (EFT and LFT) from the network calculations. The EST is the time at which it *can* begin, and the LST is the time at which it *must* commence if the minimum project duration is to be achieved. The EFT and LFT are simply the appropriate start times *plus* the duration of the activity.

The relationships between the various types of float are indicated in Figure 16.3.

Table 16.1 shows the results of all the time and float calculations for the network considered in Figure 16.1. The essential time relationships between the activities may now be analyzed, and decisions can be made concerning the complete timing of the construction works. As a visual aid, the CPM bar chart shown in Figure 16.4 may be useful. It will be immediately apparent that the CPM bar chart is similar to the conventional bar chart model, but that it shows, in addition, the critical activities and the essential information on float times. One can see at a glance which are the critical activities (no float) that must not

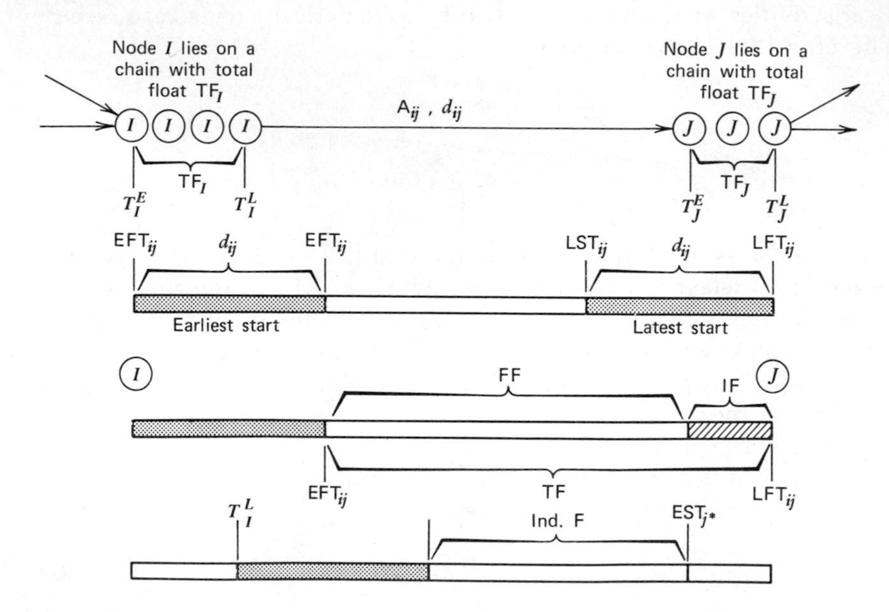

FIGURE 16.3 Relationships between floats.

Table 16.1 SCHEDULING OF ACTIVITIES (SEE FIGURE 16.1)

Activity										
Item	Arrow	Duration	EST	LST	EFT	LFT	TF	FF	IF	Remarks
A	1-2	10	0	0	10	10	0	0	0	Critical
B	1-3	10	0	6	10	16	6	0	6	—
C	2-4	23	10	10	33	33	0	0	0	Critical
I	2-6	13	10	29	23	42	19	0	19	—
G	3-4	15	10	18	25	33	8	8	0	—
H	3-5	17	10	16	27	33	6	6	0	—
Dummy	4-5	0	33	33	33	33	0	0	0	Critical
D	4-7	12	33	42	45	54	9	2	7	—
E	5-7	14	33	40	47	54	7	0	7	—
F	5-9	30	33	33	63	63	0	0	0	Critical
J	6-7	12	23	42	35	54	19	12	7	—
L	6-8	1	23	54	24	55	31	0	31	—
K	7-9	9	47	54	56	63	7	7	0	—
M	8-9	8	24	55	32	63	31	31	0	—

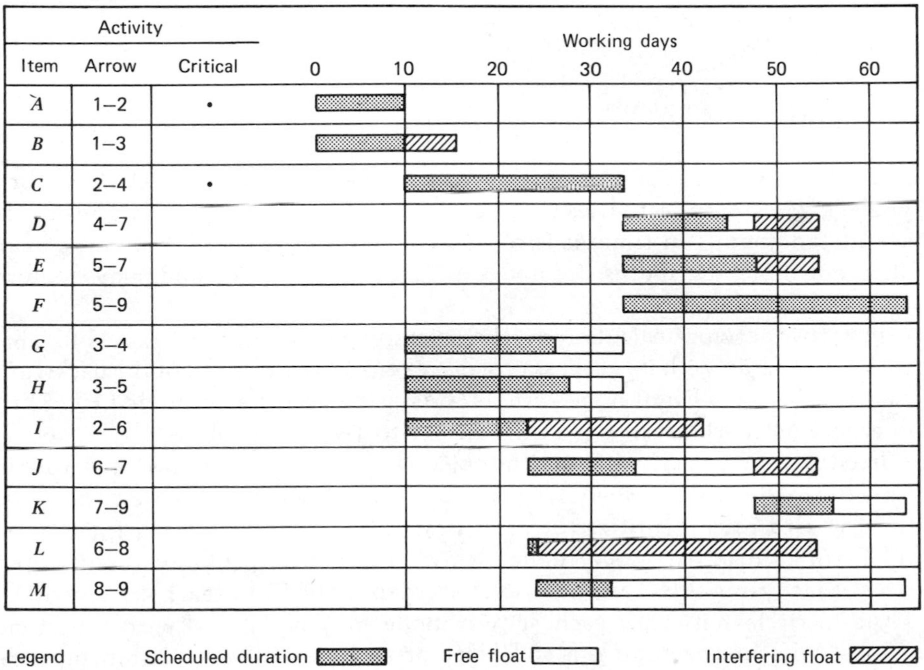

FIGURE 16.4 Critical path bar chart program.

be delayed if the project is to finish on time; and also how much delay can be tolerated in other activities. If any of these run late by an amount greater than its free float (if any), their lateness will interfere with the start of subsequent activities in the chain; but they (and subsequent activities) may be delayed by an amount of time not exceeding the total available float without prolonging the completion of the project. Float time is therefore a safety margin that may be used to offset unforseen or deliberate delays in activities along noncritical paths.

The information obtained at the scheduling stage is of great importance to the construction manager in controlling the project in the field, and it is also equally useful to the planner. Knowledge of available float time enables activities to be shifted about the program, within their float limits, to help smooth out labor and plant requirements. This is another of the major advantages that the construction industry has found in CPM: by judicious manipulation of free and interfering floats, a smoother construction plan can be achieved logically and mathematically, without extending the total project time. This aspect is discussed later, and is known as resource leveling. After this manipulation of noncritical activities, the revised arrangement is re-scheduled with fixed dates to become the final project schedule for the specific duration required.

16.5 CIRCLE NETWORKS

In circle networks, nodes model activities, whereas in arrow networks they model events. Since activities require a time period for completion, and events are instantaneous in time, it is obvious that nodes in circle networks have different time attributes to nodes in arrow networks and may require different labeling.

In arrow network calculations, the objective of the forward pass is to label each event node with its earliest possible event time EFT, and of the backward pass to label each event node with its latest permissible event time LFT. Thus in arrow networks each node is labeled with two time values.

In circle network calculations, the objective of the forward pass is to locate in time the earliest time span during which the activity may be carried out. Thus the forward pass identifies the start time (EST) and the earliest finish time (EFT) for each activity node. In the backward pass, the calculations identify the latest finish time (LFT) and the latest start time (LST) for each activity node. Thus in circle networks each activity node may be labeled with four time values. Whether each node is so labeled or whether known relationships are used to reduce the labeling to two time values is immaterial since all four values must be computed in any case.

16.6 RESOURCE LEVELING AND SCHEDULING

Once the network diagram has been analyzed and all the event times and activity floats established, the scheduling of all the project activities may proceed. It is important to realize, at this stage, that the individual activity durations used in the critical path calculations imply a commitment to working each activity with sufficient resources to ensure compatibility between the work volume involved in the activity and the productivity and production rate achievable by these resources. In other words, inherent in the initial approach to critical path network analysis is the assumption that the project duration must be minimized and the resource requirements for the various activities can be met, whenever required, from the available project resources, independently of the requirements of other concurrent activities. In this sense the normal CPM calculations imply infinite resource availabilities, since resource constraints are not considered. Therefore, at the scheduling stage, it is important to schedule activities within resource availabilities. In many cases this simply means evaluating that minimum level of resources required to maintain the project schedule as given by the CPM calculations. If these minimum levels of resources are not available then the basis for the determination of the critical path is invalid. When resources are strictly limited, the situation can arise where chains that are otherwise noncritical must exceed their total float while waiting for their special resources to become available, thus delaying the entire project. Limitations of manpower, finance, equipment, or other resources frequently occur in practice. Sometimes these resource limitations can be incorporated into the network logic by means of dummies or allowed for by skilled resource leveling. When this is not possible, the project duration is increased and the usual critical path (based on unlimited resources) may be radically altered. It is therefore obvious that, in some realistic networks, the critical path, although still being the most time-consuming path through the network, is not easily determined, since it depends on the availability of a wide variety of resources and on the network characteristics.

The limited resource problem arises naturally in the CPM process because of the positions adopted during the planning and the manner in which the network model is developed. The emphasis on the separation of the planning and scheduling functions ensures an initial concentration on the development of the construction logic and the deferment of project resource considerations. Frequently, the deferment is total and tantamount to a blind acceptance of project resource requirements as dictated by the CPM schedule based on earliest starts. Often this attitude is discovered only the first time a schedule is generated, which leads to a conflict between the resources that can be made available to a project and those apparently needed as determined from the network model.

The prime emphasis on the planning function in the CPM process means that there is a strong tendency to formulate the construction logic independent of project resource requirements. This tendency, coupled with the initial assignment of resources (implied or otherwise) to each individual activity during the activity utility data definition stage, gives little scope for resource considerations during the formative stages of the construction plan and the network model. In fact, the CPM process has been described as being based on the assumption of infinite resource availability.

The CPM process leads to two classical limited resource problems, the resource leveling problem and the resource scheduling problem. The *resource leveling problem* adopts the network model and project duration based on the assumption of infinite resources and individually defined activity resource requirements, and searches for that schedule of the project activities that removes peak requirements of limited resources, or minimizes the number of certain specific limited resource types. The resulting resource requirements are then often adopted as being a natural attribute of the project when, in fact, they are really a consequence of the specific construction plan and model developed independently of resource requirements.

The *resource-scheduling problem* commences with the network model and a project duration based on the assumption of infinite resources and individual activity resource requirements, and seeks that minimum time extension of the project duration so that a schedule of the activities becomes possible, yielding resource requirements that can be met from a defined resource availability while still maintaining the construction logic. This approach does not question the resources originally assigned to the activities nor their durations.

16.7 A RESOURCE-LEVELING PROBLEM

In order to illustrate resource leveling and scheduling techniques, consider the simple project shown in Figure 16.5 in which only labor resources are involved. The project comprises nine activities each tagged with its duration and predetermined labor crew size. Thus, for example, activity 1-2 has a duration of 8 days and a 10-man crew implying an 80-man day work content. Figure 16.5 also indicates the results of CPM calculations showing event times, activity total floats and the time critical path as passing through activities 1-2, 2-5, and 5-7.

Examination of Figure 16.5 shows that many possibilities exist for scheduling the noncritical activities within the limits required by the network duration (25 days) and labor resources. Each scheduling possibility implies a definite rate of application of resources and required maximum work force level and of consumption of the necessary materials.

One obvious schedule is to begin every activity as soon as possible. This *earliest start schedule* is shown in Figure 16.6 and Table 16.2. In the table a separate

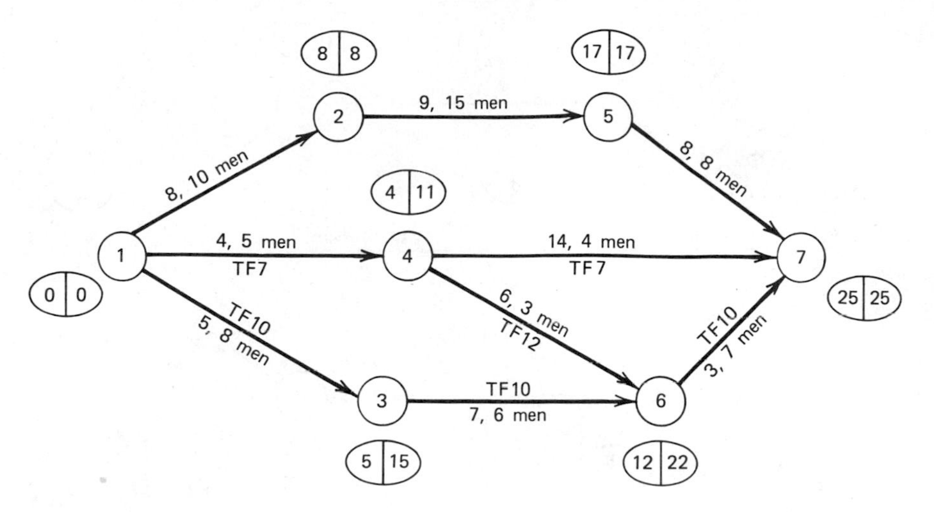

FIGURE 16.5 Project model showing critical path calculations.

row is used for each activity. All activities are scheduled for their normal durations; no float is scheduled so that free and total floats only become future safety margins. This schedule requires a specific daily work force as summarized at the bottom of the tabulation. The required work force varies from 8 men to a maximum of 28 (for 2 days only) with an average of about 19 men.

Another schedule is starting each activity as late as possible. This *latest start schedule* appears both in Figure 16.7 and Table 16.3. The required work force varies from 10 men to a maximum of 28 men (for 2 isolated days), the average again being 19 men. It will be seen from the table and the figure that all floats have been discarded and hence all chains and activities are critical—with this schedule a delay in any activity delays the project completion.

Comparison of the two schedules shows that the earliest start schedule requires a heavy initial rate of investment (application of resources), which decreases considerably later in the project. The latest start schedule, on the other hand, permits a small initial rate of investment but requires a considerable increase later. It is therefore apparent that the free float available with the former schedule is purchased, in the form of a safety margin, at the cost of a heavy initial rate of expenditure; and that the low initial rate of expenditure for the latter schedule is obtained at the cost of discarding all float and accepting the risk of a completely critical network.

It will be obvious that a compromise between these two schedules (earliest and latest start) for the project should result in a more constant work force (and hence a more uniform rate of investment) over the whole project duration. The development of such a compromise schedule (for the given project duration) is known as *resource leveling*.

The resource-leveled schedule is shown in Figure 16.8 and Table 16.4, where the required work force varies from a maximum of 25 men (for 1 day) to a

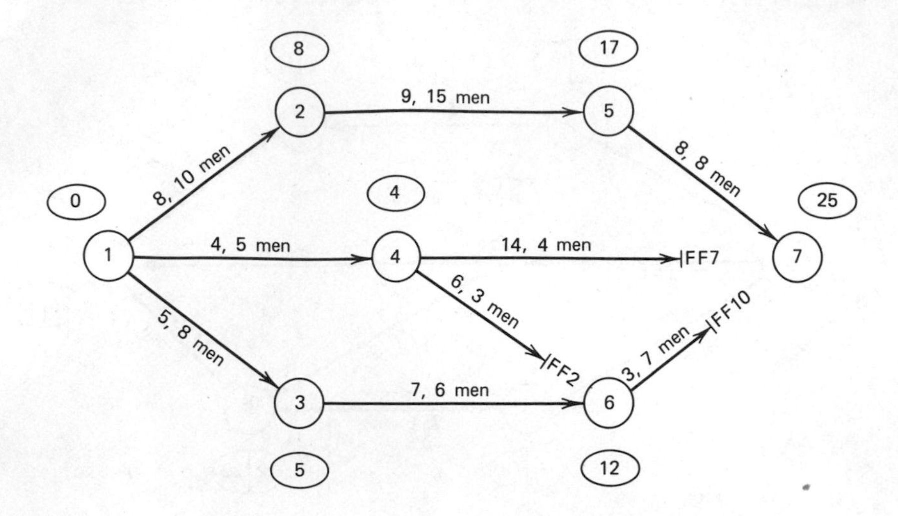

FIGURE 16.6 Earliest start schedule.

TABLE 16.2 Earliest Start Schedule: Maximum Work Force 28 Men.

Activity		Working Days (0, 10, 20, 30)
Arrow	Critical	
1–2	*	10 men/day; 0–8
1–3	TF10	8 men; 0–5; TF10; 15
1–4	TF7	5 men; 0–4; TF7; 11
2–5	*	② ①; 15 men/day; 8–17
3–6	TF10	③; 6 men; 5–12; TF10; 22
4–6	TF12	3 men; 4–10; TF12; 22
4–7	TF7	④ ⑦; 4 men; 4–18; TF7; 23
5–7	*	⑤ ⑥; 8 men/day; 17–25
6–7	TF10	7 men; 12–15; TF10; 25
Daily work force		23 23 25 23 (28) 25 26 26 19 8 8 8 8 23 23 23 23 (28) 25 26 19 12 8 8 8

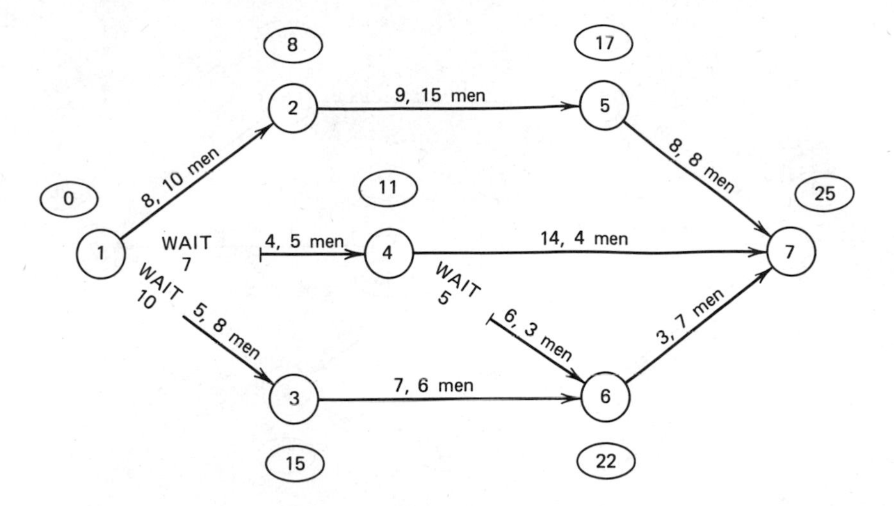

FIGURE 16.7 Latest start schedule.

TABLE 16.3 Latest Start Schedule: Maximum Work Force 28 Men.

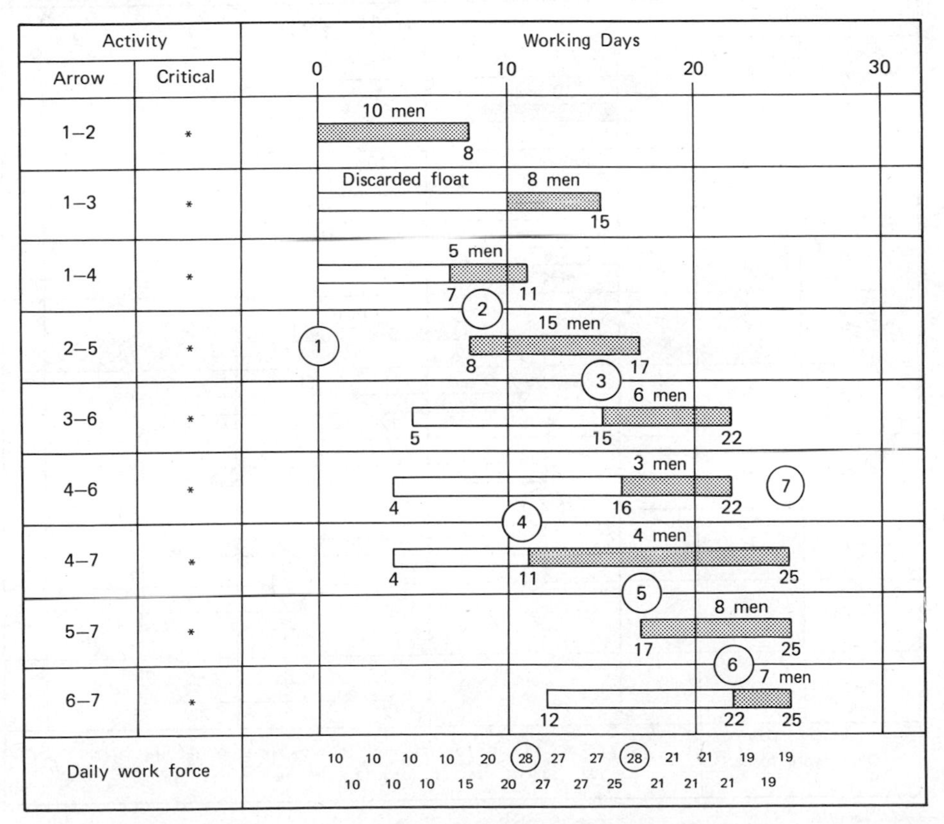

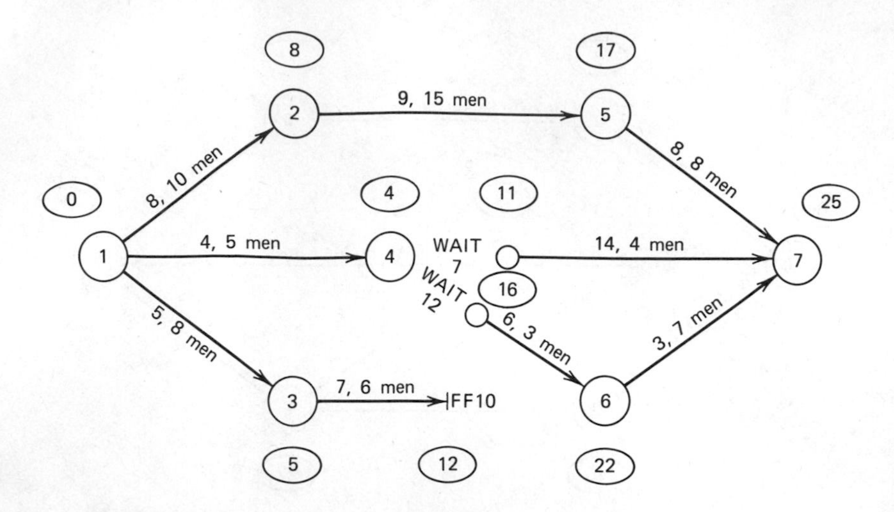

FIGURE 16.8 Resource-leveling schedule.

TABLE 16.4 Resource Leveling: Maximum Work Force 25 Men.

Activity		Working Days (0, 10, 20, 30)
Arrow	Critical	
1–2	*	10 men; 0 – 8
1–3	TF10	8 men; 5; TF10; 15
1–4		5 men; 4; 11
2–5	*	(1) (2) 15 men; 8 – 17
3–6	TF10	(3) 6 men; 5; 12; TF10; 22
4–6	*	4; 3 men; 16; 22
4–7	*	(4) (7) 4 men; 4; 11; 25
5–7	*	(5) 8 men; 17 – 25
6–7	*	(6) 7 men; 12; 22; 25
Daily work force		23 23 18 16 21 21 19 19 22 15 15 19 19 23 23 16 16 21 (25) 19 19 15 15 15 19

minimum of 15 men. Thus the leveled schedule has produced a much more uniform and desirable work force usage for the project. Notice that this has been achieved by shifting activities 4-6, 4-7, and 6-7 to their latest starts and the schedule of activities 1-3, 1-4, and 3-6 to their earliest starts.* In this way the maximum work force requirements for 28 men in the previous schedules have been reduced to 25 men for this schedule.

Resource leveling therefore involves the careful shifting of activities within their available floats in order to reduce maximum resource usage requirements. In situations involving different nodes it is helpful to tabulate labor requirements and plot the manpower required for each trade against time; this may be done when the original schedule is reviewed and amended. A similar plot may be prepared for each type of equipment. The resource-leveling procedure becomes complicated because of the resource interrelations between activities; but the effort is amply rewarded by the production of a detailed works schedule in which peak resource levels and fluctuations in demand are minimized.

In chains of activities involving numerous operations, with various labor requirements and skills and different types of equipment, the most satisfactory way of carrying out detailed resource leveling is to begin with the labor craft (or major equipment item) having the greatest fluctuations. Smoothing this out first, without regard to other skills or equipment, one then proceeds with a second craft (or piece of equipment), and so on, one skill (or machine) at a time, until a satisfactory overall labor and equipment force is attained. With complex chains, the same craft or machine may have to be reviewed several times and relevant activities shifted and reshifted before an acceptable solution appears. As pointed out, the best approach is first to shift those activities with small float and then those with large float.

In drawing up the tabulation, before beginning resource leveling, it is often helpful to sketch in (as on a critical path bar chart) the time ranges available for all the noncritical activities; this will ensure not making any incompatible decisions during the activity-shifting procedure, especially when reshifting is found necessary. An alternative aid, preferred by some planners, is to block out all times not available for the execution of each activity. Finally, once the ultimate works schedule is determined, financial expenditure and income may also be added to the resource-leveling tabulation; in this way the estimator can present a complete picture of all the essential resources in strict conformity with the network model.

*The deliberate delay to the start of an activity, or to its continuous execution, introduces into the network diagram a number of artificial "delay" activities (having time but no cost) within the length of the original activities. These are seen in activities 4-6 and 4-7 in Figure 16.8. On completing activity-shifting and resource-leveling procedures, the final network diagram thus contains a variety of artificial activities and artificial events, and the nodes must be renumbered from start to finish of the project before tabulating the final works schedule.

16.8 THE RESOURCE-SCHEDULING METHOD

An alternative resource-leveling procedure, and in fact the basis for a resource-scheduling approach, is to examine the project in successive steps of one time unit at a time, shifting chains of activities as required at each step until a valid solution appears for each successive time unit examined. Activity-shifting begins with the chain having the largest total float and proceeds on this basis through the project.

The resource-scheduling procedure suggested above can be generalized and formulated as follows: Consider two activities A and B as shown in Figure 16.9 in general CPM bar chart form. Assuming that a resource conflict occurs for A and B only when both activities are working during any specific time interval and not otherwise, then the conflict can be resolved by either scheduling B to follow A, or A to follow B. Obviously, it is better to select the sequence that increases the project duration the least.

If B is to follow A, then the best strategy is to schedule A as soon as possible (i.e., at EST_A) and B as soon as A is finished (i.e., at EFT_A). The increase in project duration T_p then becomes.

$$\begin{aligned} \Delta T_p\,(A \text{ then } B) &= (EFT_A - LST_B) && \text{if } LST_B < EFT_A \\ &= 0 && \text{if } LST_B \geq EFT_A \end{aligned} \tag{16.5}$$

The second equation means that no project extension results since B can be shifted after A within the float available to B.

Similarly, if A is to follow B, then the increase in project duration becomes

$$\begin{aligned} \Delta T_p\,(B \text{ then } A) &= (EFT_B - LST_A) && \text{if } LST_A < EFT_B \\ &= 0 && \text{if } LST_A \geq EFT_B \end{aligned} \tag{16.6}$$

The best scheduling sequence is then readily determined.

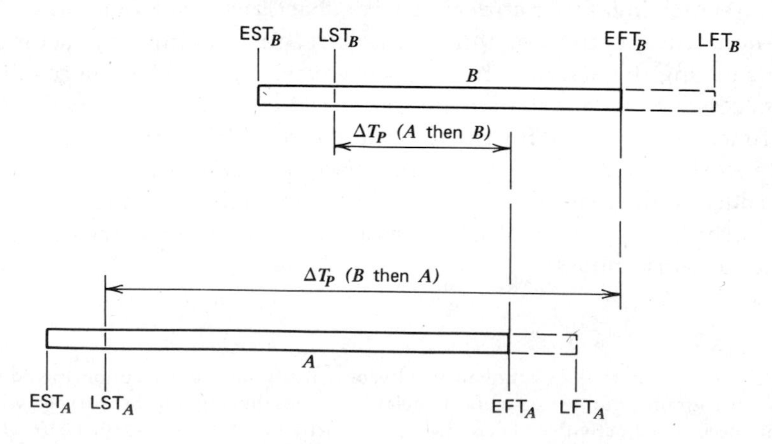

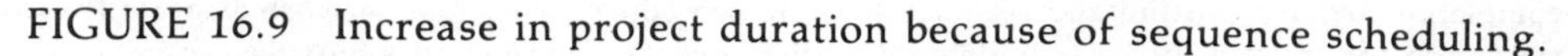
FIGURE 16.9 Increase in project duration because of sequence scheduling.

If more than two activities are simultaneously involved in a resource conflict, a systematic evaluation, similar to that indicated above, becomes necessary for each possible scheduling sequence. Since the result of each search is to order two activities only, the process may have to be repeated again and again until the resource conflict has been removed. Before proceeding to the next time interval, the normal CPM calculations may have to be repeated on the new network model to obtain updated values of the total floats for the network activities. The method then continues on through all the remaining time intervals in the project.

It is interesting to note that if the resource scheduling strategy is adopted of not shifting "in progress" activities, then equations 16.5 and 16.6 shift resource-conflict activities in the total float order. Thus if activities B and C simultaneously enter in resource conflict with in progress activity A, then equations 16.5 and 16.6 yield:

$$\begin{aligned} \Delta T_p\,(A \text{ then } B) &= (\mathrm{EFT}_A - \mathrm{LST}_B) \\ \Delta T_p\,(A \text{ then } C) &= (\mathrm{EFT}_A - \mathrm{LST}_C) \end{aligned} \qquad (16.7)$$

and since $\mathrm{EST}_B = \mathrm{EST}_C$

and $\mathrm{TF}_B = \mathrm{LST}_B - \mathrm{EST}_B$

$\mathrm{TF}_C = \mathrm{LST}_C - \mathrm{EST}_C$

there results

$$\begin{aligned} \Delta T_p\,(A \text{ then } B) &= \mathrm{EFT}_A - \mathrm{EST}_B - \mathrm{TF}_B = \text{constant} - \mathrm{TF}_B \\ \Delta T_p\,(A \text{ then } C) &= \mathrm{EFT}_A - \mathrm{EST}_C - \mathrm{TF}_C = \text{constant} - \mathrm{TF}_C \end{aligned} \qquad (16.8)$$

Thus the activity with largest total float is shifted in sequence first. On this basis resource scheduling can be carried out expeditiously by focusing on total float chains and very carefully revising total floats as the chains of rescheduled activities impact on the various event times in the network.

16.9 A RESOURCE-SCHEDULING PROBLEM

As an illustration of the resource-scheduling problem consider the problem encountered for the project of Figure 16.5 when for some reason (e.g., a management constraint) it is necessary to restrict the maximum work force to 20 men.

The resource-scheduled network shown in Figure 16.10 and Table 16.5 is obtained from a sequence of resource conflict resolution decisions as follows:

The T = *0 decision*

Available activities	1-2	TF = 0	10 men
	1-3	TF = 10	8 men
	1-4	TF = 7	5 men
			23 men

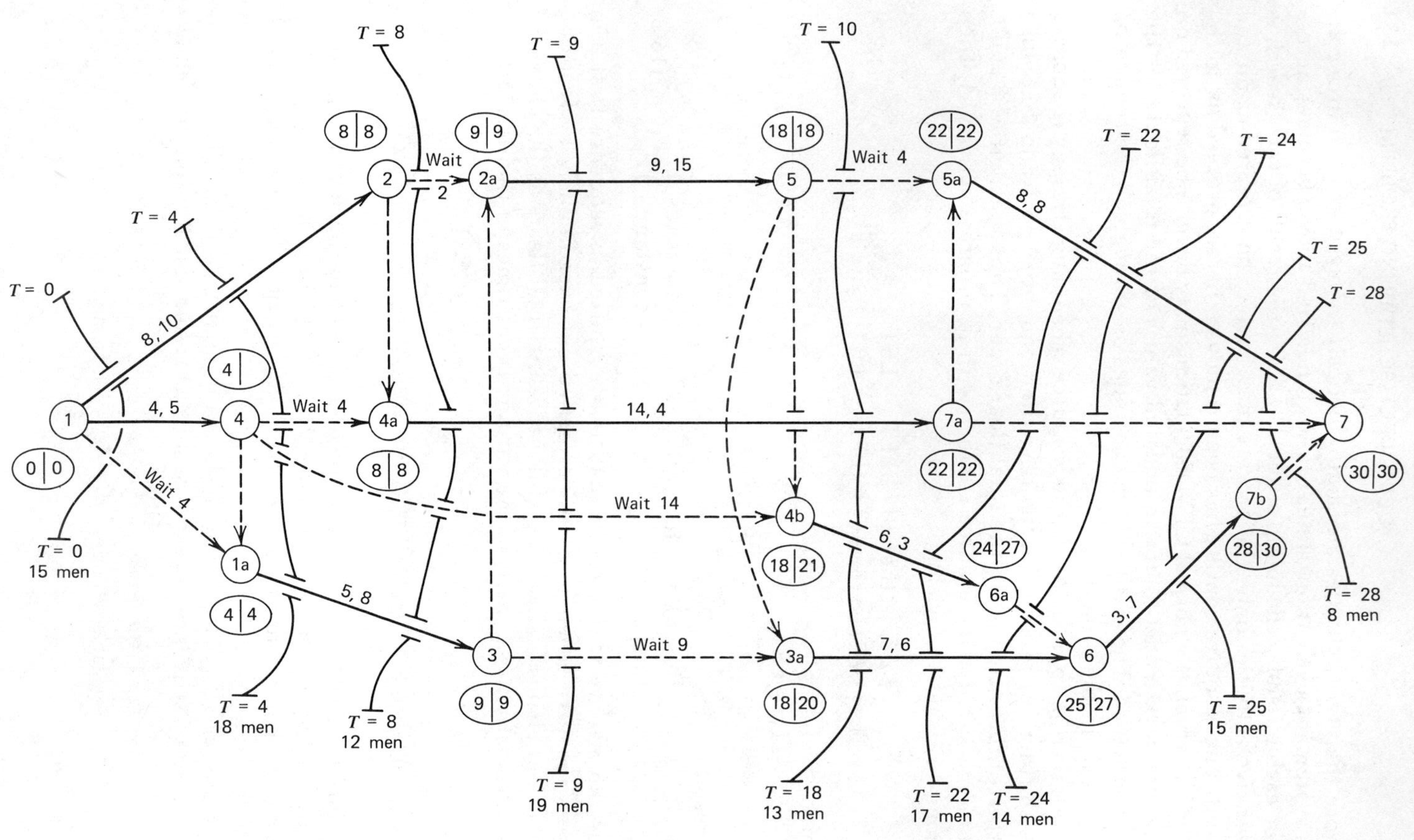

FIGURE 16.10 Resource-scheduled network.

Table 16.5 RESOURCE SCHEDULING: 20-MAN WORK FORCE LIMIT

Activity		Working Days																														
Arrow	Critical	0					5					10					15					20					25					30
1-2			10	10	10	10	10	10	10	10																						
1-3							8	8	8	8	8																					
1-4			5	5	5	5																										
2-5												15	15	15	15	15	15	15	15	15												
3-6																					6	6	6	6	6	6	6					
4-6																					3	3	3	3	3	3						
4-7											4	4	4	4	4	4	4	4	4	4	4	4	4	4								
5-7																									8	8	8	8	8	8	8	8
6-7																												7	7	7		
Daily Workforce			15	15	15	15	18	18	18	18	12	19	19	19	19	19	19	19	19	19	13	13	13	13	17	17	14	15	15	15	8	8

Resolve conflict by working activities 1-2 and 1-4 (for a total of 15 men), and delay activity 1-3 by 4 days. At completion of activity 1-4, review position. Referring to Figure 16.10 this decision requires the insertion of dummy node 1a and artificial delay activity 1-1a WAIT 4 with the consequent reduction in total float of activities 3-6 and 6-7 to six days.

The T = 4 *decision*

Available activities	1-2	*TF* = 0	10 men (in progress)
	1-3	*TF* = 6	8 men
	4-6	*TF* = 12	3 men
	4-7	*TF* = 7	4 men
			25 men

Resolve conflict by working activities 1-2 and 1-3 (for a total of 18 men), and delay activities 4-6 and 4-7. Review situation at completion of activity 1-2 that is, at time 8. Referring to Figure 16.10 this decision requires the insertion of an artificial delay activity 4-4a WAIT 4 with the consequent reduction in total floats of activities 4-7 to 3 days and of 4-6 to 8 days.

The remaining steps require decisions at times 8, 9, 18, 22, 24, 25, and 28 days and are left to the reader as an exercise. This resource-scheduling method leads to the solution indicated in Figure 16.10 and Table 16.5. Notice carefully the resource contours of Figure 16.10 where, for example, that labeled for $T = 18$ as 13 men involves the working of activities 4-7, 4-6, and 3-6, shown as 4a-7a, 4b-6a, and 3a-6 on Figure 16.10.

In summary then the enforcement of a 20-man maximum work force cannot be satisfied in the 25 days of the CPM (infinite resources) calculation but requires the project to be extended to 30 days. Notice, however, that if each activity could be worked with any size crew intermittently then the one 20-man crew could finish activity 1-2 in 4 days. On this basis the project contains 476 man-days of work and could theoretically be finished in 24 days. In most practical cases crew sizes are fixed by union agreement and vary from trade to trade so that the problem and solutions posed above are purely for illustrative purposes only.

Currently, no formal mathematical model exists that considers the interaction between the various possible resource allocations per activity, the construction logic, and the defined acceptable resource availabilities. Many heuristic models exist based on arbitrary criteria, which obtain feasible solutions by considering a limited view of the problem area and taking advantage of a particular structuring of the limited-resources problem. None consider trade-offs between different resource allocations for an activity and its effect on project resource requirements, and all require a new start (human initiated) to provide feedback if unacceptable solutions result. Possibly the most successful attempts to date are those that incorporate repeated subjective input, in an on-line real time mode on computers. This approach permits reevaluation of activity status, construction logic. and acceptable resource availabilities. In this way the user can repeatedly make trade-offs between planning and scheduling. Although optimal solutions are rarely obtained, at least the solutions so derived are feasible, useful, and understandable to the construction planner.

REVIEW PROBLEMS AND EXERCISES

16.1 Draw a precedence (circle notation) network for the arrow notation network of Figure 16.1 and recompute the network calculations. Check your calculations against Table 16.1 values.

16.2 Check the network event times of Figure 15.9. Hence determine all activity floats. Identify all critical path activities.

16.3 Determine the arrow and circle network diagrams for the following project, and hence calculate the total float, free float, and interfering float for each activity.

Activity	Duration	Immediately Following Activities
a	22	*dj*
b	10	*cf*
c	13	*dj*
d	8	—
e	15	*cfg*
f	17	*hik*
g	15	*hik*
h	6	*dj*
i	11	*j*
j	12	—
k	20	—

16.4 From the following network data, determine the critical path, starting and finishing times, and total and free floats.

Activity	Description	Duration
1-2	Excavate stage 1	4
1-8	Order and deliver steelwork	7
2-3	Formwork stage 1	4
2-4	Excavate stage 2	5
3-4	Dummy	0
3-5	Concrete stage 1	8
4-6	Formwork stage 2	2
5-6	Dummy	0
5-9	Backfill stage 1	3
6-7	Concrete stage 2	8
7-8	Dummy	0
7-9	Dummy	0
8-10	Erect steel work	10
9-10	Backfill stage 2	5

16.5 Given the following network data, carry out resource leveling within the minimum project duration for the smoothest use of equipment and smallest number of men, if only one piece of equipment A is available and all activities are continuous.

Activity	Duration	Men	Equipment (including operator)
1-2	2	3	B
1-4	2	6	A
1-7	1	4	B
2-3	4	—	A
3-6	1	4	A
4-5	5	—	A
4-8	8	4	—
5-6	4	2	B
6-9	3	4	A
7-8	3	5	B
8-9	5	2	B

16.6 Using the information derived in Problem 16.5 determine the shortest project durations for the following cases of resource limitations:

(*a*) Not more than 8 men, one A and one B, where the continuous operation of B is essential, whereas that of A is not.

(*b*) Not more than 6 men, one *A* and one *B*, where continuity of equipment operation is not essential, but is more desirable for *A* than for *B*.

16.7 In the rock-fill dam example of Figure 15.9, suppose start on the bank construction is delayed 30 days. Redraw the project network and hence recompute the critical path and activity floats.

16.8 A new road deviation with concrete pavement, shown below* in longitudinal section, is 11,600 ft long. It is to be constructed in accordance with the following conditions:

(*a*) The balanced earthworks from chainage 00 to 58(00) may be done at the same time as the balanced earthworks from chainage 58(00) to 116(00) using two separate independent crews.

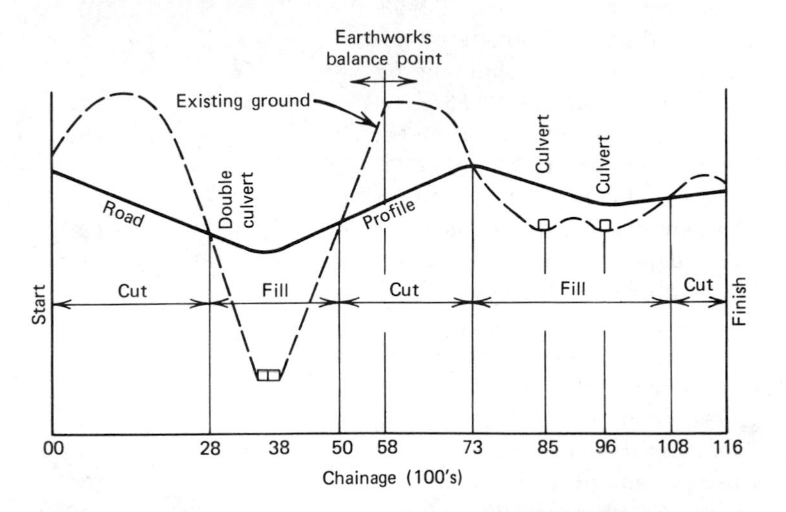

(*b*) The double box culvert will be built by one crew, and another crew will build the two small culverts. Concrete may be supplied either from the Paving Batch Plant or from small independent mixers at the culvert sites, whichever is expedient.

(*c*) One small slip-form paver will do all the concrete paving work, and all the shouldering will then follow with one crew after the concrete pavement is cured.

(*d*) Seeding the embankments with grass must be left as late as possible.

(1) Prepare a network diagram and determine the minimum possible project duration.

*Taken with acknowledgments from Antill and Woodhead, *Critical Path Methods in Construction Practice*, John Wiley & Sons, Inc., New York, 1970.

(2) If independent concrete mixers are used for the culverts, what is the latest day for delivery of the Paving Batch Plant to the site, so that the paving crew may have continuity of work (no idle time at all)?

Activity Description	Duration
Del. rebars—double box culvert	10
Move-in equipment	3
Del. rebars—small culverts	10
Set up paving batch plant	8
Order and del. paving mesh	10
Build and cure double box culvert, chainage 38	40
Clear and grub, chainage 00-58	10
Clear and grub, chainage 58-116	8
Build small culvert, chainage 85	14
Move dirt, chainage 00-58	27
Move part dirt, chainage 58-116	16
Build small culvert, chainage 96	14
Cure small culvert, chainage 85	10
Cure small culvert, chainage 96	10
Move balance dirt, chainage 58-116	5
Place subbase, chainage 00-58	4
Place subbase, chainage 58-116	4
Order and stockpile paving materials	7
Pave, chainage 58-116	5
Cure pavement, chainage 58-116	10
Pave, chainage 00-58	5
Cure pavement, chainage 00-58	10
Shoulders, chainage 00-58	2
Shoulders, chainage 58-116	2
Guardrail on curves	3
Seeding embankments with grass	4
Move out and open road	3

16.9 Visit several local building and heavy construction contractors and ascertain their attitude concerning networking methods. Are they actively committed to network planning, compelled by contract clauses to its use, or considering whether network methods are necessary?

16.10 Examine several project networks and record the level of detail used, the number of activities portrayed, and the modeling form adopted. Are these project networks used in the field and updated from time to time?

CHAPTER SEVENTEEN

Safety

17.1 NEED FOR SAFE PRACTICE

A disabling injury or fatal accident on the job site has negative impact on operations at many levels. Accidents cost money and affect worker morale. Because of the type of work involved in construction, many dangers exist both for the workers and for the public. For this reason, the subject of safety offers one area of noncontroversial mutual interest between management and the work force. The necessity of safe operations and of protecting and conserving lives by preventing accidents is understood by all.

Statistics maintained by the National Safety Council verify the fact that construction work is extremely hazardous and subject to high accident incidence. According to Robert L. Jenkins:

> The construction industry ranks fortieth in injury experience among the 42 industries for which records are compiled by the National Safety Council. This has been the approximate injury-experience ranking for construction during the past forty years.... The Safety Council reports that injuries occur in construction 16 times more frequently than in the automobile industry, eight times more often than in the chemical and steel industries, seven times more often than in the cement industry, and four times more often than in the shipbuilding industry.*

Although the fatality rate in construction has been reduced within recent years, the improvement in safety record achieved by the construction industry still lags seriously behind that achieved in other hazardous industries.

*See Chapter 11 in O'Brien and Zilly (eds.), *Contractor's Management Handbook*, McGraw-Hill Book Company, New York, 1971.

It is the contractor's responsibility to see that he does everything within his power to provide a safe working environment for his work force and the public in general. The factors that motivate safe practices at the job site are generally identified as follows:

1. Humanitarian concern.
2. Economic costs and benefits.
3. Legal and regulatory considerations.

Society has taken the position that because of the high health and accident potential intrinsic to the construction industry, the contractor must accept the liabilities associated with this hazardous environment and make an appropriate commitment to safe practice and accident prevention.

17.2 HUMANITARIAN CONCERN

It is normally accepted that day-to-day living has intrinsic risks that may result in members of the society being subjected to mental and physical hardship. One of the functions of society is to minimize pain and suffering. Particularly at the level of the work site, society has defined the principle that the employer is responsible for providing a safe environment for the work force. This is based on humanitarian concern. If, for instance, a worker loses a leg because of a job-related accident and is confined to a wheel chair, the worker is, in a sense, a casualty of the work place. Through his desire to be a participating member of society and support members of his family, the worker is injured. Society has traditionally shouldered the responsibility for this limitation on his abilities. Over the past century, the principle of employer liability for death and injury resulting from accidents or health hazards occuring at the work place has been firmly established in common law. The courts have further charged the employer with the following five responsibilities:*

1. To provide a reasonably safe work place.
2. To provide reasonably safe appliances, tools, and equipment.
3. To use reasonable care in selecting employees.
4. To enforce reasonable safety rules.
5. To provide reasonable instructions regarding the dangers of employment.

Mandatory requirements for the employer to make formal provision for injuries and deaths on the job resulted in the enactment of Workmen's Compensation laws at the end of the nineteenth century.

*Lee E. Knack, *Handbook of Construction Management and Organization*, Chapter 25, Bonny and Frein (eds.), Van Nostrand Reinhold Co., New York. 1973.

In 1884, Germany enacted the first Workmen's Compensation Act, followed by Austria in 1887 and England in 1897. The federal government passed the first American compensation act in 1908 covering government employees. Following several legal battles, the Supreme Court in 1917, declared "that states could enact and enforce compulsary Workmen's Compensation Laws under its power to provide for the public health, safety and welfare . . . ."

17.3 ECONOMIC COSTS AND BENEFITS

The *Manual of Accident Prevention** published by the Associated General Contractors of America (AGC) breaks safety costs into three categories as follows:

1. Direct cost of previous accidents
   Insurance premiums and ratings
   Mandatory accident prevention methods
   Records, safety personnel
2. Direct cost of each accident occurence
   Delay to project
   Uninsured damages
   Lost production
3. Indirect cost
   Investigation
   Loss of skilled men
   Loss of equipment

Direct costs from previous accidents come primarily in the form of insurance premiums, which have a significant effect on a contractor's operating expense. Workmen's compensation and liability insurance premiums can be calculated using either *manual* or *merit* rating systems. Manual rating is based on the past losses of the industry as a whole. The premium rate for compensation is normally set by the individual state Compensation Rating Bureaus. Many states are guided by or actually have their rates set by the National Council on Compensation Insurance. The premium rates are based on factors such as classification of operations, rates of pay, the frequency and severity of accidents in a particular classification, increases in the cost of cases, and the attitudes of various industrial compensation commissions. The rates as set and approved by each state insurance commissioner are known as the manual (standard) rates. These manual rates are published periodically in the *Engineering News Record* (ENR) Quarterly Cost Roundup issues (see Figure 17.1).

**Manual of Accident Prevention in Construction*, 6th Edition, The Associated General Contractors of America, Washington, D. C., 1971.

Compensation insurance base rates for construction workers

RATE IS PER $100 PAYROLL. COMPILED BY HERBERT L. JAMISON CO., INSURANCE ADVISERS AND AUDITORS, NEW YORK, N.Y. THESE RATES ARE SUBJECT TO CHANGE ACCORDING TO EXPERIENCE RATING.

Effective July 1, 1978

CLASSIFICATION OF WORK	Ala.	Ak.	Ariz.	Ark.	Cal.	Colo.	Conn.	Del.	D.C.	Fla.	Ga.	Hi.	Idaho	Ill.	Ind.	Iowa	Kan.	Ky.	La.	Me.	Md.	Mass.	Mich.	Minn.	Miss.	Mo.
Carpentry—1, 2 family residence	3.16	6.32	10.81	5.54	8.83	5.72B	6.78	4.55	12.12	10.22	4.45	9.50	7.07	5.51	2.17	3.95	4.98	8.76	8.06	4.32	4.98	4.94	7.15	6.30	3.04	3.29
Carpentry—3 stories or less	3.93	7.69	10.22	5.65	8.83	7.43B	6.78	4.55	12.12	10.22	4.45	9.22	7.07	7.02	2.54	3.85	4.98	7.37	11.01	4.32	4.98	6.32	7.15	6.30	3.04	3.29
Carpentry—interior cab wk	2.18	5.51	11.03	3.60	8.83	2.98	3.77	4.55	9.24	7.47	2.59	5.80	3.72	3.42	1.51	2.46	3.11	3.53	5.03	3.33	6.22	3.37	4.72	6.30	2.40	2.31
Carpentry—general	3.92	7.12	20.76	6.23	11.65	5.97	8.66	4.55	15.46	14.53	5.36	21.72	10.06	8.27	2.55	5.83	4.34	10.94	10.05	10.91	4.96	12.72	9.14	9.80	5.33	2.92
Chimney—construction brick con	9.93	21.86	21.09	20.05	22.80	12.23	30.86	A	65.76	35.25	10.39	30.48	18.44	21.36	6.91	14.87	14.77	18.33	44.65	22.56	26.86	14.37	21.04	26.08	13.17	16.86
Concrete work—bridges culverts-C	4.80	4.87	16.41	9.14	22.80	7.45	8.40	5.50	21.47	18.29	7.00	9.53	7.82	8.50	2.81	8.48	8.94	11.77	9.35	14.76	8.31	6.60	12.37	4.99	5.28	3.50
Concrete work—dwelling 1-2 family	2.44	7.36	7.17	3.10	5.90B	4.40	5.57	5.50	19.92	7.39	1.85	9.25	4.12	3.76	1.77	2.65	3.33	5.92	4.27	4.46	5.34	4.92	9.08	3.92	1.99	2.40
Concrete work—N.O.C.	4.14	8.14	14.67	5.06	9.69	6.35	4.93B	5.50	20.99	14.42	4.28	10.93	5.34	9.67	2.72	4.44	5.32	5.51	9.52	7.70	8.63	7.27	13.94	7.65	4.40	3.47
Concrete floors, sidewalks	2.94	5.96	8.34	2.91	4.76	3.81	4.47	5.50	10.89	10.02	3.43	6.23	3.37	3.64	6.53	4.74	3.41	5.78	5.44	4.46	4.87	3.69	9.31	4.86	2.16	1.93
Electrical wiring—inside	2.05	4.83	6.50	3.00	4.13	3.03	2.66	1.35	8.99	6.03	2.32	7.37	3.21	3.32	1.20	2.01	2.67	3.29	3.56	2.61	4.22	2.30	4.56	2.59	2.29	1.96
Excavation earth N.O.C.	5.00	7.02	7.40	5.54	7.04	4.80	4.77	6.65	14.31	10.58	5.47	20.40	9.96	5.31	2.16	4.42	4.65	8.28	7.50	6.78	5.58	4.62	9.95	5.18	3.81	4.51
Excavation—rock	5.94	7.62	16.51	6.18	13.86	6.91	6.41	6.65	40.64	15.86	8.85	14.83	6.50	28.17	3.90	8.81	6.39	9.37	30.53	11.11	9.86	5.08	14.70	6.94	8.10	5.15
Glaziers	4.79	7.80	10.42	6.54	8.88	6.23	5.83	4.25	17.52	9.88	4.94	11.32	5.80	8.34	2.19	4.01	3.96	6.72	8.41	7.23	9.42	5.32	7.18	6.74	5.42	3.44
Insulation work	5.27	7.20	12.63	4.55	7.97	5.66	3.99B	4.55	11.67	10.64	3.29	6.41	6.89	7.24	1.76	3.89	4.12	7.35	9.90	6.26	5.63	4.69	8.42	3.97	3.17	3.28
Lathing	2.21	4.86	7.79	3.32	6.53B	3.64	3.40	4.35	9.36	7.56	2.69	5.48	3.57	2.68	0.98	2.81	2.73	4.56	4.05	3.76	3.93	3.60	4.46	3.26	2.22	1.11
Masonry	2.81	9.31	10.56	3.50	8.75	5.00	6.30	5.15	20.15	9.41	3.14	10.12	4.79	6.24	1.70	2.96	4.97	7.11	5.20	5.66	4.97	5.62	8.78	4.49	2.03	2.88
Painting and decorating	3.18	7.21	7.96	4.49	9.51	3.65	6.05	7.20	12.51	8.10	3.78	8.34	6.06	6.31	1.86	5.53	3.76	7.74	9.11	7.05	4.60	6.15	8.18	5.84	2.77	2.55
Pile driving	11.42	21.83	36.30	16.86	20.31	15.67	4.81	7.85	56.73	26.44	7.31	26.36	15.32	14.56	5.06	11.08	11.49	25.62	29.30	13.52	15.44	12.10	27.04	10.93	14.95	7.69
Plastering	4.01	6.94	16.97	7.22	12.18	5.13	3.99	4.35	10.79	9.42	3.27	8.22	5.41	4.38	1.64	3.77	4.06	6.63	8.24	6.04	5.63	4.32	8.45	4.10	3.09	2.73
Plumbing	3.64	6.15	6.63	3.34	5.65B	3.70	2.47	2.55	10.14	6.70	3.32	5.92	3.36	4.18	1.40	3.09	2.63	3.12	6.20	3.22	4.15	3.00	5.36	5.04	2.26	2.43
Roofing	6.21	17.83	22.15	11.41	21.23	14.13	10.55	6.35	28.44	22.08	5.85	25.95	14.05	14.40	3.72	8.88	8.42	17.99	12.68	15.03	17.52	A	20.54	17.12	7.39	8.79
Sheet metal work	2.96	5.59	9.58	3.70	5.93B	3.17	4.49	6.35	11.03	7.46	3.78	5.59	4.00	4.46	1.79	3.08	3.80	4.27	6.64	5.12	5.31	4.74	7.12	3.65	4.78	2.82
Steel erection—doors and sash	2.84	7.73	8.99	4.34	6.94	4.80	3.52	10.95	18.10	6.43	3.49	5.06	5.08	5.28	1.72	3.72	3.53	8.74	6.04	5.49	5.92	4.65	10.72	4.93	2.88	2.89
Steel erection—interior ornament	2.84	7.73	8.99	4.34	6.94	9.80	3.52	10.95	18.10	6.43	3.49	5.06	5.08	5.28	1.72	3.72	3.53	8.74	6.04	5.49	5.92	4.65	10.72	4.93	2.88	2.89
Steel erection—structure	6.73	32.21	43.63	[illegible]	24.02	15.56	15.90	10.95	49.95	30.10	10.58	28.06	14.11	22.90	7.82	30.72	18.25	22.40	14.25	19.94	20.39	20.22	24.78	22.16	11.47	10.99
Steel erection—dwelling 2 stories	7.81	22.97	35.68	10.92	24.22	9.51	10.92	10.95	49.84	20.27	9.20	26.12	16.89	25.13	4.85	13.17	11.27	17.91	16.61	14.92	14.46	16.69	17.06	11.12	8.94	8.05
Steel erection—N.O.C.	8.44	19.75	19.87	12.05	15.17	15.31	15.52	10.95	61.87	21.19	10.19	35.78	9.87	23.68	6.51	11.19	12.79	17.16	20.65	21.40	16.73	16.69	16.51	28.37	8.58	14.60
Tile work—interior	2.87	2.66	8.25	3.20	5.27	3.23	2.71	3.05	27.74	5.94	2.04	6.39	2.92	2.52	1.37	2.17	2.20	3.69	4.93	3.15	3.39	3.74	5.39	2.96	2.19	1.47
Timekeepers and watchmen	3.23	7.15	6.66	4.50	6.65	4.75	4.46	A	18.15	9.64	5.16	8.42	5.14	5.91	2.16	4.14	4.10	6.85	9.29	5.84	6.26	6.62	6.76	5.80	2.95	6.58
Waterproofing interior (brush)	3.18	7.21	7.46	4.49	9.51	3.55	6.00	A	12.51	8.10	3.78	8.34	6.06	6.31	1.86	5.53	3.76	7.74	9.11	7.05	4.60	6.15	A	5.84	2.77	2.55
Waterproofing—trowel interior	4.21	6.94	16.97	4.22	12.18	5.13	3.99	A	10.79	9.42	3.29	8.22	5.41	4.38	1.64	3.77	4.06	6.63	8.24	6.04	5.63	4.32	8.45	4.10	3.09	2.73
Waterproofing—trowel exterior	2.81	9.31	10.56	3.50	8.75	5.00	6.30	A	20.15	9.41	3.14	10.12	4.79	6.24	1.70	2.96	4.97	7.11	5.20	5.66	4.97	5.62	8.78	4.49	2.03	2.88
Waterproofing (pressure gun)	4.14	8.14	14.67	5.00	9.69	6.35	4.93	A	20.99	14.42	4.28	10.93	5.34	9.67	2.72	4.44	5.32	5.51	9.52	7.70	8.63	7.27	13.94	7.65	4.40	3.47
Wrecking	12.84	27.39	66.78	20.75	A	23.45	30.40	15.65	43.41	52.27	21.52	58.71	40.07	21.74	7.32	25.65	18.86	24.95	59.50	31.61	28.13	19.25	A	13.55	15.30	23.45

CLASSIFICATION OF WORK	Mont.	Neb.	(C) Nev.	N.H.	N.J.	N.M.	N.Y.	N.C.	(C) N.D.	(C) Ohio	Okla.	Ore.	Penn.	R.I.	S.C.	S.D.	Tenn.	Tex.	Utah	Vt.	Va.	(C) Wash.	(C) W.Va.	Wisc.	(C) Wyo
Carpentry—1, 2 family residence	6.13	2.93	7.80	8.67	6.43	4.70	6.09	3.21	6.95	3.71	5.70	16.02	6.25	5.39	4.03	2.89	4.18	7.15	A	4.16	4.00	D	3.01	4.15	E
Carpentry—3 stories or less	6.13	3.03	7.80	8.67	5.05	5.85	6.92	3.21	6.95	3.71	7.41	11.27	6.25	6.87	4.03	2.89	4.18	8.62	A	4.16	3.63	D	3.01	1.90	E
Carpentry—interior cab wk	3.92	1.69	7.80	2.73	5.05	2.82	A	2.97	6.95	3.71	3.50	9.12	6.25	3.59	2.02	1.78	2.17	5.14	A	2.28	2.97	D	3.01	2.75	E
Carpentry—general	8.17	3.86	7.80	7.23	6.04	7.50	7.91	3.81	6.95	4.21	6.62	24.97	6.25	7.76	4.94	4.27	4.90	9.57	4.23	5.17	5.00	D	3.01	5.03	E
Chimney—construction brick con	25.24	9.58	15.42	25.43	21.46	11.14	26.75	10.72	11.65	2.68	24.00	36.37	5.25	25.43	12.41	9.12	11.74	27.86	16.55	15.75	15.43	D	3.01	13.90	E
Concrete work—bridges culverts-C	10.67	4.58	7.80	11.00	7.38	10.29	8.40	4.95	6.90	4.28	18.47	13.02	7.10	8.79	6.21	4.90	7.02	7.56	6.07	7.05	6.45	D	4.54	7.64	E
Concrete work dwelling 1-2 family	4.97	1.99	7.80	4.66	5.65	3.78	A	1.97	6.00	4.28	3.36	8.53	7.10	4.82	2.38	2.04	2.73	6.46	3.07	2.95	2.23	D	3.01	2.40	E
Concrete work—N.O.C.	5.97	3.09	7.80	6.20	6.48	6.66	8.40	2.78	6.00	4.28	7.45	15.00	7.10	6.70	2.83	3.78	4.00	9.92	4.71	4.52	5.73	D	3.01	3.69	E
Concrete or cement work—floors, sidewalks	6.26	2.67	7.80	3.62	4.79	3.92	4.41	2.16	6.90	4.28	4.36	9.15	7.10	5.93	2.43	2.25	2.45	6.46	3.90	2.68	2.62	D	3.01	2.19	E
Electrical wiring—inside	3.04	1.29	4.01	2.96	3.11	2.75	3.46	2.02	2.65	1.72	2.77	5.59	2.50	3.59	2.09	1.37	2.48	3.99	1.77	1.80	2.75	D	3.01	1.93	E
Excavation earth N.O.C.	9.51	2.36	5.90	8.83	4.79	4.87	5.73	3.40	6.00	4.02	6.96	14.97	7.40	5.10	3.51	3.80	4.54	5.83	3.99	4.03	5.84	D	3.80	4.70	E
Excavation—rock	13.34	6.45	5.90	10.02	6.32	7.12	6.16	5.48	6.00	4.02	3.94	13.95	7.40	12.34	6.82	4.69	5.04	9.71	6.16	5.89	4.25	D	3.80	8.77	E
Glaziers	8.33	3.51	3.96	8.16	6.25	5.56	7.73	4.13	6.95	4.91	5.07	7.27	3.90	6.80	4.69	3.75	4.52	6.03	2.93	5.42	4.16	D	3.01	3.72	E
Insulation work	6.48	3.21	7.80	5.23	6.48	7.34	3.97	3.47	6.95	2.61	5.64	11.62	6.25	5.22	3.56	2.65	3.41	7.21	4.07	4.06	4.08	D	3.01	2.94	E
Lathing	3.58	1.73	7.80	4.30	3.55	2.95	4.56	1.88	3.80	2.61	2.11	14.92	3.80	3.99	2.16	1.70	2.50	5.58	3.01	2.56	3.38	D	3.01	2.11	E
Masonry	7.36	2.05	7.80	4.66	6.71	6.37	6.81	1.71	3.80	2.68	4.74	15.02	5.25	5.58	2.36	2.68	3.40	6.18	3.09	2.93	3.63	D	3.01	3.32	E
Painting and decorating	9.03	2.95	7.62	7.76	7.24	4.62	9.00	3.15	5.20	4.91	3.91	11.27	8.30	6.13	3.47	2.48	3.35	5.11	4.56	3.71	4.00	D	3.01	3.18	E
Pile driving	20.92	10.01	5.90	12.35	9.31	14.87	10.64	6.93	18.00	4.02	16.27	31.40	13.35	16.28	9.45	10.04	21.48	14.41	7.51	10.52	11.46	D	3.01	9.14	E
Plastering	6.89	2.99	7.80	5.23	3.55	6.98	8.78	2.98	3.65	2.61	4.52	11.32	3.80	5.53	3.02	3.07	3.39	9.50	4.07	3.84	3.89	D	3.01	2.48	E
Plumbing	3.25	1.76	3.92	3.73	3.23	3.77	4.90	2.12	5.10	2.09	3.85	7.67	3.55	3.03	1.94	1.78	2.29	4.93	2.18	2.07	3.07	D	1.43	2.48	E
Roofing	14.21	9.17	7.80	20.48	17.01	12.93	A	6.61	10.75	11.02	15.66	30.06	8.15	16.81	7.96	7.16	6.69	15.73	2.00	9.15	9.17	D	3.01	8.55	E
Sheet metal work—erection—inst. & repair	4.73	3.04	3.92	3.98	3.78	4.39	7.03	2.74	5.10	3.31	4.54	8.90	8.15	4.83	4.20	2.47	4.10	8.32	3.01	3.53	4.98	D	6.91	3.04	E
Steel erection—doors and sash	5.31	2.61	7.80	6.05	A	4.34	3.89	2.48	18.00	4.21	3.36	16.31	15.75	4.89	2.29	2.73	3.14	6.71	3.04	3.65	2.87	D	6.91	3.98	E
Steel erection—interior ornament	5.31	2.61	7.80	6.05	5.00	4.34	3.89	2.48	18.00	4.21	3.36	16.31	15.75	4.89	2.29	2.73	3.14	6.71	3.04	3.65	2.87	D	6.91	3.98	E
Steel erection—structure	25.72	9.26	15.42	24.68	16.39	8.33	15.55	16.02	18.00	17.41	38.65	29.99	15.75	26.71	10.27	10.80	14.75	24.44	A	18.91	11.24	D	6.91	9.73	E
Steel erection—dwelling 2 stories	17.39	7.37	15.42	14.14	9.54	16.21	8.76	8.02	18.00	17.41	17.50	23.10	15.75	20.86	8.54	7.23	9.07	24.44	9.19	13.71	10.57	D	6.91	7.43	E
Steel erection—N.O.C.	12.27	6.01	15.42	17.67	8.60	12.90	13.86	6.28	18.00	17.41	15.45	28.23	15.75	33.29	7.78	6.16	9.14	8.42	7.03	14.78	12.68	D	6.91	6.84	E
Tile work—interior	3.07	1.64	3.96	2.88	5.14	3.03	5.21	1.72	2.75	1.87	3.34	7.24	2.85	3.18	1.75	1.53	2.46	4.35	2.28	2.11	2.55	D	3.01	2.05	E
Timekeepers and watchmen	4.10	2.66	7.80	6.37	3.43	4.89	4.05	2.82	4.05	3.71	5.08	18.21		6.08	2.94	2.79	3.75	3.09	4.51	3.92	2.91	D	3.01	3.47	E
Waterproofing interior (brush)	9.03	2.95	7.62	7.76	7.24	4.62	9.00	3.15	5.20	4.28	3.91	11.27	5.25	6.13	3.47	2.48	3.35	5.11	4.56	3.71	4.00	D	3.01	3.18	E
Waterproofing—trowel interior	6.89	2.99	7.80	5.23	3.55	6.98	8.78	2.98	10.75	11.02	4.52	11.32	5.25	5.53	3.02	3.07	3.39	9.50	4.07	3.84	3.89	D	3.01	2.48	E
Waterproofing—trowel exterior	7.36	2.05	7.80	4.66	6.71	6.37	6.81	1.71	10.75	11.02	4.74	15.02	5.25	5.58	2.36	2.68	3.40	6.18	3.09	2.93	3.63	D	3.01	3.32	E
Waterproofing (pressure gun)	5.97	3.09	7.80	6.20	6.48	6.66	8.40	2.78	10.75	11.02	7.45	15.00	5.25	6.70	2.83	3.78	4.00	9.92	4.71	4.52	5.73	D	3.01	3.69	E
Wrecking	70.70	10.96	7.80	32.82	37.41	7.42	—	9.43	10.75	27.49	35.58	47.51	3.65	22.81	14.37	14.34	19.78	17.17	27.71	12.12	14.70	D	6.91	11.44	E

• A Specialty rated-refer to company. • B Rates indicated are published by the National Council on Compensation Insurance, New York City. • C Subscription to State Fund Mandatory Industrial Commission, Carson City, Nev.; Workman's Compensation Bureau, Bismarck, N.D.; The Industrial Commission of Ohio, Columbus, Ohio; Dept. of Labor and Industry, Olympia, Wash.; State Compensation Commission, Charleston, W. Va.; Workmen's Compensation, State Treasurer's Office, Cheyenne, Wyo. • General contractor pays composite $1.01 rate per working manhour for all workers. • E Rate equals 5% of gross salary each month.

FIGURE 17.1 Compensation insurance base rates for construction workers.

The merit rating system bases premiums on a particular company's safety record. High-risk (high-accident-rated) companies are therefore penalized with higher premiums than those paid by companies with low accident rates. In this way, a good safety program can result in substantial financial savings to a company. Higher returns on jobs can be realized, and the ability to bid lower and win more jobs is greatly enhanced.

Once the premiums reach a value of $1000, the contractor is eligible for a merit system rating. That is, the cost of his premium will be individually calculated with his safety record being the critical consideration. Under the merit system, there are two basic methods utilized to incorporate the safety record into the final cost of the premium. These are referred to as the *experience* rating and *retrospective* rating methods.

Most insurance carriers use the experience rating method, which is based on the company's record for the past three years. In this system, an *experience modification factor* is applied to the manual rate to establish the premium for a given firm. Data on losses, the actual project being insured, and other variables are considered in deriving the experience modification factor. If the company has an experience modification factor credit of 25%, it will pay only 75% of the manual premium. Good experience ratings and credit modification factors can lead to significant savings. Clough (Wiley, 1975) illustrates this with the following example:

> Assume that a building contractor does an annual volume of $10 million worth of work a year. Considering a typical amount of subcontracting and the cost of materials, this general contractor's annual payroll will be of the order of magnitude of $2.5 million. If his present workmen's compensation rate averages about 8 percent, his annual premium cost will be about $200,000. Now assume that an effective accident prevention program results in an experience-modification rate reduction for this contractor of 30 percent below the workmen's compensation insurance rates formerly paid. Annual savings of the order of $60,000 are thereby realized on the cost of this one insurance coverage alone.

Retrospective rating is basically the same as experience rating except for one point. It utilizes the loss record of the contractor for the previous year or other defined retrospective period to compute the premium. This can raise or lower the premium cost based on performance during the retrospective period. The starting point or basis for this method is again the manual premium. A percentage (usually 20%) of the standard premium resulting from applying the experience modification factor to the manual rate is used to obtain the basic premium. The retrospective rate is then calculated as:

$$\begin{matrix}\text{Retrospective}\\ \text{rate}\end{matrix} = \begin{matrix}\text{tax}\\ \text{multiplier}\end{matrix} \times \left[\begin{matrix}\text{Basic}\\ \text{premium}\end{matrix} + \left(\begin{matrix}\text{Incurred}\\ \text{loss}\end{matrix} \times \begin{matrix}\text{Loss}\\ \text{conversion}\\ \text{factor}\end{matrix}\right)\right]$$

The incurred loss is the amount paid out to settle claims over the retrospective period. The loss conversion factor is a percentage loading used to weight the incurred losses to cover general claims investigation and adjustment expenses. The tax multiplier covers premium taxes that must be paid to the state. If the data for a given company are as follows,

Manual premium	$25,000
Experience modification factor (25% Credit)	0.75

Then

Standard premium = 0.75($25,000) = $18,750
Basic premium @ 20% of standard = 0.20($18,750) = $3750
Loss conversion factor = 1.135 (derived from experience)
Tax multiplier = 1.03 (based on state tax)
Incurred losses = $10,000

Then:

Retrospective premium = 1.03 [(3750 + (1.135 × $10,000))]
= $15,553

This is a nice savings over the standard premium of $18,750 and provides the contractor with a clear incentive to minimize incurred losses. By so doing, the contractor can expect a large premium rebate at the end of the year.

17.4 UNINSURED ACCIDENT COSTS

In addition to the cost of insurance premiums, additional direct costs for things such as the salary of the safety engineer and his staff as well as costs associated with the implementation of a good safety program can be identified. The precise amount of the costs associated with the *Manual for Accident Prevention* categories (2) and (3) is more difficult to assess, and these costs can be thought of as additional uninsured costs resulting from accidents. Typical uninsured costs associated with an accident as identified by Lee E. Knack are shown in Table 17.1.

Although varying slightly from source to source, hidden losses of this variety have been estimated to be as much as nine times the amount spent on comprehensive insurance. In addition to the costs noted in Table 17.1, another cost is that of paying an injured employee to show up for work even if he cannot perform at his best. This is common practice for minor injuries. This is done to avoid recording a lost time accident which might impact the insurance premium.

The following situation illustrates the additional losses resulting from hidden costs. At a large industrial construction site, the survey party chief was

Table 17.1 UNINSURED COSTS

Injuries	Associated Costs
1. First-aid expenses	1. Difference between actual losses and amount recovered
2. Transportation costs	2. Rental of equipment to replace damaged equipment
3. Cost of investigations	3. Surplus workers for replacement of injured workmen
4. Cost of processing reports	4. Wages or other benefits paid to disabled workers
	5. Overhead costs while production is stopped
	6. Loss of bonus or payment of forfeiture of delays
Wage Losses	**Off the Job Accidents**
1. Idle time of workers whose work is interrupted	1. Cost of medical services
2. Man-hours spent in cleaning up accident area	2. Time spent on injured workers' welfare
3. Time spent repairing damaged equipment	3. Loss of skill and experience
4. Time lost by workers receiving first aid	4. Training replacement worker
	5. Decreased production of replacement
	6. Benefits paid to injured worker or dependents
Production Losses	**Intangibles**
1. Product spoiled by accident	1. Lowered employee morale
2. Loss of skill and experience	2. Increased labor conflict
3. Lowered production of worker replacement	3. Unfavorable public relations
4. Idle machine time	

Source. From Lee. E. Knack, *Handbook of Construction Management and Organization* ,Chapter 25, in Bonny and Frein (eds.), Van Nostrand Reinhold Co., New York, 1973.

on the way to the office to get a set of plans. The survey crew was to lay out four machine foundations that morning. The wooden walkway beneath the party chief collapsed. He was in the hospital for five weeks with a shattered pelvis. Another party chief who was unfamiliar with the site was assigned to the surveying crew. As a result, the four machine foundations were constructed 2 ft farther west than called for in the plans. After this was discovered, six laborers worked for 20 hours removing the reinforced concrete. The survey crew of four spent another five hours laying out the foundations. Four carpenters worked 20 more hours preparing new forms. Five more hours were required for the ironworkers to place the steel reinforcement. The total indirect cost was approximately $1700. Although this activity was not on the critical path, if it had been, liquidated damages might have been charged to the contractor. Still, the accident resulted in costs amounting to one week's pay for the employees affected and the cost of material that had to be replaced.

17.5 FEDERAL LEGISLATION AND REGULATION

The federal government implemented a formal program of mandatory safety practices in 1969 with the passage of the Construction Safety Act as an amendment to the Contract Work Hours Standard Act. This legislation requires contractors working on federally funded projects to meet certain requirements to protect the worker against health and accident hazards. In addition, certain reporting and training provisions were established. This program of required procedures has been refered to as a *physical* approach to achieving safety. That is, regulations are prescribed that are designed to minimize the possibility of an unsafe condition arising. A typical physical measure of this type is the requirement to install guard rails around all open upper floors of a multistory building during construction. Furthermore, physical measures are implemented to minimize injury in the event of an accident. An example of this is the requirement to wear a safety belt when working high steel and the installation of safety nets to protect a man who slips and falls. This physical approach is in contrast to the behavioural approach that is designed to make all levels of the work force from top management to the laborer think in a safe way and thus avoid unsafe situations. Research on the behavioural approach is discussed in detail in Barrie and Paulson, *Professional Construction Management* (1978).

Shortly after the passage of the Construction Safety Act, a more comprehensive approach to mandatory safe practices was adopted in the form of the Williams-Steiger Occupational Safety and Health Act (OSHA) passed by the Congress in 1970. This act established mandatory safety and health procedures to be followed by all firms operating in interstate commerce.

Under this act, all employers are required to provide:

> . . . employment and a place of employment which are free from recognized hazards that are causing or are likely to cause death or serious physical harm to his employees.*

The provisions of the Construction Safety Act were included in the act by reference. The provisions of OSHA fall within the jurisdiction of the Secretary of Labor. In 1971, he implemented the law by publishing Section 1926 that specifically refers to the construction industry and Section 1910 that pertains to General Standards. Many existing standards issued by various standards organizations including the American National Standards Institute (ANSI) were included in the basic law by reference. The schematic development of the legislation is shown in Figure 17.2. The jumble of regulations and references was reconciled to some degree in 1974 with the issuance of the "Construction Safety and Health Regulations" in the *Federal Register*.† OSHA has a service for providing update information on standards that is designed to aid in keeping the five volumes of regulations current. The regulations are divided as follows:

Volume I: General Industry Standards.
Volume II: Maritime Industry Standards.
Volume III: Construction Industry Standards.
Volume IV: OSHA Regulations.
Volume V: Compliance Operations Manual.

Under the OSHA legislation, the assistant Secretary of Labor for occupational safety and health administers and enforces OSHA through the labor department's Occupational Safety and Health Administration with its 10 offices around the country. The Occupational Safety and Health Review Commission is designated by OSHA as the review body to which citations for alleged violations and proposed penalties can be appealed. Research in this area is under the control of the National Institute for Occupational Safety and Health (NIOSH), which is part of the Department of Health, Education and Welfare.

17.6 OSHA REQUIREMENTS

Employers must "make, keep and preserve, and make available to representatives of the Secretaries of Labor, and Health Education and Welfare" records of recordable occupational injuries and illnesses. Any fatal or serious accidents must be reported to OSHA within 48 hours. Certain records of

Engineering News Record, August 24, 1972, p. 23.

†"Construction Safety and Health Regulations," U.S. Department of Labor, Occupation Safety and Health Administration, *Federal Register*, Vol. 39, No. 122, June 24, 1974, Washington, D.C.

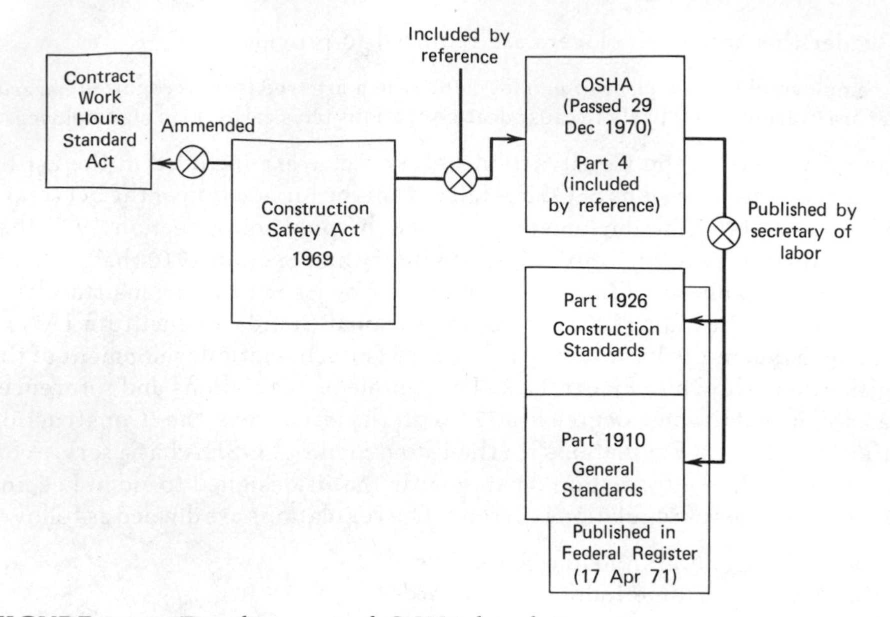

FIGURE 17.2 Development of OSHA legislation.

job-related fatalities, injuries, and illnesses must be maintained by firms having eight or more employees. The two key forms that must be available for review when a compliance officer makes an inspection are the:

1. *OSHA 200.* This is a log that summarizes each reportable case as a single line entry and must be posted for employee inspection (see Figure 17.3).
2. *OSHA 101.* A supplementary form with details on each line entry in the OSHA 200 (see Figure 17.4).

These records must be preserved for five years.

The employer is also required to post at the work site records of citations and notices of employees' rights. There have been strong drives to change OSHA so that employers with scattered work sites, as in construction, can be allowed to maintain the required records at a central location (i.e., home office).

17.7 HOW THE LAW IS APPLIED

The 10 OSHA regional offices employ inspectors whose duties include visits to active projects to determine if the builders are conforming to the regulations. An inspection can be initiated at random by OSHA or state safety inspectors, or by an employee (or his union) who submits a written statement to the labor department that he believes there is a violation that threatens

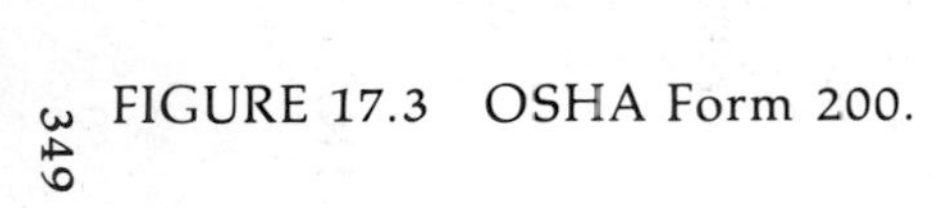

Bureau of Labor Statistics
Log and Summary of Occupational
Injuries and Illnesses

U.S. Department of Labor

For Calendar Year 19 ______ Page ___ of ___

Form Approved
O.M.B. No. 44R 1453

NOTE: This form is required by Public Law 91-596 and must be kept in the establishment for *5 years.* Failure to maintain and post can result in the issuance of citations and assessment of penalties. *(See posting requirements on the other side of form.)*

RECORDABLE CASES: You are required to record information about every occupational **death**; every nonfatal occupational **illness**; and those nonfatal occupational **injuries** which involve one or more of the following: loss of consciousness, restriction of work or motion, transfer to another job, or medical treatment (other than first aid). *(See definitions on the other side of form.)*

Company Name
Establishment Name
Establishment Address

Case or File Number	Date of Injury or Onset of Illness	Employee's Name	Occupation	Department	Description of Injury or Illness
Enter a nonduplicating number which will facilitate comparisons with supplementary records.	Enter Mo./day.	Enter first name or initial, middle initial, last name.	Enter regular job title, not activity employee was performing when injured or at onset of illness. In the absence of a formal title, enter a brief description of the employee's duties.	Enter department in which the employee is regularly employed or a description of normal workplace to which employee is assigned, even though temporarily working in another department at the time of injury or illness.	Enter a brief description of the injury or illness and indicate the part or parts of body affected. Typical entries for this column might be: Amputation of 1st joint right forefinger; Strain of lower back; Contact dermatitis on both hands; Electrocution—body.
(A)	(B)	(C)	(D)	(E)	(F)
					PREVIOUS PAGE TOTALS →
					TOTALS (Instructions on other side of form.) →

OSHA No. 200

Extent of and Outcome of INJURY					
Fatalities	Nonfatal Injuries				
Injury Related	Injuries With Lost Workdays				Injuries Without Lost Workdays
Enter DATE of death. Mo./day/yr.	Enter a CHECK if injury involves days away from work, or days of restricted work activity, or both.	Enter a CHECK if injury involves days away from work.	Enter number of DAYS away from work.	Enter number of DAYS of restricted work activity.	Enter a CHECK if no entry was made in columns 1 or 2 but the injury is recordable as defined above.
(1)	(2)	(3)	(4)	(5)	(6)

Type, Extent of, and Outcome of ILLNESS												
Type of Illness							Fatalities	Nonfatal Illnesses				
CHECK Only One Column for Each Illness *(See other side of form for terminations or permanent transfers.)*							Illness Related	Illnesses With Lost Workdays				Illnesses Without Lost Workdays
Occupational skin diseases or disorders	Dust diseases of the lungs	Respiratory conditions due to toxic agents	Poisoning (systemic effects of toxic materials)	Disorders due to physical agents	Disorders associated with repeated trauma	All other occupational illnesses	Enter DATE of death. Mo./day/yr.	Enter a CHECK if illness involves days away from work, or days of restricted work activity, or both.	Enter a CHECK if illness involves days away from work.	Enter number of DAYS away from work.	Enter number of DAYS of restricted work activity.	Enter a CHECK if no entry was made in columns 8 or 9.
(7)												
(a)	(b)	(c)	(d)	(e)	(f)	(g)	(8)	(9)	(10)	(11)	(12)	(13)

INJURIES

ILLNESSES

FOLD

Certification of Annual Summary Totals By ______ Title ______ Date ______

OSHA No. 200

POST ONLY THIS PORTION OF THE LAST PAGE NO LATER THAN FEBRUARY 1.

FIGURE 17.3 OSHA Form 200.

OSHA No. 101
Case or File No. __________

Form approved
OMB No. 44R 1453

Supplementary Record of Occupational Injuries and Illnesses

EMPLOYER

1. Name ____________________
2. Mail address ____________________
   (No. and street) (City or town) (State)
3. Location, if different from mail address ____________________

INJURED OR ILL EMPLOYEE

4. Name ____________________ Social Security No. __________
   (First name) (Middle name) (Last name)
5. Home address ____________________
   (No. and street) (City or town) (State)
6. Age __________ 7. Sex: Male__________ Female__________ (Check one)
8. Occupation ____________________
   (Enter regular job title, *not* the specific activity he was performing at time of injury.)
9. Department ____________________
   (Enter name of department or division in which the injured person is regularly employed, even though he may have been temporarily working in another department at the time of injury.)

THE ACCIDENT OR EXPOSURE TO OCCUPATIONAL ILLNESS

10. Place of accident or exposure ____________________
    (No. and street) (City or town) (State)
    If accident or exposure occurred on employer's premises, give address of plant or establishment in which it occurred. Do not indicate department or division within the plant or establishment. If accident occurred outside employer's premises at an identifiable address, give that address. If it occurred on a public highway or at any other place which cannot be identified by number and street, please provide place references locating the place of injury as accurately as possible.
11. Was place of accident or exposure on employer's premises? __________ (Yes or No)
12. What was the employee doing when injured? ____________________
    (Be specific. If he was using tools or equipment or handling material,
    ____
    name them and tell what he was doing with them.)
    ____
13. How did the accident occur? ____________________
    (Describe fully the events which resulted in the injury or occupational illness. Tell what
    ____
    happened and how it happened. Name any objects or substances involved and tell how they were involved. Give
    ____
    full details on all factors which led or contributed to the accident. Use separate sheet for additional space.)

OCCUPATIONAL INJURY OR OCCUPATIONAL ILLNESS

14. Describe the injury or illness in detail and indicate the part of body affected. __________
    (e.g.: amputation of right index finger
    ____
    at second joint; fracture of ribs; lead poisoning; dermatitis of left hand, etc.)
15. Name the object or substance which directly injured the employee. (For example, the machine or thing he struck against or which struck him; the vapor or poison he inhaled or swallowed; the chemical or radiation which irritated his skin; or in cases of strains, hernias, etc., the thing he was lifting, pulling, etc.)
    ____
    ____
16. Date of injury or initial diagnosis of occupational illness ____________________
    (Date)
17. Did employee die? __________ (Yes or No)

OTHER

18. Name and address of physician ____________________
19. If hospitalized, name and address of hospital ____________________
    ____

Date of report __________ Prepared by ____________________
Official position ____________________

FIGURE 17.4 OSHA Form 101.

physical harm or imminent danger. All inspections must be on an unannounced basis during the working day. Recent rulings by the Supreme Court, however, require that an inspection warrant be obtained from proper authority. This essentially changes the "surprise" nature of the inspection and allows the work site supervisor to prepare for the inspection.

The inspection is divided into four parts:*

1. An opening conference with the employer.
2. Selection of a representative of the employees and of the employer to accompany the inspector on a tour of the work place.
3. The walk around inspection. The inspector is allowed to talk with any employees.
4. The closing conference during which the inspector discusses the conditions and practices he observed which might be safety or health violations. He will explain that citations may be issued and that a reasonable time will be allowed to correct the situation. Fines may be proposed with the citations. The employer is allowed 15 days to appeal a penalty. A copy of the citation must be posted near the site of the violation.

The appeals of citations, penalties, or abatement periods are made through a procedure to the Occupational Safety and Health Review Commission. The Commission, after a hearing and review, can affirm, modify, or vacate the citation, proposed penalty, and abatement period.

During the first fiscal year of the law, OSHA conducted 32,701 inspections in 29,505 establishments employing 5.9 million persons. About 25% of the work places were in compliance. Citations were issued for 102,861 violations with proposed penalties of $2,291,147. The incidence of inspections in construction was high, partly because its roofing and sheet metal segments were among the first target industries singled out for inspection.

Table 17.2 is a list of OSHA standards most commonly cited for violations during July to October 1972.† It shows in descending order the most frequently alleged violations.

If, during the course of such an inspection, a violation is noted a written citation is given to the employer and the area where the violation occurs will be posted. A reasonable length of time shall be granted the employer for correction of the violation.

These violations, and failure to abate in the given time, incur monetary violations up to $10,000.00. Serious violations incur a mandatory fine of $1000. Failure to abate within the given time period can result in a fine of $1000 a day for the period the violation persists.

*See Vincent Bush, *Safety in the Construction Industry*, Reston, 1975.

†*Engineering News Record*, 1 Feb. 1973, p. 67.

Table 17.2 OSHA STANDARDS MOST COMMONLY CITED FOR VIOLATIONS

Section	Subject	Section	Subject
1926.500	Guardrails, handrails, covers	1926.100	Head protection
.451	Scaffolding	.552	Materials hoists, personnel hoists, elevators
.450	Ladders	.50	Medical services, first aid
.350	Gas welding and cutting	.501	Stairways
.401	Grounding and bending	.300	General requirements, hand and power tools
.550	Cranes and derricks	.651	Excavation
.25	Housekeeping	.51	Sanitation
.152	Flammable and combustible liquids	.28	Personal protective equipment
.400	General electrical	.102	Eye and face protection
.402	Electrical equipment installation and maintenance	.302	Power-operated hand tools
.150	Fire protection	.351	Arc welding and cutting
.652	Trenching	.105	Safety nets
.601	Motor Vehicles		

17.8 SAFETY RECORD KEEPING

Documentation under the Williams-Steiger Act is required as follows:

> . . . every employer who is covered under this act must keep occupational injury and ilness records for his employees in the establishment in which his employees usually report to work.

The OSHA laws require employers to keep both a log of recordable occupational injuries and illnesses and a supplementary record of each injury or illness. These records must be kept up to date and should be available to government representatives.

These records are also used to compile an annual accident report (OSHA 200) that must be posted in a prominent place in the establishment available to the employees. Also the poster entitled "Safety and Health Protection on the Job" shall be posted in a similar manner.

The only employers excluded from this portion of the act are those who are already reporting this material under the Federal Coal Mine Health and Safety Act or the Federal Metal and Nonmetallic Mine Safety Act.

Recordable occupational illnesses and injuries are those that result from a work accident or from exposure to the work environment and lead to fatalities, lost workdays, transfer to another job (temporary or permanent), or termination of employment. Also those cases involving loss of consciousness or restriction of work or motion are recordable.

Reporting at the job site level breaks into six reporting levels as follows:

1. First aid log.
2. First report of injury.
3. Supervisor's accident investigation report.
4. Project accident report.
5. OSHA required Injury or Illness Reports (OSHA 200 and 101).
6. Fatality or major accident report.

The *First aid log* is kept on the job and lists every treatment given. The *First report of injury* is required by the Workmen's Compensation Laws in most states. It is prepared to record every personal injury that requires off-site medical treatment regardless of whether the employee lost time from work or not. The *Supervisor's accident investigation report* is prepared by the foreman for each recordable accident and places special emphasis on identifying methods by which the accidents can be prevented in the future. A typical project accident report form is shown in Figure 17.5. It is a monthly summary of disabling injuries and lost time sent to the home office. The form shown is a report of information on each disabling injury required by OSHA and kept at the job site. Finally, as noted above, any fatality or accident that hospitalizes five or more employees must be reported to the OSHA area director within 48 hours.

17.9 THE SAFETY PROGRAM

A good job site safety program should be founded on:

1. Safety indoctrination of all new personnel arriving on the site.
2. Continuous inspection for possible safety hazards.
3. Regular briefings to increase the safety awareness of personnel at all levels.

If workers or supervisors flagrantly neglect safety rules and regulations, warnings should be issued. If this is not effective, disciplinary action or discharge should be considered.

It is good practice to personally brief each employee arriving on site regarding job procedures. A briefing sheet such as the one shown in Figure 17.6 is an effective aid for conducting this type of briefing. This focuses the worker's attention at the outset on the importance of safety and indicates management's interest in this phase of the job. Safety rules and regulations such as those shown in Figure 17.7 should be available in "handout" form and be conspicuously posted around the job site.

JOB NAME Peachtree Shopping Mall JOB NO. 10-100 LOCATION Atlanta, Georgia MONTH April 1974

THIS REPORT SHOULD BE COMPLETED AND MAILED TO THE SAFETY BRANCH OF THE INDUSTRIAL RELATIONS DEPARTMENT IN THE ATLANTA OFFICE BY THE FIFTH DAY OF THE MONTH.

PROJECT SUPERINTENDENT ____________

THIS FIGURE MAY BE TAKEN FROM PAYROLL RECORDS. IN THE CASE OF FRACTIONS USE THE NEAREST WHOLE NUMBER. DO NOT INCLUDE SUBCONTRACTORS OR OTHERS.

1. AVERAGE NUMBER OF EMPLOYEES ____________

FIGURE ACTUAL HOURS WORKED WHETHER STRAIGHT TIME OR OVERTIME. INCLUDES ONLY THOSE ON OUR PAYROLL.

2. TOTAL HOURS WORKED BY ALL EMPLOYEES ______

RECORD ONLY THOSE INJURIES THAT CAUSE DEATH, PERMANENT DISABILITY (LOSS OF A FINGER, ETC.), OR LOSS OF TIME BEYOND THE DAY ON WHICH THE ACCIDENT OCCURRED. NO MATTER WHAT TIME OF DAY THE INJURY MAY OCCUR, IF THE EMPLOYEE RETURNS TO HIS REGULAR JOB AT THE START OF HIS NEXT REGULAR SHIFT, THE INJURY IS NOT COUNTED. IF HE DOES NOT RETURN AT THAT TIME, IT MUST BE COUNTED AS A DISABLING INJURY.

3. NUMBER OF:
   TEMPORARY DISABLING INJURIES ____________
   PERMANENT DISABLING INJURIES ____________
   DEATHS ____________
   TOTAL DISABLING INJURIES
   FOR THIS MONTH ____________

FOR TEMPORARY INJURIES, COUNT THE ACTUAL CALENDAR DAYS LOST, EXCLUDING THE DAY OF INJURY. IF THE INJURED EMPLOYEE HAS NOT RETURNED BY THE END OF THE MONTH, MAKE AN ESTIMATE OF PROJECTED NUMBER OF LOST DAYS. FOR DEATHS AND PERMANENT INJURIES, USE THE NUMBER OF DAYS SPECIFIED IN THE STANDARD TABLE.

4. NUMBER OF DAYS LOST AS A RESULT OF:
   TEMPORARY DISABLING INJURIES ______________
   TEMPORARY DISABLING INJURIES ______________
   PERMANENT DISABLING INJURIES ______________
   DEATH ______________
   TOTAL DAYS LOST ATTRIBUTABLE TO THIS MONTH ______________

FIGURE 17.5 Project accident report.

August 2, 1973

PEACHTREE SHOPPING MALL
Atlanta, Georgia

Welcome to the job! ABC Construction Company is interested in you and during your employment with us, we will exert every effort to make this a pleasant job, with a good working atmosphere. On the other hand, your skills, ability and performance are most important and essential to the successful completion of the project. To set up and complete a good job, certain rules and regulations must be established. For our mutual benefit, these rules and regulations are as follows:

WORKING RULES AND REGULATIONS

Employment

The Project Manager, or his duly authorized representative, will do all the hiring on the job.

Identification

Employees shall wear a company badge at all times, in full view, above the waist, on an outer garment. Badge numbers will be used in gate clearance, payroll and timekeeping identification.

Hours of Work

The regular workday will begin as per individual instructions, with a lunch period of one-half hour at a designated time. The work week shall be five days, Monday through Friday.

All employees will be at their work locations, ready to start work at work time. All employees are expected to remain at work until the authorized quitting time, at which time they may put up their tools and leave their place of work. Loitering in the change rooms and/or other places during working hours, or late starting of work and early quitting of work will be subject to proper disciplinary measures.

Checking In and Out

Employees are to check in and out at starting and quitting time. Infractions of this rule will be treated with appropriate disciplinary measures. Employees authorized to leave the project during regular working hours must check out with the timekeeper.

Issuing, Care and Use of Tools

Certain company tools will be issued to journeymen and apprentices, or the foreman on a check or receipt system. Tools (while issued) must be properly used and maintained. A toolroom clearance will be required on termination. Loss of or damage to tools will be noted on the employee's record.

A Day's Work

Each employee on the job is expected to perform a full day's work. Your willingness, cooperation and right attitudes will go a long way in accomplishing this objective.

Conduct on the Job

Good conduct on the job is essential to the overall welfare of all employees and the daily progress of the job. Therefore, conduct including, but not limited to, the following violations will be subject to appropriate disciplinary action or discharge.

Theft of company's or employees' property
Recurring tardiness
Leaving company's premises without proper authorization
Possession and/or use of intoxicants and/or narcotics on company's premises
Willful damage to company's materials, tools, and/or equipment
Engaging in horseplay (including shouting to passers-by)
Insubordination
Gambling
Fighting on company's premises
Sleeping on the job
Failure to observe established safety rules and regulations

Housekeeping

Good housekeeping is essential to the safe and efficient construction of the job and is the responsibility of each employee. Work areas, stairways, walkways, and change rooms shall be kept clean at all times.

Safety Rules

Established safety rules and regulations will be observed and followed by all employees in the best interest of accident-free operations.

All unsafe working conditions should be reported to your immediate foreman, who in turn reports it to the company Safety Engineer.

All employees will be required to wear proper clothing above and below the waist. Hard hats must be worn by all employees and visitors while on the construction site.

Pay Period

Wednesday thru Tuesday is the pay period, with pay day on Friday of each week.

Use of First Aid Facilities

First Aid facilities are available at the jobsite and direct contacts have been established with local doctors, hospitals and emergency crews for accidents of a serious nature. All injuries, regardless of severity, must be reported to the employee's supervisor, Field Safety Supervisor and/or First Aid immediately upon occurence. Insurance regulations make this requirement mandatory.

Sanitary Facilities

Adequate sanitary facilities are provided on the jobsite and are to be used by all employees. We request your cooperation in maintaining these facilities in a clean and orderly condition.

Raincoats and Boots

Raincoats and boots are supplied to employees where the conditions of the job being performed require them.

Remaining in Work Areas

Each employee must remain on the jobsite and at his work location at all times during regular working hours, unless authorized to leave by his supervisor.

Absenteeism

Unauthorized absenteeisms will result in termination of employment. An employee who must be absent or late should call 999-9000 and report to timekeeper.

Your cooperation in observing the rules and regulations for the job will show proper consideration for other employees and will be appreciated by the company.

If you agree to and will abide by the above, please sign and return to our Field Safety Supervisor, Charlie Hoarse.

____

cc: Employee File

FIGURE 17.6 Job briefing sheet.

ABC CONTRACTORS AND ENGINEERS

760 SPRING STREET, N. W., ATLANTA, GEORGIA 30308
(404) 999-9000

July 30, 19

Re: OCCUPATIONAL SAFETY & HEALTH ACT 1970 (Construction) (OSHA)

Employers, Owner, Contractors, Sub-Contractors, Superintendents or Foremen in charge shall not direct or permit an employee to work under conditions which are not in compliance with the above code.

Where one Contractor is selected to execute the work of the project he shall assure compliance with the requirements of this code from his employees as well as all Subcontractors.

Every employee shall observe all provisions of the above codes which directly concern or affect his conduct. He shall use the safety devices provided for his personal protection and he shall not tamper with or render ineffective any safety device or safeguard.

1. *Overhead Hazards* — All employees shall be provided with *HARD HATS* and shall use HARD HATS.
2. *Falling Hazards* — Every hole or opening in floors, roofs, platforms, etc., into or through which a person may fall shall be guarded by a barrier sufficient to PREVENT FALLS.
3. *Slipping Hazards* — Scaffolds, platforms or other elevated working surfaces covered with ice, snow, grease or other substances causing slippery footing shall be removed, turned, sanded, etc., to ensure safe footing.
4. *Tripping* — Areas where employees must work shall be kept *reasonably free* from accumulations of dirt, debris, scattered tools, materials and sharp projections.
5. *Projecting Nails* — Projecting nails in boards, planks and timbers shall be *removed*, hammered in or *bent over* in a safe way.
6. *Riding of Hoisting Equipment* — No employee shall ride on or in the load bucket, sling, platform, ball or hook.
7. *Lumber & Nail Fastenings* — Lumber used for temporary structures must be sound. Nails shall be driven full length and shall be of the proper size, length and number. The proper use of double-headed nails is not prohibited.
8. *Guard Rail* or *Safety Rail* — Should be 2 x 4 at a height of 42″ plus a midrail of 1 x 4. The hand rail shall be smooth and free from splinters and

protruding nails. Other material or construction may be used provided the assembly *assures* equivalent *safety.*

9. *Toe Boards* — Shall extend 6″ above platform level and shall be installed *where needed* for the safety of those working below.
10. *Protective Eye Equipment* — Eye protection shall be provided by employers and *shall be used* for cutting, chipping, drilling, cleaning, buffing, grinding, polishing, shaping or surfacing masonry, concrete, brick, metal or similar substances. Also for the use and handling of corrosive substances.
11. *Protective Apparel* — *Waterproof Boots* where required shall have safety insoles unless they are the overshoe type.
    Waterproof Clothing shall be supplied to the employee required to work in the rain.
12. *Safety Belts & Lines* — Shall be *arranged* so that a free fall of no more than 6′ will be allowed.
13. *Stairways* — Temporary stairways shall be not less than *3 feet* in *width* and shall have treads of not less than *2 inch x 10 inch* plank. Must have hand rails. (See #8).
14. *Smoking* — Prohibited in areas used for gasoline dispensing and fueling operations or other *high hazard fire areas.*
15. *Flammable* — *Flammable liquids shall be kept in safety cans* or approved use and storage containers.
    Suitable grounding to prevent the build-up of static charges shall be provided on all flammable liquid transfer systems.
16. *Sanitation* — *Toilet facilities* shall be provided and made available in sufficient number to accomodate all employees.
17. *Drinking Water* — A supply of *clean* and *cool* potable water shall be provided in readily accessible locations on all projects.
18. *Salt Tablets* — Shall be made available at *drinking stations* when required.
19. *Excavations* — Material and other superimposed loads shall be placed at least three feet back from the edge of any excavation and shall be piled or retained so as to prevent them from falling into the excavation. Sides and slopes of excavations shall be stripped of loose rocks or other material. Slopes shall be at an angle of 45 degrees or less (*1 on 1 slope*).
20. *Structural Steel Erection* — When erection connections are made 20% of the bolts in each connection must be drawn up wrench tight. At least *2 bolts* must be used at each end of the member.

    No loads shall be placed on a frame work until the permanent bolting is complete.

    Only employees of the structural steel erector engaged in work directly involved in the steel erection shall be permitted to work under any single story structural steel framework which is not in true alignment and *permanently bolted.*

21. *Use of Ladders* — Ladders shall be provided to give access to floors, stagings or platforms. Ladders shall be maintained in a safe condition at all times. Ladders shall be securely *fastened top* and *bottom* as well as braced where required.

Ladders leading to floors, roofs, stagings or platforms shall extend at least 3 feet above the level of such floors, roofs, stagings or platforms.

22. *Scaffolds* — All scaffolding shall be constructed so as to *support* 4 times the anticipated working load, and shall be braced to prevent lateral movement.

Planks shall overhang their end supports not less than 6″ or more than 12″.

2″ planking may span up to and including 10′. The minimum *width* of any planked platform shall be 18 inches.

Guard rails and *toe rails* shall be provided on the open sides and ends of all scaffold platforms more than 8′ high (See #8).

23. *Rigging, Ropes and Chains* — All rope, chains, sheaves and blocks shall be of sufficient strength, condition, and size to safely raise, lower or *sustain* the imposed load *in any position.*

Wire rope shall be used with power-driven hoisting machinery.

No rope shall be used when visual inspection of the rope shows marked signs of corrosion, misuse or *damage.*

All load hooks shall have *safety clips.*

Loads which tend to swing or turn during hoisting shall be controlled by a *tag line* whenever practicable.

24. *Welding and Cutting* — Oxygen from a cylinder or torch shall never be used for *ventilation.*

Shields or goggles must be worn where applicable. *Cradles* shall be used for lifting or lowering cylinders.

25. *Cranes & Derricks* — All cranes and derricks shall be equipped with a properly operating boom angle *indicator* located within the normal view of the operator.

Every derrick and crane shall be operated by a designated person.

A copy of the *signals in use* shall be posted in a conspicuous place on or near each derrick or crane. Cranes and derricks shall have a *fire extinguisher* attached.

26. *Trucks* — Trucks shall not be *backed* or dumped in places where men are working nor backed into a hazardous location unless guided by a person so stationed on the side where he can see the truck driver and the space in back of the vehicle.

THE ABOVE ITEMS DO NOT ENCOMPASS ALL THE CONSTRUCTION SAFETY REGULATIONS AS THEY PERTAIN TO "OSHA," BUT ARE INTENDED AS A GUIDE TO THE EVER PRESENT HAZARDS AND PRIMARY CAUSES OF ACCIDENTS IN OUR INDUSTRY.

FIGURE 17.7 Job safety rules and regulations.

General safety meetings conducted by the safety engineer should be held at least once a week with supervisors at the foreman and job steward level. A typical report for such a meeting is shown in Figure 17.8. The objective of these meetings is primarily to heighten the safety awareness of supervisors directly in charge of workers. These foreman level personnel in turn should hold at least one "tool box" safety meeting each week to transfer this awareness to workmen and discuss safety conditions with their crew. The report format includes a record of those in attendance, the first aid report, and a description of the safety topics discussed. In addition to the general safety meetings, each job should have a designated safety committee that meets regularly. The members of the safety committee should include key supervisory personnel and craftsmen with an alertness to potential danger and a genuine desire to prevent accidents and injuries. One of the purposes of the safety committee should be to make suggestions as to how to improve overall job safety. Therefore, the members appointed should be sensitive to safety and innovative in devising safe methods.

REVIEW QUESTIONS

17.1 What factors should motivate a contractor to have a safe operation and a good safety program?

17.2 What factors influence the rate assigned to a contractor for workmen's compensation insurance?

17.3 What are two major economic benefits of a good construction program?

17.4 Explain organizing for safety.

17.5 What actions could you as the contractor take to instill a sense of safety among your workers?

17.6 Observe several construction sites and ascertain details of their safety program. If possible attend a tool box safety meeting. Then prepare a list of both good and bad examples of safety practice.

17.7 Using OSHA Regulations as a guide, determine what are the accepted safety standards for:
(a) Guard rails
(b) Exposed reinforcing steel.
(c) Protection of openings.
(d) Man hoists.

17.8 Many construction workers resist the use of safety helmets, goggles, and protective mittens and clothing despite the fact they are designed to protect them. Give several reasons why this practice persists.

ABC CONSTRUCTION COMPANY
Job-10-100
Peachtree Shopping Mall
Atlanta, Georgia
Sept. 1, 19

GENERAL SAFETY MEETING #7

SAFETY SLOGAN FOR THE WEEK:

'BE ALERT, DON'T GET HURT."

C. HOARSE — Safety Supervisor
A. APPLE — Carpenter Foreman

D. DUCK — Surveyor
M. MAUS — Laborer
D. HALPIN — Field Engineer
R. WOODHEAD — Tool Room

SUB-CONTRACTORS PRESENT:

Live Wire Electric
Henry Purcell
James Wallace

THE FIRST AID REPORT FOR AUGUST 15 TO AUGUST 31 WAS GIVEN. THERE WERE:

First Aid	7
Doctor's Cases	0
Lost Time Injuries	0

SHORTCUTS

All of us, supposedly, at one time or another, have been exposed to possible injury by short cutting when a few extra steps would have meant the safe way. We did so as kids when we jumped the fence instead of using the gate and we do so as men when we cross streets by jaywalking instead of using the intersection. Accident statistics plainly indicate the fact that people disregard the fact that minor safety violations may have very serious results.

FIGURE 17.8 Safety meeting minutes.

In construction work, short cutting can be deadly. All of us know of cases in which this kind of thoughtless act resulted in a serious injury. For instance, an iron worker tried to cross an opening by swinging on reinforcing rods, his hands slipped and he fell about 20 feet to a concrete floor. If he had bothered to take a few moments to walk around the building, he would still be tying rods.

The safe way is not always the shortest way and choosing the safe way is your *Personal Responsibility*. When you are told to go to work in a particular area, you are expected to take the safe route; not an unsafe short route. We can not be your guardian angel, that is one thing you will have to do for yourself.

If you are told to go to work in some place that has no safe access, report this fact to your Foreman so that necessary means of access can be provided.

Ladders and scaffolds are provided for high work, use them. Even though a high job may take only a few moments, DO NOT CLIMB ON FALSE WORK, or on some improvised platform.

Your first responsibility is to yourself. Remember that ladders, steps and walk ways have been built to save you trouble and to save your neck, too. Use them always.

Gambling a few minutes and a little energy against a possible lifetime of pain and misery is a poor bet.

GENERAL DISCUSSION:

FLAGMEN MUST CONTROL ALL THE BACK-UP OPERATIONS ON THIS JOB.

TRAFFIC- BE ON THE ALERT FOR MOVING VEHICLES, OUR AREA IS SLIPPERY. *DON'T* WALK BESIDE MOVING EQUIPMENT.

INJURIES- REPORT ALL INJURIES TO YOUR FOREMAN IMMEDIATELY.

____

C. Hoarse, Safety Supervisor

CHAPTER EIGHTEEN

The Construction Management Approach

18.1 CONSTRUCTION MANAGEMENT CONTRACTS

Although there is a wide spectrum of opinion regarding what construction management is, most professionals and firms operating as professional construction managers are providing management services carried out during the predesign, design, and construction phases. These services provide control of time and cost in the construction of a new facility. The professional construction manager, then, is the individual or firm who binds himself to an owner in a professional arrangement and applies the proper combination of management tools to a construction project to achieve time and cost control.

In construction management type contracts, one firm is retained to coordinate all activities from concept design through acceptance of the facility. The firm represents the owner in all *construction management* activities. In this type of contract, construction management is defined as:

> . . . that group of management activities related to a construction program, carried out during the pre-design, design, and construction phases, that contributes to the control of time and cost in the construction of a new facility.*

The construction management firm's position in the classical relationship linking owner, contractor, and architect engineer is as shown in Figure 18.1.

This firm has the function of a traffic cop or enforcer, controlling the flows of information among all parties active on the project. The

*George T. Heery, "Construction Management Defined," *The Military Engineer*, No.430, March-April 1974, p.85.

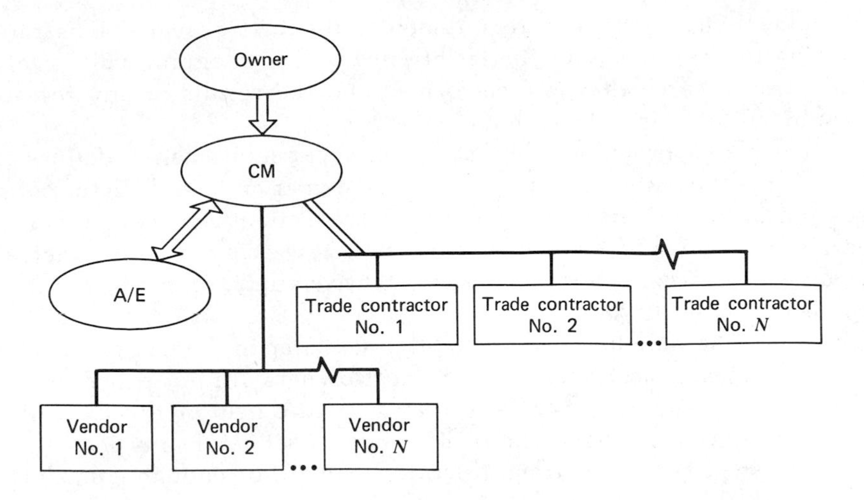

FIGURE 18.1 Construction management relationships to other principal parties on the project.

CM establishes the procedures for award of all contracts to architect/ engineers, principal vendors, and the so-called trade contractors. Once contractual relationships are established, the CM controls not only the prime or major contractor but all subcontractors as well as major vendors and off-site fabricators. Major and minor contractors on the site are referred to as trade contractors. In this control or management function, the CM firm utilizes the project schedule as a road map or flight plan to keep things moving forward in a timely and cost effective manner. The major functions carried out by the CM firm vary depending on whether the project is in the (1) predesign, (2) design, or (3) construction phase.

18.2 CM FUNCTIONS DURING PREDESIGN PHASE

During the predesign phase, the CM is occupied with what is referred to as *programming*. Programming consists of the development of the owner's facility requirements. The requirements may be wide and varied, and so may the program. The program may be used for funding projections, site selection, design, and so on. Programming may also include research and development of design criteria.

In some cases, an owner will get an idea of his needs from an in-house staff. However, if the owner's construction requirements are only periodic, this leads to the maintenance of unneeded staff space and equipment. The CM works closely with the owner in determining his actual needs and either supplements existing staff or preempts the need to maintain permanent staff.

Along with listing facility requirements, the CM provides a narrative describing the philosophy of the facility needs. The program will normally include any research that was performed and the results of any research shown in a design-oriented fashion.

The completed program fully lists all project requirements, defines the project philosophy, and attempts to give the owner an overall picture of his needs and an analysis showing trade-offs and reductions. This comprehensive program goes as far as is possible at this time toward giving the owner cost control. This is accomplished through extensive study, research, and interviews with owner personnel.

Once the programming is complete, the next step in the CM process is to set a reasonable project budget. Since most owners are not experienced in budgeting and estimating facilities, they need the help of the CM. Many projects can have their maximum realistic budget set before any design work begins. Most projects are straightforward enough, or enough comparative data are available, so that a budget may be set. These projects include most schools, office buildings, hospitals, apartment buildings, and many others.

Many projects have their budgets set by economic conditions such as public bonds, loan availability, and economic requirements. An owner of a new apartment complex must know what rents he can get in his area before he designs. If he overdesigns an apartment by using a high budget, he soon finds that no one will rent his apartment. This is an economic requirement. The CM provides help to the owner in this area by determining the project's income based on applicable statistical data. These income projections are based on the rentable floor area, rentable dwelling units, number of parking places, and so on, depending on the type of building. With the total number of income-generating units, the CM and owner develop the maximum realistic budget, depending on the availability of money from the funding source.

Once the project budget is established, it must be broken down into categories such as the award prices for the various contracts (if separate contracts are being used), budget for the design and engineering, budget for the construction management services, and budget for miscellaneous expenses and factors to allow the project budget to prepare for in-progress contingencies and cost index adjustments.

Once the total project budget is broken down, the owner sees exactly what he is getting for his money. It may then be determined that his desired program is too large or too small for the budget. This is when the owner turns to the CM for help. The CM may provide answers such as using lower-cost construction materials that bring the budget into line while still fulfilling the owner's needs. Another answer he might find may be the use of trade-offs, changing from one system to another, or reducing other elements. By careful cooperation, the CM and the owner reach a budget that fits the owner's pocket and a program that fits his needs.

18.3 SELECTION OF DESIGNER

The next step during the predesign phase of the project is the selection of the design firm. The owner may have already commissioned an architect before the CM is selected. It is more desirable to have the CM participate in this decision. If the CM is brought on board after the architect has been commissioned, there is no guarantee that the two parties can work together with a minimum of conflict. However, if the CM has a say in who is commissioned to design the project, he will chose a firm with which he can work and that is reputable.

The first step in the selection of the design firm is the preparation of lists of suitable firms. In some cases, the CM is associated with a design firm and will contract with this in-house firm. This is a very desirable situation, since the CM has an excellent working base and is familiar with the people and procedures of the architect/engineer. If the in-house firm is not desirable as a designer, the list of prospective architect/engineers should start with firms that have expertise in a certain type of project. Some A/E firms are particularly experienced with schools, others with sports complexes, and others with shopping centers. In some cases, the CM is limited to a local architect because of local political constraints. This is most frequent with government work.

Once a list of suitable firms is worked up, the CM coordinates interviews between the owner and representatives of the design firms. A briefing of the owner before each interview is desirable. The information presented can include the size of the firm, their professional reputation, and their experience in work of this particular nature. It should be the job of the CM to take an active part in leading the interview, since the owner may not be familiar with conducting meetings of this type.

At the end of all interviews, the CM prepares his recommendations to be presented to the owner. He states which firm he recommends and presents the reasons for his recommendation. If none of the firms satisfy the CM, he might recommend that all firms be rejected and a new list be drawn up.

When the owner agrees to a firm, it is the job of the CM to prepare the contract documents. He includes in the contract, whenever possible, the agreed-on program, the project budget, and a project time schedule. These are needed to ensure the time and cost control of the project and also to give the CM a certain hold on the architect. If a schedule is not included, the architect could delay the project considerably by producing tardy or poor designs and by continually running over the budget. This could lead to the owner continually rejecting the drawings and could serve to delay the project. The only other hold the owner has over the architect is to withhold his pay requests. This action will serve to create ill feelings between the owner and architect and can be disastrous to the continuity of the project. Therefore, it is up to the CM to see that the project is continued with a mutual feeling of cooperation.

18.4 PREDESIGN PROJECT ANALYSIS

The last, and probably most important step of the predesign phase is predesign project analysis. It is during this period that the design concept (i.e., the general direction in which the design should proceed) is developed. This period is considered to be part of the predesign function, since it should normally be completed before any actual design or engineering work takes place. During predesign project analysis, a team approach is used. This team consists of the project architect, construction manager, the architectural designer(s), the structural, mechanical, electrical, and civil engineers, the landscape planner, the CM's estimator and the owner or the owner's representative.

Initially, the construction manager leads team discussions to make sure the sessions don't turn into committee meetings or simple briefings. This would destroy the team analysis approach. The construction manager attempts to organize the sessions into a type of "brainstorming" session with all members of the team injecting their opinions and points of view.

Once the sessions are on course and become productive, the CM should turn leadership of the sessions over to the project architect. He should direct the discussion into areas such as budgeting, design concepts, systems utilization, and scheduling. It may be initially difficult making these sessions truly productive, since 10 to 12 people are working with each other for the first time and they may be reluctant to present ideas subject to criticism by others. It is the job of the CM to tear down any personality barriers and create a harmonious working atmosphere, where ideas can be freely discussed, reviewed, and criticized.

The format of the sessions is unstructured, allowing for the development of new concepts as they relate to the subject under consideration. If discussion centers on foundations and bay size, structural systems should also be discussed, since all systems are based on certain bay sizes. Discussion of interfacing systems should not be delayed, since decisions on one system impact decisions regarding others.

The period of time set aside for these sessions should be two or three days. If a week or longer is taken, the sessions tend to become unproductive. These sessions are not intended to yield a detail plan for the entire project, but instead to give all persons involved in the project the same planning base. The following areas should be developed during predesign project analysis:

1. Design concept development.
2. Elimination of design blind alleys.
3. Identification of constraints.
4. Systems utilization analysis.
5. Systems development potentials analysis—in the event available industrialized building systems are not applicable.

6. Owner evaluation.
7. Construction management plan.
8. Critical date schedule.
9. Budget analysis.
10. Tentative selection of major engineering systems.

Depending on the size and scope of the project, there is a point of diminishing return in predesign project analysis. If the project is small, the analysis might progress as far as selecting the type of structural frame, bay size, and sidewall construction. If the project is very large (e.g., a high-rise apartment building-office complex), the project analysis should not proceed past establishing the basic design direction.

When identifying constraints, care should be taken to ensure that identified constraints actually apply to the project being considered. Constraints that should be reviewed are:

Structural frame fabrication, delivery, and erection.

Site availability and zoning.

Utility, street, and site access work.

Electric switch gear delivery and other major electric equipment.

Labor contract terminations.

Major mechanical machinery delivery.

Elevator installation.

Laboratory equipment delivery.

Furniture delivery.

Special equipment.

The construction management plan is the flight plan that the construction manager sets for his use throughout the remainder of the project. He attempts to develop a plan that will be most suitable for the owner. His plan should include:

1. *Bidding strategy.* This consideration guides selection of the type of contract to be used for construction. If the owner is a developer who would be in the market for repeat business from a contractor, he might do well to negotiate a contract. However, if the owner is in a business where he would not expect any more dealings with a contractor, he would consider contracting in a competitively bid format.
2. *Plans for early bidding or negotiations.* Depending on analysis of the project constraints, it might be in the owner's best interest to bid portions of the work early. Staggering of the bid packages to get construction work going as early as possible may be desirable.
3. *Plans to transfer contracts.* It has been found through experience that in

order to effectively monitor cost and time, any separate contract still incomplete at the time of awarding the general contract should be transferred to the general contract.

At the conclusion of the predesign project analysis, all operating parties should have a clear understanding of project goals that will enable them to proceed faster and more productively. The team should develop a feeling of mutual respect and cooperation. The most important accomplishment of the analysis is, however, the formation and initiation of a team that is capable of realizing project goals and remaining within budget and schedule.

18.5 CM ACTIVITIES DURING THE DESIGN PHASE

From the time the designer and the CM are satisfied with the predesign project analysis, the CM is continually feeding the designer with input. This will continue until the final design work and specifications are complete.

During this period, the designer (architect) completes a series of phases as defined in his contract. Usually these phases are (1) the schematic design phase, (2) the design development phase, and (3) the contract document preparation phase. A description of these phases plus the architect's responsibilities to the owner may be found on AIA Document B-131 (see Appendix G). These are the standard phases for the design, which may vary in duration based on the specific project.

Design is of primary importance to the CM. He must be able to advise the architect and make professional suggestions in areas of cost, methods of construction, systems, and scheduling. The design of the project is the "personification" of the project program set up during the predesign phase. Because of this, the design has a direct relationship between both time and cost. Therefore, design becomes the basic tool available for cost control.

The responsibility for design is designated to the architect, where it properly belongs. The extent to which the CM should advise the architect depends on what type of project the owner wants, an aesthetic monument, or an economical functioning building. The CM's responsibility is based on the owner's desire to control time and cost.

It is important for the CM to understand how early in the design certain fundamentals such as the structural system are established. He should be prepared to study trade-offs and advise the architect and the owner regarding the cost and scheduling impacts of selecting various systems. If the building is a one-story school building and the engineer has designed it using conventional structural steel, the construction manager should be well prepared to show the time, cost, and functional differences using a framing system, a precast system, a cast-in-place system, or some combination.

Throughout the design phase, the estimator must work closely with both the CM and the architect. Time spent with the architect should range from full-time on large projects to two or three days a week for smaller projects.

The secret to effective cost control lies with the estimator. The estimate is only as good as the individual who makes it, and the estimator must know how to use the available tools properly. The estimator's first task is to compile and examine the continuously changing cost data that apply to the project. He should know where to find this information, how to analyze it with respect to his project, and how to compile it in such a manner that its use will be extended to other projects. The careful presentation of comparative data is extremely important to the initial project budget.

Once all comparative data and any other predesign information, such as the program for the project, are assembled, the estimator makes a preliminary project budget without use of any designs. The budget would be broken down into components such as sitework, masonry, and HVAC. This is the initial project budget on which the predesign project analysis is based.

The estimator's next major function occurs during the predesign project analysis. He presents a more detailed budget, breaking it down by components. He should also be available for advice on construction sequencing and be knowledgeable in areas such as labor wages, local unions, projected work stoppages, and wage increases. As soon as the design process is underway, the estimator further breaks his components down until, at the end of the design, the budget is subdivided into unit prices and small components (e.g., footing steel and interior finishes, etc.). The first major breakdown comes during the schematic design. Here the sitework may be broken into earthmoving, drainage, and landscaping. As the design progresses, other major areas are detailed. The final components are broken into unit prices. This method of development adds to the flexibility of cost control, since the impact of design changes can be reflected in the budget on a unit price basis.

Along with budget preparation, the estimator provides the CM and the architect with three additional services:

1. He provides a list of areas of potential cost savings. These may be worked up with the help of the CM and should include results of take-off studies and evaluations.
2. He develops a list of areas of potential cost problems. These might include items that are in short supply and items with unstabilized prices, such as steel.
3. He establishes budget targets for major component systems of the building, such as electrical and structural frame.

As each estimate is worked up and revised, it should be brought to the attention of the owner for approval. The purpose of the approvals is not to let the owner know how much the project will cost him now, but hopefully to reassure him that costs are being controlled.

The continual process of checking and adjusting the budget is not to change the final budget with each change in design, but instead to check the design, bringing it in line with the budget rather than the budget in line with the design.

Throughout the design phase, the CM must be continually aware of the progress of the design, present level of the budget, and current design philosophy. He updates the owner regarding these facts on a periodic basis (e.g., monthly, weekly). If a change is made, the CM should review the technical and economic aspects of the change and make his recommendation to the owner. Conversely, he must take the wishes of the owner and make recommendations to the architects. It is also during the design phase that the project schedule should be finalized and a network scheduling consultant should be retained.

18.6 PREPARATION OF CONTRACT DOCUMENTS

The contract for each individual project should be tailored to fit the needs of the project. This is done by having a set of general conditions that may be applicable to any project and modifying them with a set of supplementary conditions that suit the needs of a particular project.

The time control contract provisions are generally of four types:

1. Completion dates.
2. Liquidated damages.
3. Grants of contract time extension.
4. Progress payments.

The master network schedule establishes a projected completion date. This schedule should include the bid date, notice to proceed date, access to site date (if required), any equipment delivery dates if purchased under separate contract, and dates for substantial and final completion. The substantial completion date establishes the date at which the project may be fully occupied by the owner except for any corrective work under punch lists and any final cleanup. The contract should also specify provisions that may be taken if these dates are not met. One such provision normally requires that the contractor work overtime to complete the work.

Liquidated damages have been discussed previously in Chapter 3: if the liquidated damages are written properly and the amounts are substantial, a competent contractor will do all he can to comply with the project completion dates in order to avoid being assessed liquidated damages. When a contractor submits a request for a time extension, the CM should:

1. Acknowledge the receipt of the request.

2. If it is immediately known that the request will be rejected, reject it immediately by written communication to the contractor.

3. If an immediate decision cannot be made, so inform the contractor and answer him as soon as a decision is reached. All time extensions should be carefully reviewed and analyzed as to how they affect the critical path. Owner approval should be given on all time extensions.

Progress payments are the simplest and yet the most effective way of controlling time. Dates are stipulated in the contract as to when the contractor may be paid. The last date is usually within 30 days of final completion. If the contractor is running late, he must realize that no payments will be made past the last progress payment, until the project has reached final completion. Progress payments can be thought of as a complement to liquidated damages and other time-controlling contract provisions.

18.7 CONSTRUCTION MANAGEMENT DURING THE BIDDING PROCESS

Once the plans and specifications are complete and contract documents are prepared, the work is ready to be bid. This is a crucial step, since the cost and time controls are wasted if a low bidder is well above the estimated contract price. Careful planning must be utilized to ensure competitive or well-negotiated bids. Whether to negotiate or competitively bid a contract is normally determined during the predesign project analysis in the CM's plan.

To ensure that competition is keen, the CM must have an up-to-date listing of all contractors who might be qualified to bid the job. These contractors should be of an appropriate size, be capable of efficiently mobilizing into the area of the job, and have experience on similar projects. It is helpful for the contractor to have qualifications such as a good industry-wide reputation, a good credit rating (e.g., Dun and Bradstreet), and a good record of consistency in bringing projects in on time under budget. All of these qualities are helpful in obtaining a competitive bid and good craftsmanship.

After a list of suitable bidders has been compiled, each one should be contacted to make sure they are interested. They should be informed of the date of issuance of contract documents and the tentative bid date. As the time to issue documents draws near, they should again be contacted to confirm their interest.

During the two weeks preceding issuance of contract documents, notices to bidders should be mailed. Notices should also be given to construction reporting services and construction periodicals such as *Engineering News Record*. For all public projects, it is required that projects be properly advertised. Prequalification is the only limiting factor on public projects.

About a week into the bid period, after the contractors have had the opportunity to review the bid documents, a prebid conference should be held. All contractors should be encouraged to attend along with the architect, engineer, construction manager, scheduling consultant, and owner's representative. There are four major purposes of the prebid conference. They are:

1. To evaluate the level of competitive interests. If a prebid conference draws only two bidders, there is a need to create more interest.
2. To brief prospective bidders and other interested parties, such as subcontractors and suppliers, on the project schedule.
3. To brief prospective bidders on project scheduling fundamentals in general, and to make appointments for each bidder with the network (CPM)* consultant for private consultation and provisional preliminary network modification prior to the date of bid receipt.
4. To bring out questions regarding the contract documents early in the bid phase, in order to avoid extensive last-minute changes in the form of addenda.

At the completion of the prebid conference, the people in attendance should be allowed a question-and-answer session to clear up any questions arising from the drawings and documents.

As the bid date approaches, final estimates should be made and compared to original estimates. Unless the project is publicly funded, the formality of simultaneous bid opening can be dispensed with and comparisons made immediately between each bid and final owner's estimate. If all bids submitted are above the estimate, a decision may be made to negotiate with the lowest, or some other, bidder. Once a good bid or best available bid is accepted, all bidders should be notified and the award presented.

18.8 CM FUNCTIONS DURING THE CONSTRUCTION PHASE

The construction phase of the project begins when the operations of the CM are transferred from an office to an on-site location — usually in a job trailer. This normally occurs concurrently with the movement of the first contractor onto the site to commence site preparation. The specifications, drawings, and a final budget define the flight plan in terms of which the CM must ensure project completion. His primary function is to control time, cost, and quality for the owner. The CM must perform the services of an inspector, expeditor, quality-control man, arbitrator, and any other job that may be necessary to make sure the project runs smoothly and efficiently.

*See Chapters Fifteen and Sixteen for a discussion of Critical Path Method (CPM).

One of the primary tasks carried out by the CM in the field is the maintenance of all necessary paperwork. Some of the more important types of documents that the field team must supervise or generate are discussed below.

Daily Field Log

This is a historical document filled out for each day the contractor performs work on the site. Data on the reports must include the date, weather conditions, amount of equipment in use for the day, number of people working, and major areas of work performed. A "remarks" section should also be included to specify any verbal instructions given by the architect, any problems arising from the day's work, and anything else along these lines that should be documented. The purpose of the daily log is two-fold:

1. It keeps the owner and all interested parties aware of the status of the project.

2. It documents events on a daily basis. If a contractor wants a time extension because he says he was rained out for two days, all that is needed is to check the daily logs for those two days.

Field Orders

A field order is a written authorization from the architect to the contractor to change certain item(s) in either the specifications or the drawings, when the changes involved do not involve major design changes or any change in contract price. The field order is processed from the architect to the contractor through the CM who reviews the change. If the change is one that would not require the authorization of the owner, the CM logs the field order and then passes it on to the contractor who effects the change. A field order is usually a written authorization confirming verbal instructions.

Change Order

A change order is a written authorization from both the architect and the owner to the contractor that involves a change in design or contract price or both. In order to expedite the processing of a change order, the entire process should be handled through the construction manager. If any party (architect, owner. contractor, or CM) wishes to initiate a change order, the CM should be notified. He then proceeds to issue a proposal to the contractor for the work involved. When a reply is received on the proposal, the CM reviews the proposal, makes a recommendation to the owner for action, and forwards both the proposal and recommendation to the owner for review. If the owner agrees with the proposal, he notifies the architect to issue a change order that all parties sign. Once again careful documentation is needed on both the proposal and the change order. The CM must also watch the time to make sure the process does not endanger the project schedule.

Other documents that must be maintained include logs and copies of transmittals, shop drawings, change orders, test reports, concrete pours, and equipment deliveries. The point to stress in making any kind of report and keeping any kind of log is to take note of all dates, decisions, verbal agreements, and anything else that may come back to haunt the owner if it is not properly documented.

During the construction phase, in addition to the documentation function, the CM must coordinate all trades on the job. This also means coordinating the efforts of the architect/engineer, as well as prime and trade contractors into a working team. If a contractor has a question, the CM must develop the answer as quickly as possible. This enables the contractor to continue his work. If two contractors on the site have a coordination problem, the CM must become the arbitrator in resolving the problem.

The CM should avoid, if at all possible, becoming involved in interpreting the drawings. He must consult the architect or engineer, depending on which would be applicable. The architect/engineer, and not the CM, is solely responsible to the owner for the correct application of the drawings.

Coordination of trades also involves keeping all superintendents on the job apprised of their progress as it pertains to the master schedule. This should take place about once a week through the vehicle of a weekly construction progress meeting. At these weekly project meetings, the progress of the project with reference to the schedule should be evaluated. If one trade is holding up another, this delay should be brought up and reconciled. If the entire project is behind, the reasons should be discussed. Careful criticism should be given where it is needed. However, the cooperation of everyone is needed to make the project a success. These meetings also provide a good forum for discussing any questions from the superintendents that require action. The theme of these meetings should be communication. The need for full communication from all parties cannot be overemphasized.

The last major function of the CM during construction is monitoring the project cost analysis. Most large projects will have a printout with daily cost figures that are developed in conjunction with the network schedule. This enables the CM to keep a careful eye on project cash flows and detect at an early date any budget overruns or underruns.

The job of the construction manager does not conclude with the final completion date of the project, but continues through tenant move-in and into the warranty period. The CM is involved in writing up instruction manuals for the use of the owner as well as presentation of any warranties for installed equipment. If during the warranty period of the building the roof leaks, the owner will contact the CM who will contact the proper party to have it repaired or replaced.

REVIEW QUESTIONS AND EXERCISES

18.1 Using the section breakout of this chapter as a guide, identify the different stages of the CM's involvement in a project. Hence develop for each stage a listing of the major functions performed by the CM.

18.2 The users of both the construction management and design-build approaches to project management claim significant reductions in the overall duration of a project and increased reliability in meeting the project schedules over that pertaining to the traditional approach. Suggest a number of reasons why this claim may be true. Identify the major reasons why the traditional construction approach may inherently take longer to complete the same construction project.

18.3 Assuming the following idealized construction work durations:

(a) Site preparation	1 time unit
(b) Foundations	2 time units
(c) Erection of structure	3 time units
(d) Services	3 time units
(e) Cladding	2 time units
(f) Interior finishing	4 time units
(g) Landscaping	1 time unit

prepare simple time schedules based on the following strategies:

(i) Purely sequential construction effort.

(ii) Simultaneous activity in (a) and (b), (d) and (e), and slight lagging of activities (d) and (e) from (c), and (f) from (d) and (e).

18.4 Develop a schematic illustration of the construction management approach (in a focus similar to that of Figure 1.2), and compare with that obtained in problem 1.4 for the design-build approach and that of Figure 1.2 for the traditional approach to the building process.

18.5 Conduct an interview with a local contractor of your selection and prepare a report of approximately 8 to 10 pages concerning the contractor's response on questions in the following areas:

(a) Organizational Considerations—Structure of firm both legal and functional (i.e., organization of functional responsibilities).
(b) Type of Work (heavy, building, sanitary, etc.).
(c) Type of Contracts (fixed price, negotiated, etc.).
(d) Source of jobs.
(e) Maximum job size.
(f) Financial Procedures (how is work financed during construction, profit margins or philosophy)
(g) Bonding limit.
(h) Equipment costing and charging.
(i) Estimating techniques.

(j) Scheduling techniques.
(k) Percent of work subcontracted.
(l) Size of staff.
(m) Labor relations.
(n) OSHA (safety) experience—safety program.
(o) Permanent work force.
(p) Inventory of materials maintained.
(q) Quality control techniques.

Appendices

APPENDIX A: General Conditions: The American Institute of Architects (AIA)*

APPENDIX B: Typical Considerations Affecting the Decision to Bid

APPENDIX C: Performance and Payment Bond Forms*

APPENDIX D: AIA Document A101 Standard Form of Agreement between Owner and Contractor—Stipulated Sum*

APPENDIX E: AIA Document A111 Standard Form of Agreement between Owner and Contractor—Cost of Work plus a Fee*

APPENDIX F: AGC Sample Specification: Implementation and Use of Network Scheduling Methods†

APPENDIX G: AIA Document B141 Standard Form of Agreement between Owner and Architect*

APPENDIX H: AGC Builder's Association of Chicago: Typical Job Descriptions**

APPENDIX I: AGC Standard Subcontract Agreement†

APPENDIX J: Interest Tables

APPENDIX K: AGC-CSI List of Cost Accounts†

APPENDIX L: Plans for Small Gas Station

APPENDIX M: Site Reconnaissance Checklist

*Reproduced with the permission of the American Institute of Architects under application number 80015. Further reproduction, in part or in whole, is prohibited. Because AIA Documents are revised from time to time, users should ascertain from the AIA the current edition(s) of the Document(s) reproduced herein.

**Builder's Association of Chicago, Inc.

†Reproduced with the permission of the Associated General Contractors of America.

APPENDIX A

General Conditions: The American Institute of Architects (AIA)

THE AMERICAN INSTITUTE OF ARCHITECTS

AIA Document A201

General Conditions of the Contract for Construction

THIS DOCUMENT HAS IMPORTANT LEGAL CONSEQUENCES; CONSULTATION WITH AN ATTORNEY IS ENCOURAGED WITH RESPECT TO ITS MODIFICATION

1976 EDITION

TABLE OF ARTICLES

This document has been approved and endorsed by The Associated General Contractors of America.

INDEX

GENERAL CONDITIONS OF THE CONTRACT FOR CONSTRUCTION

ARTICLE 1

CONTRACT DOCUMENTS

1.1 DEFINITIONS

1.1.1 THE CONTRACT DOCUMENTS

The Contract Documents consist of the Owner-Contractor Agreement, the Conditions of the Contract (General, Supplementary and other Conditions), the Drawings, the Specifications, and all Addenda issued prior to and all Modifications issued after execution of the Contract. A Modification is (1) a written amendment to the Contract signed by both parties, (2) a Change Order, (3) a written interpretation issued by the Architect pursuant to Subparagraph 2.2.8, or (4) a written order for a minor change in the Work issued by the Architect pursuant to Paragraph 12.3. The Contract Documents do not include Bidding Documents such as the Advertisement or Invitation to Bid, the Instructions to Bidders, sample forms, the Contractor's Bid or portions of Addenda relating to any of these, or any other documents, unless specifically enumerated in the Owner-Contractor Agreement.

1.1.2 THE CONTRACT

The Contract Documents form the Contract for Construction. This Contract represents the entire and integrated agreement between the parties hereto and supersedes all prior negotiations, representations, or agreements, either written or oral. The Contract may be amended or modified only by a Modification as defined in Subparagraph 1.1.1. The Contract Documents shall not be construed to create any contractual relationship of any kind between the Architect and the Contractor, but the Architect shall be entitled to performance of obligations intended for his benefit, and to enforcement thereof. Nothing contained in the Contract Documents shall create any contractual relationship between the Owner or the Architect and any Subcontractor or Sub-subcontractor.

1.1.3 THE WORK

The Work comprises the completed construction required by the Contract Documents and includes all labor necessary to produce such construction, and all materials and equipment incorporated or to be incorporated in such construction.

1.1.4 THE PROJECT

The Project is the total construction of which the Work performed under the Contract Documents may be the whole or a part.

1.2 EXECUTION, CORRELATION AND INTENT

1.2.1 The Contract Documents shall be signed in not less than triplicate by the Owner and Contractor. If either the Owner or the Contractor or both do not sign the Conditions of the Contract, Drawings, Specifications, or any of the other Contract Documents, the Architect shall identify such Documents.

1.2.2 By executing the Contract, the Contractor represents that he has visited the site, familiarized himself with the local conditions under which the Work is to be performed, and correlated his observations with the requirements of the Contract Documents.

1.2.3 The intent of the Contract Documents is to include all items necessary for the proper execution and completion of the Work. The Contract Documents are complementary, and what is required by any one shall be as binding as if required by all. Work not covered in the Contract Documents will not be required unless it is consistent therewith and is reasonably inferable therefrom as being necessary to produce the intended results. Words and abbreviations which have well-known technical or trade meanings are used in the Contract Documents in accordance with such recognized meanings.

1.2.4 The organization of the Specifications into divisions, sections and articles, and the arrangement of Drawings shall not control the Contractor in dividing the Work among Subcontractors or in establishing the extent of Work to be performed by any trade.

1.3 OWNERSHIP AND USE OF DOCUMENTS

1.3.1 All Drawings, Specifications and copies thereof furnished by the Architect are and shall remain his property. They are to be used only with respect to this Project and are not to be used on any other project. With the exception of one contract set for each party to the Contract, such documents are to be returned or suitably accounted for to the Architect on request at the completion of the Work. Submission or distribution to meet official regulatory requirements or for other purposes in connection with the Project is not to be construed as publication in derogation of the Architect's common law copyright or other reserved rights.

ARTICLE 2

ARCHITECT

2.1 DEFINITION

2.1.1 The Architect is the person lawfully licensed to practice architecture, or an entity lawfully practicing architecture identified as such in the Owner-Contractor Agreement, and is referred to throughout the Contract Documents as if singular in number and masculine in gender. The term Architect means the Architect or his authorized representative.

2.2 ADMINISTRATION OF THE CONTRACT

2.2.1 The Architect will provide administration of the Contract as hereinafter described.

2.2.2 The Architect will be the Owner's representative during construction and until final payment is due. The Architect will advise and consult with the Owner. The Owner's instructions to the Contractor shall be forwarded through the Architect. The Architect will have authority to act on behalf of the Owner only to the extent provided in the Contract Documents, unless otherwise modified by written instrument in accordance with Subparagraph 2.2.18.

2.2.3 The Architect will visit the site at intervals appropriate to the stage of construction to familiarize himself generally with the progress and quality of the Work and to determine in general if the Work is proceeding in accordance with the Contract Documents. However, the Architect will not be required to make exhaustive or continuous on-site inspections to check the quality or quantity of the Work. On the basis of his on-site observations as an architect, he will keep the Owner informed of the progress of the Work, and will endeavor to guard the Owner against defects and deficiencies in the Work of the Contractor.

2.2.4 The Architect will not be responsible for and will not have control or charge of construction means, methods, techniques, sequences or procedures, or for safety precautions and programs in connection with the Work, and he will not be responsible for the Contractor's failure to carry out the Work in accordance with the Contract Documents. The Architect will not be responsible for or have control or charge over the acts or omissions of the Contractor, Subcontractors, or any of their agents or employees, or any other persons performing any of the Work.

2.2.5 The Architect shall at all times have access to the Work wherever it is in preparation and progress. The Contractor shall provide facilities for such access so the Architect may perform his functions under the Contract Documents.

2.2.6 Based on the Architect's observations and an evaluation of the Contractor's Applications for Payment, the Architect will determine the amounts owing to the Contractor and will issue Certificates for Payment in such amounts, as provided in Paragraph 9.4.

2.2.7 The Architect will be the interpreter of the requirements of the Contract Documents and the judge of the performance thereunder by both the Owner and Contractor.

2.2.8 The Architect will render interpretations necessary for the proper execution or progress of the Work, with reasonable promptness and in accordance with any time limit agreed upon. Either party to the Contract may make written request to the Architect for such interpretations.

2.2.9 Claims, disputes and other matters in question between the Contractor and the Owner relating to the execution or progress of the Work or the interpretation of the Contract Documents shall be referred initially to the Architect for decision which he will render in writing within a reasonable time.

2.2.10 All interpretations and decisions of the Architect shall be consistent with the intent of and reasonably inferable from the Contract Documents and will be in writing or in the form of drawings. In his capacity as interpreter and judge, he will endeavor to secure faithful performance by both the Owner and the Contractor, will not show partiality to either, and will not be liable for the result of any interpretation or decision rendered in good faith in such capacity.

2.2.11 The Architect's decisions in matters relating to artistic effect will be final if consistent with the intent of the Contract Documents.

2.2.12 Any claim, dispute or other matter in question between the Contractor and the Owner referred to the Architect, except those relating to artistic effect as provided in Subparagraph 2.2.11 and except those which have been waived by the making or acceptance of final payment as provided in Subparagraphs 9.9.4 and 9.9.5, shall be subject to arbitration upon the written demand of either party. However, no demand for arbitration of any such claim, dispute or other matter may be made until the earlier of (1) the date on which the Architect has rendered a written decision, or (2) the tenth day after the parties have presented their evidence to the Architect or have been given a reasonable opportunity to do so, if the Architect has not rendered his written decision by that date. When such a written decision of the Architect states (1) that the decision is final but subject to appeal, and (2) that any demand for arbitration of a claim, dispute or other matter covered by such decision must be made within thirty days after the date on which the party making the demand receives the written decision, failure to demand arbitration within said thirty days' period will result in the Architect's decision becoming final and binding upon the Owner and the Contractor. If the Architect renders a decision after arbitration proceedings have been initiated, such decision may be entered as evidence but will not supersede any arbitration proceedings unless the decision is acceptable to all parties concerned.

2.2.13 The Architect will have authority to reject Work which does not conform to the Contract Documents. Whenever, in his opinion, he considers it necessary or advisable for the implementation of the intent of the Contract Documents, he will have authority to require special inspection or testing of the Work in accordance with Subparagraph 7.7.2 whether or not such Work be then fabricated, installed or completed. However, neither the Architect's authority to act under this Subparagraph 2.2.13, nor any decision made by him in good faith either to exercise or not to exercise such authority, shall give rise to any duty or responsibility of the Architect to the Contractor, any Subcontractor, any of their agents or employees, or any other person performing any of the Work.

2.2.14 The Architect will review and approve or take other appropriate action upon Contractor's submittals such as Shop Drawings, Product Data and Samples, but only for conformance with the design concept of the Work and with the information given in the Contract Documents. Such action shall be taken with reasonable promptness so as to cause no delay. The Architect's approval of a specific item shall not indicate approval of an assembly of which the item is a component.

2.2.15 The Architect will prepare Change Orders in accordance with Article 12, and will have authority to order minor changes in the Work as provided in Subparagraph 12.4.1.

2.2.16 The Architect will conduct inspections to determine the dates of Substantial Completion and final completion, will receive and forward to the Owner for the Owner's review written warranties and related documents required by the Contract and assembled by the Contractor, and will issue a final Certificate for Payment upon compliance with the requirements of Paragraph 9.9.

2.2.17 If the Owner and Architect agree, the Architect will provide one or more Project Representatives to assist the Architect in carrying out his responsibilities at the site. The duties, responsibilities and limitations of authority of any such Project Representative shall be as set forth in an exhibit to be incorporated in the Contract Documents.

2.2.18 The duties, responsibilities and limitations of authority of the Architect as the Owner's representative during construction as set forth in the Contract Documents will not be modified or extended without written consent of the Owner, the Contractor and the Architect.

2.2.19 In case of the termination of the employment of the Architect, the Owner shall appoint an architect against whom the Contractor makes no reasonable objection whose status under the Contract Documents shall be that of the former architect. Any dispute in connection with such appointment shall be subject to arbitration.

ARTICLE 3

OWNER

3.1 DEFINITION

3.1.1 The Owner is the person or entity identified as such in the Owner-Contractor Agreement and is referred to throughout the Contract Documents as if singular in number and masculine in gender. The term Owner means the Owner or his authorized representative.

3.2 INFORMATION AND SERVICES REQUIRED OF THE OWNER

3.2.1 The Owner shall, at the request of the Contractor, at the time of execution of the Owner-Contractor Agreement, furnish to the Contractor reasonable evidence that he has made financial arrangements to fulfill his obligations under the Contract. Unless such reasonable evidence is furnished, the Contractor is not required to execute the Owner-Contractor Agreement or to commence the Work.

3.2.2 The Owner shall furnish all surveys describing the physical characteristics, legal limitations and utility locations for the site of the Project, and a legal description of the site.

3.2.3 Except as provided in Subparagraph 4.7.1, the Owner shall secure and pay for necessary approvals, easements, assessments and charges required for the construction, use or occupancy of permanent structures or for permanent changes in existing facilities.

3.2.4 Information or services under the Owner's control shall be furnished by the Owner with reasonable promptness to avoid delay in the orderly progress of the Work.

3.2.5 Unless otherwise provided in the Contract Documents, the Contractor will be furnished, free of charge, all copies of Drawings and Specifications reasonably necessary for the execution of the Work.

3.2.6 The Owner shall forward all instructions to the Contractor through the Architect.

3.2.7 The foregoing are in addition to other duties and responsibilities of the Owner enumerated herein and especially those in respect to Work by Owner or by Separate Contractors, Payments and Completion, and Insurance in Articles 6, 9 and 11 respectively.

3.3 OWNER'S RIGHT TO STOP THE WORK

3.3.1 If the Contractor fails to correct defective Work as required by Paragraph 13.2 or persistently fails to carry out the Work in accordance with the Contract Documents, the Owner, by a written order signed personally or by an agent specifically so empowered by the Owner in writing, may order the Contractor to stop the Work, or any portion thereof, until the cause for such order has been eliminated; however, this right of the Owner to stop the Work shall not give rise to any duty on the part of the Owner to exercise this right for the benefit of the Contractor or any other person or entity, except to the extent required by Subparagraph 6.1.3.

3.4 OWNER'S RIGHT TO CARRY OUT THE WORK

3.4.1 If the Contractor defaults or neglects to carry out the Work in accordance with the Contract Documents and fails within seven days after receipt of written notice from the Owner to commence and continue correction of such default or neglect with diligence and promptness, the Owner may, after seven days following receipt by the Contractor of an additional written notice and without prejudice to any other remedy he may have, make good such deficiencies. In such case an appropriate Change Order shall be issued deducting from the payments then or thereafter due the Contractor the cost of correcting such deficiencies, including compensation for the Architect's additional services made necessary by such default, neglect or failure. Such action by the Owner and the amount charged to the Contractor are both subject to the prior approval of the Architect. If the payments then or thereafter due the Contractor are not sufficient to cover such amount, the Contractor shall pay the difference to the Owner.

ARTICLE 4

CONTRACTOR

4.1 DEFINITION

4.1.1 The Contractor is the person or entity identified as such in the Owner-Contractor Agreement and is referred to throughout the Contract Documents as if singular in number and masculine in gender. The term Contractor means the Contractor or his authorized representative.

4.2 REVIEW OF CONTRACT DOCUMENTS

4.2.1 The Contractor shall carefully study and compare the Contract Documents and shall at once report to the Architect any error, inconsistency or omission he may discover. The Contractor shall not be liable to the Owner or the Architect for any damage resulting from any such errors, inconsistencies or omissions in the Contract Documents. The Contractor shall perform no portion of the Work at any time without Contract Documents or, where required, approved Shop Drawings, Product Data or Samples for such portion of the Work.

4.3 SUPERVISION AND CONSTRUCTION PROCEDURES

4.3.1 The Contractor shall supervise and direct the Work, using his best skill and attention. He shall be solely responsible for all construction means, methods, techniques, sequences and procedures and for coordinating all portions of the Work under the Contract.

4.3.2 The Contractor shall be responsible to the Owner for the acts and omissions of his employees, Subcontractors and their agents and employees, and other persons performing any of the Work under a contract with the Contractor.

4.3.3 The Contractor shall not be relieved from his obligations to perform the Work in accordance with the Contract Documents either by the activities or duties of the Architect in his administration of the Contract, or by inspections, tests or approvals required or performed under Paragraph 7.7 by persons other than the Contractor.

4.4 LABOR AND MATERIALS

4.4.1 Unless otherwise provided in the Contract Documents, the Contractor shall provide and pay for all labor, materials, equipment, tools, construction equipment and machinery, water, heat, utilities, transportation, and other facilities and services necessary for the proper execution and completion of the Work, whether temporary or permanent and whether or not incorporated or to be incorporated in the Work.

4.4.2 The Contractor shall at all times enforce strict discipline and good order among his employees and shall not employ on the Work any unfit person or anyone not skilled in the task assigned to him.

4.5 WARRANTY

4.5.1 The Contractor warrants to the Owner and the Architect that all materials and equipment furnished under this Contract will be new unless otherwise specified, and that all Work will be of good quality, free from faults and defects and in conformance with the Contract Documents. All Work not conforming to these requirements, including substitutions not properly approved and authorized, may be considered defective. If required by the Architect, the Contractor shall furnish satisfactory evidence as to the kind and quality of materials and equipment. This warranty is not limited by the provisions of Paragraph 13.2.

4.6 TAXES

4.6.1 The Contractor shall pay all sales, consumer, use and other similar taxes for the Work or portions thereof provided by the Contractor which are legally enacted at the time bids are received, whether or not yet effective.

4.7 PERMITS, FEES AND NOTICES

4.7.1 Unless otherwise provided in the Contract Documents, the Contractor shall secure and pay for the building permit and for all other permits and governmental fees, licenses and inspections necessary for the proper execution and completion of the Work which are customarily secured after execution of the Contract and which are legally required at the time the bids are received.

4.7.2 The Contractor shall give all notices and comply with all laws, ordinances, rules, regulations and lawful orders of any public authority bearing on the performance of the Work.

4.7.3 It is not the responsibility of the Contractor to make certain that the Contract Documents are in accordance with applicable laws, statutes, building codes and regulations. If the Contractor observes that any of the Contract Documents are at variance therewith in any respect, he shall promptly notify the Architect in writing, and any necessary changes shall be accomplished by appropriate Modification.

4.7.4 If the Contractor performs any Work knowing it to be contrary to such laws, ordinances, rules and regulations, and without such notice to the Architect, he shall assume full responsibility therefor and shall bear all costs attributable thereto.

4.8 ALLOWANCES

4.8.1 The Contractor shall include in the Contract Sum all allowances stated in the Contract Documents. Items covered by these allowances shall be supplied for such amounts and by such persons as the Owner may direct, but the Contractor will not be required to employ persons against whom he makes a reasonable objection.

4.8.2 Unless otherwise provided in the Contract Documents:

.1 these allowances shall cover the cost to the Contractor, less any applicable trade discount, of the materials and equipment required by the allowance delivered at the site, and all applicable taxes;

.2 the Contractor's costs for unloading and handling on the site, labor, installation costs, overhead, profit and other expenses contemplated for the original allowance shall be included in the Contract Sum and not in the allowance;

.3 whenever the cost is more than or less than the allowance, the Contract Sum shall be adjusted accordingly by Change Order, the amount of which will recognize changes, if any, in handling costs on the site, labor, installation costs, overhead, profit and other expenses.

4.9 SUPERINTENDENT

4.9.1 The Contractor shall employ a competent superintendent and necessary assistants who shall be in attendance at the Project site during the progress of the Work. The superintendent shall represent the Contractor and all communications given to the superintendent shall be as binding as if given to the Contractor. Important communications shall be confirmed in writing. Other communications shall be so confirmed on written request in each case.

4.10 PROGRESS SCHEDULE

4.10.1 The Contractor, immediately after being awarded the Contract, shall prepare and submit for the Owner's and Architect's information an estimated progress schedule for the Work. The progress schedule shall be related to the entire Project to the extent required by the Contract Documents, and shall provide for expeditious and practicable execution of the Work.

4.11 DOCUMENTS AND SAMPLES AT THE SITE

4.11.1 The Contractor shall maintain at the site for the Owner one record copy of all Drawings, Specifications, Addenda, Change Orders and other Modifications, in good order and marked currently to record all changes made during construction, and approved Shop Drawings, Product Data and Samples. These shall be available to the Architect and shall be delivered to him for the Owner upon completion of the Work.

4.12 SHOP DRAWINGS, PRODUCT DATA AND SAMPLES

4.12.1 Shop Drawings are drawings, diagrams, schedules and other data specially prepared for the Work by the Contractor or any Subcontractor, manufacturer, supplier or distributor to illustrate some portion of the Work.

4.12.2 Product Data are illustrations, standard schedules, performance charts, instructions, brochures, diagrams and other information furnished by the Contractor to illustrate a material, product or system for some portion of the Work.

4.12.3 Samples are physical examples which illustrate materials, equipment or workmanship and establish standards by which the Work will be judged.

4.12.4 The Contractor shall review, approve and submit, with reasonable promptness and in such sequence as to cause no delay in the Work or in the work of the Owner or any separate contractor, all Shop Drawings, Product Data and Samples required by the Contract Documents.

4.12.5 By approving and submitting Shop Drawings, Product Data and Samples, the Contractor represents that he has determined and verified all materials, field measurements, and field construction criteria related thereto, or will do so, and that he has checked and coordinated the information contained within such submittals with the requirements of the Work and of the Contract Documents.

4.12.6 The Contractor shall not be relieved of responsibility for any deviation from the requirements of the Contract Documents by the Architect's approval of Shop Drawings, Product Data or Samples under Subparagraph 2.2.14 unless the Contractor has specifically informed the Architect in writing of such deviation at the time of submission and the Architect has given written approval to the specific deviation. The Contractor shall not be relieved from responsibility for errors or omissions in the Shop Drawings, Product Data or Samples by the Architect's approval thereof.

4.12.7 The Contractor shall direct specific attention, in writing or on resubmitted Shop Drawings, Product Data or Samples, to revisions other than those requested by the Architect on previous submittals.

4.12.8 No portion of the Work requiring submission of a Shop Drawing, Product Data or Sample shall be commenced until the submittal has been approved by the Architect as provided in Subparagraph 2.2.14. All such portions of the Work shall be in accordance with approved submittals.

4.13 USE OF SITE

4.13.1 The Contractor shall confine operations at the site to areas permitted by law, ordinances, permits and the Contract Documents and shall not unreasonably encumber the site with any materials or equipment.

4.14 CUTTING AND PATCHING OF WORK

4.14.1 The Contractor shall be responsible for all cutting, fitting or patching that may be required to complete the Work or to make its several parts fit together properly.

4.14.2 The Contractor shall not damage or endanger any portion of the Work or the work of the Owner or any separate contractors by cutting, patching or otherwise altering any work, or by excavation. The Contractor shall not cut or otherwise alter the work of the Owner or any separate contractor except with the written consent of the Owner and of such separate contractor. The Contractor shall not unreasonably withhold from the Owner or any separate contractor his consent to cutting or otherwise altering the Work.

4.15 CLEANING UP

4.15.1 The Contractor at all times shall keep the premises free from accumulation of waste materials or rubbish caused by his operations. At the completion of the Work he shall remove all his waste materials and rubbish from and about the Project as well as all his tools, construction equipment, machinery and surplus materials.

4.15.2 If the Contractor fails to clean up at the completion of the Work, the Owner may do so as provided in Paragraph 3.4 and the cost thereof shall be charged to the Contractor.

4.16 COMMUNICATIONS

4.16.1 The Contractor shall forward all communications to the Owner through the Architect.

4.17 ROYALTIES AND PATENTS

4.17.1 The Contractor shall pay all royalties and license fees. He shall defend all suits or claims for infringement of any patent rights and shall save the Owner harmless from loss on account thereof, except that the Owner shall be responsible for all such loss when a particular design, process or the product of a particular manufacturer or manufacturers is specified, but if the Contractor has reason to believe that the design, process or product specified is an infringement of a patent, he shall be responsible for such loss unless he promptly gives such information to the Architect.

4.18 INDEMNIFICATION

4.18.1 To the fullest extent permitted by law, the Contractor shall indemnify and hold harmless the Owner and the Architect and their agents and employees from and against all claims, damages, losses and expenses, including but not limited to attorneys' fees, arising out of or resulting from the performance of the Work, provided that any such claim, damage, loss or expense (1) is attributable to bodily injury, sickness, disease or death, or to injury to or destruction of tangible property (other than the Work itself) including the loss of use resulting therefrom, and (2) is caused in whole or in part by any negligent act or omission of the Contractor, any Subcontractor, anyone directly or indirectly employed by any of them or anyone for whose acts any of them may be liable, regardless of whether or not it is caused in part by a party indemnified hereunder. Such obligation shall not be construed to negate, abridge, or otherwise reduce any other right or obligation of indemnity which would otherwise exist as to any party or person described in this Paragraph 4.18.

4.18.2 In any and all claims against the Owner or the Architect or any of their agents or employees by any employee of the Contractor, any Subcontractor, anyone directly or indirectly employed by any of them or anyone for whose acts any of them may be liable, the indemnification obligation under this Paragraph 4.18 shall not be limited in any way by any limitation on the amount or type of damages, compensation or benefits payable by or for the Contractor or any Subcontractor under workers' or workmen's compensation acts, disability benefit acts or other employee benefit acts.

4.18.3 The obligations of the Contractor under this Paragraph 4.18 shall not extend to the liability of the Architect, his agents or employees, arising out of (1) the preparation or approval of maps, drawings, opinions, reports, surveys, change orders, designs or specifications, or (2) the giving of or the failure to give directions or instructions by the Architect, his agents or employees provided such giving or failure to give is the primary cause of the injury or damage.

ARTICLE 5

SUBCONTRACTORS

5.1 DEFINITION

5.1.1 A Subcontractor is a person or entity who has a direct contract with the Contractor to perform any of the Work at the site. The term Subcontractor is referred to throughout the Contract Documents as if singular in number and masculine in gender and means a Subcontractor or his authorized representative. The term Subcontractor does not include any separate contractor or his subcontractors.

5.1.2 A Sub-subcontractor is a person or entity who has a direct or indirect contract with a Subcontractor to perform any of the Work at the site. The term Sub-subcontractor is referred to throughout the Contract Documents as if singular in number and masculine in gender and means a Sub-subcontractor or an authorized representative thereof.

5.2 AWARD OF SUBCONTRACTS AND OTHER CONTRACTS FOR PORTIONS OF THE WORK

5.2.1 Unless otherwise required by the Contract Documents or the Bidding Documents, the Contractor, as soon as practicable after the award of the Contract, shall furnish to the Owner and the Architect in writing the names of the persons or entities (including those who are to furnish materials or equipment fabricated to a special design) proposed for each of the principal portions of the Work. The Architect will promptly reply to the Contractor in writing stating whether or not the Owner or the Architect, after due investigation, has reasonable objection to any such proposed person or entity. Failure of the Owner or Architect to reply promptly shall constitute notice of no reasonable objection.

5.2.2 The Contractor shall not contract with any such proposed person or entity to whom the Owner or the Architect has made reasonable objection under the provisions of Subparagraph 5.2.1. The Contractor shall not be required to contract with anyone to whom he has a reasonable objection.

5.2.3 If the Owner or the Architect has reasonable objection to any such proposed person or entity, the Contractor shall submit a substitute to whom the Owner or the Architect has no reasonable objection, and the Contract Sum shall be increased or decreased by the difference in cost occasioned by such substitution and an appropriate Change Order shall be issued; however, no increase in the Contract Sum shall be allowed for any such substitution unless the Contractor has acted promptly and responsively in submitting names as required by Subparagraph 5.2.1.

5.2.4 The Contractor shall make no substitution for any Subcontractor, person or entity previously selected if the Owner or Architect makes reasonable objection to such substitution.

5.3 SUBCONTRACTUAL RELATIONS

5.3.1 By an appropriate agreement, written where legally required for validity, the Contractor shall require each Subcontractor, to the extent of the Work to be performed by the Subcontractor, to be bound to the Contractor by the terms of the Contract Documents, and to assume toward the Contractor all the obligations and responsibilities which the Contractor, by these Documents, assumes toward the Owner and the Architect. Said agreement shall preserve and protect the rights of the Owner and the Architect under the Contract Documents with respect to the Work to be performed by the Subcontractor so that the subcontracting thereof will not prejudice such rights, and shall allow to the Subcontractor, unless specifically provided otherwise in the Contractor-Subcontractor agreement, the benefit of all rights, remedies and redress against the Contractor that the Contractor, by these Documents, has against the Owner. Where appropriate, the Contractor shall require each Subcontractor to enter into similar agreements with his Sub-subcontractors. The Contractor shall make available to each proposed Subcontractor, prior to the execution of the Subcontract, copies of the Contract Documents to which the Subcontractor will be bound by this Paragraph 5.3, and identify to the Subcontractor any terms and conditions of the proposed Subcontract which may be at variance with the Contract Documents. Each Subcontractor shall similarly make copies of such Documents available to his Sub-subcontractors.

ARTICLE 6

WORK BY OWNER OR BY SEPARATE CONTRACTORS

6.1 OWNER'S RIGHT TO PERFORM WORK AND TO AWARD SEPARATE CONTRACTS

6.1.1 The Owner reserves the right to perform work related to the Project with his own forces, and to award separate contracts in connection with other portions of the Project or other work on the site under these or similar Conditions of the Contract. If the Contractor claims that delay or additional cost is involved because of such action by the Owner, he shall make such claim as provided elsewhere in the Contract Documents.

6.1.2 When separate contracts are awarded for different portions of the Project or other work on the site, the term Contractor in the Contract Documents in each case shall mean the Contractor who executes each separate Owner-Contractor Agreement.

6.1.3 The Owner will provide for the coordination of the work of his own forces and of each separate contractor with the Work of the Contractor, who shall cooperate therewith as provided in Paragraph 6.2.

6.2 MUTUAL RESPONSIBILITY

6.2.1 The Contractor shall afford the Owner and separate contractors reasonable opportunity for the introduction and storage of their materials and equipment and the execution of their work, and shall connect and coordinate his Work with theirs as required by the Contract Documents.

6.2.2 If any part of the Contractor's Work depends for proper execution or results upon the work of the Owner or any separate contractor, the Contractor shall, prior to proceeding with the Work, promptly report to the Architect any apparent discrepancies or defects in such other work that render it unsuitable for such proper execution and results. Failure of the Contractor so to report shall constitute an acceptance of the Owner's or separate contractors' work as fit and proper to receive his Work, except as to defects which may subsequently become apparent in such work by others.

6.2.3 Any costs caused by defective or ill-timed work shall be borne by the party responsible therefor.

6.2.4 Should the Contractor wrongfully cause damage to the work or property of the Owner, or to other work on the site, the Contractor shall promptly remedy such damage as provided in Subparagraph 10.2.5.

6.2.5 Should the Contractor wrongfully cause damage to the work or property of any separate contractor, the Contractor shall upon due notice promptly attempt to settle with such other contractor by agreement, or otherwise to resolve the dispute. If such separate contractor sues or initiates an arbitration proceeding against the Owner on account of any damage alleged to have been caused by the Contractor, the Owner shall notify the Contractor who shall defend such proceedings at the Owner's expense, and if any judgment or award against the Owner arises therefrom the Contractor shall pay or satisfy it and shall reimburse the Owner for all attorneys' fees and court or arbitration costs which the Owner has incurred.

6.3 OWNER'S RIGHT TO CLEAN UP

6.3.1 If a dispute arises between the Contractor and separate contractors as to their responsibility for cleaning up as required by Paragraph 4.15, the Owner may clean up and charge the cost thereof to the contractors responsible therefor as the Architect shall determine to be just.

ARTICLE 7

MISCELLANEOUS PROVISIONS

7.1 GOVERNING LAW

7.1.1 The Contract shall be governed by the law of the place where the Project is located.

7.2 SUCCESSORS AND ASSIGNS

7.2.1 The Owner and the Contractor each binds himself, his partners, successors, assigns and legal representatives to the other party hereto and to the partners, successors, assigns and legal representatives of such other party in respect to all covenants, agreements and obligations contained in the Contract Documents. Neither party to the Contract shall assign the Contract or sublet it as a whole without the written consent of the other, nor shall the Contractor assign any moneys due or to become due to him hereunder, without the previous written consent of the Owner.

7.3 WRITTEN NOTICE

7.3.1 Written notice shall be deemed to have been duly served if delivered in person to the individual or member of the firm or entity or to an officer of the corporation for whom it was intended, or if delivered at or sent by registered or certified mail to the last business address known to him who gives the notice.

7.4 CLAIMS FOR DAMAGES

7.4.1 Should either party to the Contract suffer injury or damage to person or property because of any act or omission of the other party or of any of his employees, agents or others for whose acts he is legally liable, claim shall be made in writing to such other party within a reasonable time after the first observance of such injury or damage.

7.5 PERFORMANCE BOND AND LABOR AND MATERIAL PAYMENT BOND

7.5.1 The Owner shall have the right to require the Contractor to furnish bonds covering the faithful performance of the Contract and the payment of all obligations arising thereunder if and as required in the Bidding Documents or in the Contract Documents.

7.6 RIGHTS AND REMEDIES

7.6.1 The duties and obligations imposed by the Contract Documents and the rights and remedies available thereunder shall be in addition to and not a limitation of any duties, obligations, rights and remedies otherwise imposed or available by law.

7.6.2 No action or failure to act by the Owner, Architect or Contractor shall constitute a waiver of any right or duty afforded any of them under the Contract, nor shall any such action or failure to act constitute an approval of or acquiescence in any breach thereunder, except as may be specifically agreed in writing.

7.7 TESTS

7.7.1 If the Contract Documents, laws, ordinances, rules, regulations or orders of any public authority having jurisdiction require any portion of the Work to be inspected, tested or approved, the Contractor shall give the Architect timely notice of its readiness so the Architect may observe such inspection, testing or approval. The Contractor shall bear all costs of such inspections, tests or approvals conducted by public authorities. Unless otherwise provided, the Owner shall bear all costs of other inspections, tests or approvals.

7.7.2 If the Architect determines that any Work requires special inspection, testing, or approval which Subparagraph 7.7.1 does not include, he will, upon written authorization from the Owner, instruct the Contractor to order such special inspection, testing or approval, and the Contractor shall give notice as provided in Subparagraph 7.7.1. If such special inspection or testing reveals a failure of the Work to comply with the requirements of the Contract Documents, the Contractor shall bear all costs thereof, including compensation for the Architect's additional services made necessary by such failure; otherwise the Owner shall bear such costs, and an appropriate Change Order shall be issued.

7.7.3 Required certificates of inspection, testing or approval shall be secured by the Contractor and promptly delivered by him to the Architect.

7.7.4 If the Architect is to observe the inspections, tests or approvals required by the Contract Documents, he will do so promptly and, where practicable, at the source of supply.

7.8 INTEREST

7.8.1 Payments due and unpaid under the Contract Documents shall bear interest from the date payment is due at such rate as the parties may agree upon in writing or, in the absence thereof, at the legal rate prevailing at the place of the Project.

7.9 ARBITRATION

7.9.1 All claims, disputes and other matters in question between the Contractor and the Owner arising out of, or relating to, the Contract Documents or the breach thereof, except as provided in Subparagraph 2.2.11 with respect to the Architect's decisions on matters relating to artistic effect, and except for claims which have been waived by the making or acceptance of final payment as provided by Subparagraphs 9.9.4 and 9.9.5, shall be decided by arbitration in accordance with the Construction Industry Arbitration Rules of the American Arbitration Association then obtaining unless the parties mutually agree otherwise. No arbitration arising out of or relating to the Contract Documents shall include, by consolidation, joinder or in any other manner, the Architect, his employees or consultants except by written consent containing a specific reference to the Owner-Contractor Agreement and signed by the Architect, the Owner, the Contractor and any other person sought to be joined. No arbitration shall include by consolidation, joinder or in any other manner, parties other than the Owner, the Contractor and any other persons substantially involved in a common question of fact or law, whose presence is required if complete relief is to be accorded in the arbitration. No person other than the Owner or Contractor shall be included as an original third party or additional third party to an arbitration whose interest or responsibility is insubstantial. Any consent to arbitration involving an additional person or persons shall not constitute consent to arbitration of any dispute not described therein or with any person not named or described therein. The foregoing agreement to arbitrate and any other agreement to arbitrate with an additional person or persons duly consented to by the parties to the Owner-Contractor Agreement shall be specifically enforceable under the prevailing arbitration law. The award rendered by the arbitrators shall be final, and judgment may be entered upon it in accordance with applicable law in any court having jurisdiction thereof.

7.9.2 Notice of the demand for arbitration shall be filed in writing with the other party to the Owner-Contractor Agreement and with the American Arbitration Association, and a copy shall be filed with the Architect. The demand for arbitration shall be made within the time limits specified in Subparagraph 2.2.12 where applicable, and in all other cases within a reasonable time after the claim, dispute or other matter in question has arisen, and in no event shall it be made after the date when institution of legal or equitable proceedings based on such claim, dispute or other matter in question would be barred by the applicable statute of limitations.

7.9.3 Unless otherwise agreed in writing, the Contractor shall carry on the Work and maintain its progress during any arbitration proceedings, and the Owner shall continue to make payments to the Contractor in accordance with the Contract Documents.

ARTICLE 8

TIME

8.1 DEFINITIONS

8.1.1 Unless otherwise provided, the Contract Time is the period of time allotted in the Contract Documents for Substantial Completion of the Work as defined in Subparagraph 8.1.3, including authorized adjustments thereto.

8.1.2 The date of commencement of the Work is the date established in a notice to proceed. If there is no notice to proceed, it shall be the date of the Owner-Contractor Agreement or such other date as may be established therein.

8.1.3 The Date of Substantial Completion of the Work or designated portion thereof is the Date certified by the Architect when construction is sufficiently complete, in accordance with the Contract Documents, so the Owner can occupy or utilize the Work or designated portion thereof for the use for which it is intended.

8.1.4 The term day as used in the Contract Documents shall mean calendar day unless otherwise specifically designated.

8.2 PROGRESS AND COMPLETION

8.2.1 All time limits stated in the Contract Documents are of the essence of the Contract.

8.2.2 The Contractor shall begin the Work on the date of commencement as defined in Subparagraph 8.1.2. He shall carry the Work forward expeditiously with adequate forces and shall achieve Substantial Completion within the Contract Time.

8.3 DELAYS AND EXTENSIONS OF TIME

8.3.1 If the Contractor is delayed at any time in the progress of the Work by any act or neglect of the Owner or the Architect, or by any employee of either, or by any separate contractor employed by the Owner, or by changes ordered in the Work, or by labor disputes, fire, unusual delay in transportation, adverse weather conditions not reasonably anticipatable, unavoidable casualties, or any causes beyond the Contractor's control, or by delay authorized by the Owner pending arbitration, or by any other cause which the Architect determines may justify the delay, then the Contract Time shall be extended by Change Order for such reasonable time as the Architect may determine.

8.3.2 Any claim for extension of time shall be made in writing to the Architect not more than twenty days after the commencement of the delay; otherwise it shall be waived. In the case of a continuing delay only one claim is necessary. The Contractor shall provide an estimate of the probable effect of such delay on the progress of the Work.

8.3.3 If no agreement is made stating the dates upon which interpretations as provided in Subparagraph 2.2.8 shall be furnished, then no claim for delay shall be allowed on account of failure to furnish such interpretations until fifteen days after written request is made for them, and not then unless such claim is reasonable.

8.3.4 This Paragraph 8.3 does not exclude the recovery of damages for delay by either party under other provisions of the Contract Documents.

ARTICLE 9

PAYMENTS AND COMPLETION

9.1 CONTRACT SUM

9.1.1 The Contract Sum is stated in the Owner-Contractor Agreement and, including authorized adjustments thereto, is the total amount payable by the Owner to the Contractor for the performance of the Work under the Contract Documents.

9.2 SCHEDULE OF VALUES

9.2.1 Before the first Application for Payment, the Contractor shall submit to the Architect a schedule of values allocated to the various portions of the Work, prepared in such form and supported by such data to substantiate its accuracy as the Architect may require. This schedule, unless objected to by the Architect, shall be used only as a basis for the Contractor's Applications for Payment.

9.3 APPLICATIONS FOR PAYMENT

9.3.1 At least ten days before the date for each progress payment established in the Owner-Contractor Agreement, the Contractor shall submit to the Architect an itemized Application for Payment, notarized if required, supported by such data substantiating the Contractor's right to payment as the Owner or the Architect may require, and reflecting retainage, if any, as provided elsewhere in the Contract Documents.

9.3.2 Unless otherwise provided in the Contract Documents, payments will be made on account of materials or equipment not incorporated in the Work but delivered and suitably stored at the site and, if approved in advance by the Owner, payments may similarly be made for materials or equipment suitably stored at some other location agreed upon in writing. Payments for materials or equipment stored on or off the site shall be conditioned upon submission by the Contractor of bills of sale or such other procedures satisfactory to the Owner to establish the Owner's title to such materials or equipment or otherwise protect the Owner's interest, including applicable insurance and transportation to the site for those materials and equipment stored off the site.

9.3.3 The Contractor warrants that title to all Work, materials and equipment covered by an Application for Payment will pass to the Owner either by incorporation in the construction or upon the receipt of payment by the Contractor, whichever occurs first, free and clear of all liens, claims, security interests or encumbrances, hereinafter referred to in this Article 9 as "liens"; and that no Work, materials or equipment covered by an Application for Payment will have been acquired by the Contractor, or by any other person performing Work at the site or furnishing materials and equipment for the Project, subject to an agreement under which an interest therein or an encumbrance thereon is retained by the seller or otherwise imposed by the Contractor or such other person.

9.4 CERTIFICATES FOR PAYMENT

9.4.1 The Architect will, within seven days after the receipt of the Contractor's Application for Payment, either issue a Certificate for Payment to the Owner, with a copy to the Contractor, for such amount as the Architect determines is properly due, or notify the Contractor in writing his reasons for withholding a Certificate as provided in Subparagraph 9.6.1.

9.4.2 The issuance of a Certificate for Payment will constitute a representation by the Architect to the Owner, based on his observations at the site as provided in Subparagraph 2.2.3 and the data comprising the Application for Payment, that the Work has progressed to the point indicated; that, to the best of his knowledge, information and belief, the quality of the Work is in accordance with the Contract Documents (subject to an evaluation of the Work for conformance with the Contract Documents upon Substantial Completion, to the results of any subsequent tests required by or performed under the Contract Documents, to minor deviations from the Contract Documents correctable prior to completion, and to any specific qualifications stated in his Certificate); and that the Contractor is entitled to payment in the amount certified. However, by issuing a Certificate for Payment, the Architect shall not thereby be deemed to represent that he has made exhaustive or continuous on-site inspections to check the quality or quantity of the Work or that he has reviewed the construction means, methods, techniques, sequences or procedures, or that he has made any examination to ascertain how or for what purpose the Contractor has used the moneys previously paid on account of the Contract Sum.

9.5 PROGRESS PAYMENTS

9.5.1 After the Architect has issued a Certificate for Payment, the Owner shall make payment in the manner and within the time provided in the Contract Documents.

9.5.2 The Contractor shall promptly pay each Subcontractor, upon receipt of payment from the Owner, out of the amount paid to the Contractor on account of such Subcontractor's Work, the amount to which said Subcontractor is entitled, reflecting the percentage actually retained, if any, from payments to the Contractor on account of such Subcontractor's Work. The Contractor shall, by an appropriate agreement with each Subcontractor, require each Subcontractor to make payments to his Sub-subcontractors in similar manner.

9.5.3 The Architect may, on request and at his discretion, furnish to any Subcontractor, if practicable, information regarding the percentages of completion or the amounts applied for by the Contractor and the action taken thereon by the Architect on account of Work done by such Subcontractor.

9.5.4 Neither the Owner nor the Architect shall have any obligation to pay or to see to the payment of any moneys to any Subcontractor except as may otherwise be required by law.

9.5.5 No Certificate for a progress payment, nor any progress payment, nor any partial or entire use or occupancy of the Project by the Owner, shall constitute an acceptance of any Work not in accordance with the Contract Documents.

9.6 PAYMENTS WITHHELD

9.6.1 The Architect may decline to certify payment and may withhold his Certificate in whole or in part, to the extent necessary reasonably to protect the Owner, if in his opinion he is unable to make representations to the Owner as provided in Subparagraph 9.4.2. If the Architect is unable to make representations to the Owner as provided in Subparagraph 9.4.2 and to certify payment in the amount of the Application, he will notify the Contractor as provided in Subparagraph 9.4.1. If the Contractor and the Architect cannot agree on a revised amount, the Architect will promptly issue a Certificate for Payment for the amount for which he is able to make such representations to the Owner. The Architect may also decline to certify payment or, because of subsequently discovered evidence or subsequent observations, he may nullify the whole or any part of any Certificate for Payment previously issued, to such extent as may be necessary in his opinion to protect the Owner from loss because of:

- **.1** defective work not remedied,
- **.2** third party claims filed or reasonable evidence indicating probable filing of such claims,
- **.3** failure of the Contractor to make payments properly to Subcontractors or for labor, materials or equipment,
- **.4** reasonable evidence that the Work cannot be completed for the unpaid balance of the Contract Sum,
- **.5** damage to the Owner or another contractor,
- **.6** reasonable evidence that the Work will not be completed within the Contract Time, or
- **.7** persistent failure to carry out the Work in accordance with the Contract Documents.

9.6.2 When the above grounds in Subparagraph 9.6.1 are removed, payment shall be made for amounts withheld because of them.

9.7 FAILURE OF PAYMENT

9.7.1 If the Architect does not issue a Certificate for Payment, through no fault of the Contractor, within seven days after receipt of the Contractor's Application for Payment, or if the Owner does not pay the Contractor within seven days after the date established in the Contract Documents any amount certified by the Architect or awarded by arbitration, then the Contractor may, upon seven additional days' written notice to the Owner and the Architect, stop the Work until payment of the amount owing has been received. The Contract Sum shall be increased by the amount of the Contractor's reasonable costs of shut-down, delay and start-up, which shall be effected by appropriate Change Order in accordance with Paragraph 12.3.

9.8 SUBSTANTIAL COMPLETION

9.8.1 When the Contractor considers that the Work, or a designated portion thereof which is acceptable to the Owner, is substantially complete as defined in Subparagraph 8.1.3, the Contractor shall prepare for submission to the Architect a list of items to be completed or corrected. The failure to include any items on such list does not alter the responsibility of the Contractor to complete all Work in accordance with the Contract Documents. When the Architect on the basis of an inspection determines that the Work or designated portion thereof is substantially complete, he will then prepare a Certificate of Substantial Completion which shall establish the Date of Substantial Completion, shall state the responsibilities of the Owner and the Contractor for security, maintenance, heat, utilities, damage to the Work, and insurance, and shall fix the time within which the Contractor shall complete the items listed therein. Warranties required by the Contract Documents shall commence on the Date of Substantial Completion of the Work or designated portion thereof unless otherwise provided in the Certificate of Substantial Completion. The Certificate of Substantial Completion shall be submitted to the Owner and the Contractor for their written acceptance of the responsibilities assigned to them in such Certificate.

9.8.2 Upon Substantial Completion of the Work or designated portion thereof and upon application by the Contractor and certification by the Architect, the Owner shall make payment, reflecting adjustment in retainage, if any, for such Work or portion thereof, as provided in the Contract Documents.

9.9 FINAL COMPLETION AND FINAL PAYMENT

9.9.1 Upon receipt of written notice that the Work is ready for final inspection and acceptance and upon receipt of a final Application for Payment, the Architect will promptly make such inspection and, when he finds the Work acceptable under the Contract Documents and the Contract fully performed, he will promptly issue a final Certificate for Payment stating that to the best of his knowledge, information and belief, and on the basis of his observations and inspections, the Work has been completed in accordance with the terms and conditions of the Contract Documents and that the entire balance found to be due the Contractor, and noted in said final Certificate, is due and payable. The Architect's final Certificate for Payment will constitute a further representation that the conditions precedent to the Contractor's being entitled to final payment as set forth in Subparagraph 9.9.2 have been fulfilled.

9.9.2 Neither the final payment nor the remaining retained percentage shall become due until the Contractor submits to the Architect (1) an affidavit that all payrolls, bills for materials and equipment, and other indebtedness connected with the Work for which the Owner or his property might in any way be responsible, have been paid or otherwise satisfied, (2) consent of surety, if any, to final payment and (3), if required by the Owner, other data establishing payment or satisfaction of all such obligations, such as receipts, releases and waivers of liens arising out of the Contract, to the extent and in such form as may be designated by the Owner. If any Subcontractor refuses to furnish a release or waiver required by the Owner, the Contractor may furnish a bond satisfactory to the Owner to indemnify him against any such lien. If any such lien remains unsatisfied after all payments are made, the Contractor shall refund to the Owner all moneys that the latter may be compelled to pay in discharging such lien, including all costs and reasonable attorneys' fees.

9.9.3 If, after Substantial Completion of the Work, final completion thereof is materially delayed through no fault of the Contractor or by the issuance of Change Orders affecting final completion, and the Architect so confirms, the Owner shall, upon application by the Contractor and certification by the Architect, and without terminating the Contract, make payment of the balance due for that portion of the Work fully completed and accepted. If the remaining balance for Work not fully completed or corrected is less than the retainage stipulated in the Contract Documents, and if bonds have been furnished as provided in Paragraph 7.5, the written consent of the surety to the payment of the balance due for that portion of the Work fully completed and accepted shall be submitted by the Contractor to the Architect prior to certification of such payment. Such payment shall be made under the terms and conditions governing final payment, except that it shall not constitute a waiver of claims.

9.9.4 The making of final payment shall constitute a waiver of all claims by the Owner except those arising from:

- **.1** unsettled liens,
- **.2** faulty or defective Work appearing after Substantial Completion,
- **.3** failure of the Work to comply with the requirements of the Contract Documents, or
- **.4** terms of any special warranties required by the Contract Documents.

9.9.5 The acceptance of final payment shall constitute a waiver of all claims by the Contractor except those previously made in writing and identified by the Contractor as unsettled at the time of the final Application for Payment.

ARTICLE 10

PROTECTION OF PERSONS AND PROPERTY

10.1 SAFETY PRECAUTIONS AND PROGRAMS

10.1.1 The Contractor shall be responsible for initiating, maintaining and supervising all safety precautions and programs in connection with the Work.

10.2 SAFETY OF PERSONS AND PROPERTY

10.2.1 The Contractor shall take all reasonable precautions for the safety of, and shall provide all reasonable protection to prevent damage, injury or loss to:

- **.1** all employees on the Work and all other persons who may be affected thereby;
- **.2** all the Work and all materials and equipment to be incorporated therein, whether in storage on or off the site, under the care, custody or control of the Contractor or any of his Subcontractors or Sub-subcontractors; and
- **.3** other property at the site or adjacent thereto, including trees, shrubs, lawns, walks, pavements, roadways, structures and utilities not designated for removal, relocation or replacement in the course of construction.

10.2.2 The Contractor shall give all notices and comply with all applicable laws, ordinances, rules, regulations and lawful orders of any public authority bearing on the safety of persons or property or their protection from damage, injury or loss.

10.2.3 The Contractor shall erect and maintain, as required by existing conditions and progress of the Work, all reasonable safeguards for safety and protection, including posting danger signs and other warnings against hazards, promulgating safety regulations and notifying owners and users of adjacent utilities.

10.2.4 When the use or storage of explosives or other hazardous materials or equipment is necessary for the execution of the Work, the Contractor shall exercise the utmost care and shall carry on such activities under the supervision of properly qualified personnel.

10.2.5 The Contractor shall promptly remedy all damage or loss (other than damage or loss insured under Paragraph 11.3) to any property referred to in Clauses 10.2.1.2 and 10.2.1.3 caused in whole or in part by the Contractor, any Subcontractor, any Sub-subcontractor, or anyone directly or indirectly employed by any of them, or by anyone for whose acts any of them may be liable and for which the Contractor is responsible under Clauses 10.2.1.2 and 10.2.1.3, except damage or loss attributable to the acts or omissions of the Owner or Architect or anyone directly or indirectly employed by either of them, or by anyone for whose acts either of them may be liable, and not attributable to the fault or negligence of the Contractor. The foregoing obligations of the Contractor are in addition to his obligations under Paragraph 4.18.

10.2.6 The Contractor shall designate a responsible member of his organization at the site whose duty shall be the prevention of accidents. This person shall be the Contractor's superintendent unless otherwise designated by the Contractor in writing to the Owner and the Architect.

10.2.7 The Contractor shall not load or permit any part of the Work to be loaded so as to endanger its safety.

10.3 EMERGENCIES

10.3.1 In any emergency affecting the safety of persons or property, the Contractor shall act, at his discretion, to prevent threatened damage, injury or loss. Any additional compensation or extension of time claimed by the Contractor on account of emergency work shall be determined as provided in Article 12 for Changes in the Work.

ARTICLE 11

INSURANCE

11.1 CONTRACTOR'S LIABILITY INSURANCE

11.1.1 The Contractor shall purchase and maintain such insurance as will protect him from claims set forth below which may arise out of or result from the Contractor's operations under the Contract, whether such operations be by himself or by any Subcontractor or by anyone directly or indirectly employed by any of them, or by anyone for whose acts any of them may be liable:

.1 claims under workers' or workmen's compensation, disability benefit and other similar employee benefit acts;

.2 claims for damages because of bodily injury, occupational sickness or disease, or death of his employees;

.3 claims for damages because of bodily injury, sickness or disease, or death of any person other than his employees;

.4 claims for damages insured by usual personal injury liability coverage which are sustained (1) by any person as a result of an offense directly or indirectly related to the employment of such person by the Contractor, or (2) by any other person;

.5 claims for damages, other than to the Work itself, because of injury to or destruction of tangible property, including loss of use resulting therefrom; and

.6 claims for damages because of bodily injury or death of any person or property damage arising out of the ownership, maintenance or use of any motor vehicle.

11.1.2 The insurance required by Subparagraph 11.1.1 shall be written for not less than any limits of liability specified in the Contract Documents, or required by law, whichever is greater.

11.1.3 The insurance required by Subparagraph 11.1.1 shall include contractual liability insurance applicable to the Contractor's obligations under Paragraph 4.18.

11.1.4 Certificates of Insurance acceptable to the Owner shall be filed with the Owner prior to commencement of the Work. These Certificates shall contain a provision that coverages afforded under the policies will not be cancelled until at least thirty days' prior written notice has been given to the Owner.

11.2 OWNER'S LIABILITY INSURANCE

11.2.1 The Owner shall be responsible for purchasing and maintaining his own liability insurance and, at his option, may purchase and maintain such insurance as will protect him against claims which may arise from operations under the Contract.

11.3 PROPERTY INSURANCE

11.3.1 Unless otherwise provided, the Owner shall purchase and maintain property insurance upon the entire Work at the site to the full insurable value thereof. This insurance shall include the interests of the Owner, the Contractor, Subcontractors and Sub-subcontractors in the Work and shall insure against the perils of fire and extended coverage and shall include "all risk" insurance for physical loss or damage including, without duplication of coverage, theft, vandalism and malicious mischief. If the Owner does not intend to purchase such insurance for the full insurable value of the entire Work, he shall inform the Contractor in writing prior to commencement of the Work. The Contractor may then effect insurance which will protect the interests of himself, his Subcontractors and the Sub-subcontractors in the Work, and by appropriate Change Order the cost thereof shall be charged to the Owner. If the Contractor is damaged by failure of the Owner to purchase or maintain such insurance and to so notify the Contractor, then the Owner shall bear all reasonable costs properly attributable thereto. If not covered under the all risk insurance or otherwise provided in the Contract Documents, the Contractor shall effect and maintain similar property insurance on portions of the Work stored off the site or in transit when such portions of the Work are to be included in an Application for Payment under Subparagraph 9.3.2.

11.3.2 The Owner shall purchase and maintain such boiler and machinery insurance as may be required by the Contract Documents or by law. This insurance shall include the interests of the Owner, the Contractor, Subcontractors and Sub-subcontractors in the Work.

11.3.3 Any loss insured under Subparagraph 11.3.1 is to be adjusted with the Owner and made payable to the Owner as trustee for the insureds, as their interests may appear, subject to the requirements of any applicable mortgagee clause and of Subparagraph 11.3.8. The Contractor shall pay each Subcontractor a just share of any insurance moneys received by the Contractor, and by appropriate agreement, written where legally required for validity, shall require each Subcontractor to make payments to his Sub-subcontractors in similar manner.

11.3.4 The Owner shall file a copy of all policies with the Contractor before an exposure to loss may occur.

11.3.5 If the Contractor requests in writing that insurance for risks other than those described in Subparagraphs 11.3.1 and 11.3.2 or other special hazards be included in the property insurance policy, the Owner shall, if possible, include such insurance, and the cost thereof shall be charged to the Contractor by appropriate Change Order.

11.3.6 The Owner and Contractor waive all rights against (1) each other and the Subcontractors, Sub-subcontractors, agents and employees each of the other, and (2) the Architect and separate contractors, if any, and their subcontractors, sub-subcontractors, agents and employees, for damages caused by fire or other perils to the extent covered by insurance obtained pursuant to this Paragraph 11.3 or any other property insurance applicable to the Work, except such rights as they may have to the proceeds of such insurance held by the Owner as trustee. The foregoing waiver afforded the Architect, his agents and employees shall not extend to the liability imposed by Subparagraph 4.18.3. The Owner or the Contractor, as appropriate, shall require of the Architect, separate contractors, Subcontractors and Sub-subcontractors by appropriate agreements, written where legally required for validity, similar waivers each in favor of all other parties enumerated in this Subparagraph 11.3.6.

11.3.7 If required in writing by any party in interest, the Owner as trustee shall, upon the occurrence of an insured loss, give bond for the proper performance of his duties. He shall deposit in a separate account any money so received, and he shall distribute it in accordance with such agreement as the parties in interest may reach, or in accordance with an award by arbitration in which case the procedure shall be as provided in Paragraph 7.9. If after such loss no other special agreement is made, replacement of damaged work shall be covered by an appropriate Change Order.

11.3.8 The Owner as trustee shall have power to adjust and settle any loss with the insurers unless one of the parties in interest shall object in writing within five days after the occurrence of loss to the Owner's exercise of this power, and if such objection be made, arbitrators shall be chosen as provided in Paragraph 7.9. The Owner as trustee shall, in that case, make settlement with the insurers in accordance with the directions of such arbitrators. If distribution of the insurance proceeds by arbitration is required, the arbitrators will direct such distribution.

11.3.9 If the Owner finds it necessary to occupy or use a portion or portions of the Work prior to Substantial Completion thereof, such occupancy shall not commence prior to a time mutually agreed to by the Owner and Contractor and to which the insurance company or companies providing the property insurance have consented by endorsement to the policy or policies. This insurance shall not be cancelled or lapsed on account of such partial occupancy. Consent of the Contractor and of the insurance company or companies to such occupancy or use shall not be unreasonably withheld.

11.4 LOSS OF USE INSURANCE

11.4.1 The Owner, at his option, may purchase and maintain such insurance as will insure him against loss of use of his property due to fire or other hazards, however caused. The Owner waives all rights of action against the Contractor for loss of use of his property, including consequential losses due to fire or other hazards however caused, to the extent covered by insurance under this Paragraph 11.4.

ARTICLE 12

CHANGES IN THE WORK

12.1 CHANGE ORDERS

12.1.1 A Change Order is a written order to the Contractor signed by the Owner and the Architect, issued after execution of the Contract, authorizing a change in the Work or an adjustment in the Contract Sum or the Contract Time. The Contract Sum and the Contract Time may be changed only by Change Order. A Change Order signed by the Contractor indicates his agreement therewith, including the adjustment in the Contract Sum or the Contract Time.

12.1.2 The Owner, without invalidating the Contract, may order changes in the Work within the general scope of the Contract consisting of additions, deletions or other revisions, the Contract Sum and the Contract Time being adjusted accordingly. All such changes in the Work shall be authorized by Change Order, and shall be performed under the applicable conditions of the Contract Documents.

12.1.3 The cost or credit to the Owner resulting from a change in the Work shall be determined in one or more of the following ways:

- **.1** by mutual acceptance of a lump sum properly itemized and supported by sufficient substantiating data to permit evaluation;
- **.2** by unit prices stated in the Contract Documents or subsequently agreed upon;
- **.3** by cost to be determined in a manner agreed upon by the parties and a mutually acceptable fixed or percentage fee; or
- **.4** by the method provided in Subparagraph 12.1.4.

12.1.4 If none of the methods set forth in Clauses 12.1.3.1, 12.1.3.2 or 12.1.3.3 is agreed upon, the Contractor, provided he receives a written order signed by the Owner, shall promptly proceed with the Work involved. The cost of such Work shall then be determined by the Architect on the basis of the reasonable expenditures and savings of those performing the Work attributable to the change, including, in the case of an increase in the Contract Sum, a reasonable allowance for overhead and profit. In such case, and also under Clauses 12.1.3.3 and 12.1.3.4 above, the Contractor shall keep and present, in such form as the Architect may prescribe, an itemized accounting together with appropriate supporting data for inclusion in a Change Order. Unless otherwise provided in the Contract Documents, cost shall be limited to the following: cost of materials, including sales tax and cost of delivery; cost of labor, including social security, old age and unemployment insurance, and fringe benefits required by agreement or custom; workers' or workmen's compensation insurance; bond premiums; rental value of equipment and machinery; and the additional costs of supervision and field office personnel directly attributable to the change. Pending final determination of cost to the Owner, payments on account shall be made on the Architect's Certificate for Payment. The amount of credit to be allowed by the Contractor to the Owner for any deletion or change which results in a net decrease in the Contract Sum will be the amount of the actual net cost as confirmed by the Architect. When both additions and credits covering related Work or substitutions are involved in any one change, the allowance for overhead and profit shall be figured on the basis of the net increase, if any, with respect to that change.

12.1.5 If unit prices are stated in the Contract Documents or subsequently agreed upon, and if the quantities originally contemplated are so changed in a proposed Change Order that application of the agreed unit prices to the quantities of Work proposed will cause substantial inequity to the Owner or the Contractor, the applicable unit prices shall be equitably adjusted.

12.2 CONCEALED CONDITIONS

12.2.1 Should concealed conditions encountered in the performance of the Work below the surface of the ground or should concealed or unknown conditions in an existing structure be at variance with the conditions indicated by the Contract Documents, or should unknown physical conditions below the surface of the ground or should concealed or unknown conditions in an existing structure of an unusual nature, differing materially from those ordinarily encountered and generally recognized as inherent in work of the character provided for in this Contract, be encountered, the Contract Sum shall be equitably adjusted by Change Order upon claim by either party made within twenty days after the first observance of the conditions.

12.3 CLAIMS FOR ADDITIONAL COST

12.3.1 If the Contractor wishes to make a claim for an increase in the Contract Sum, he shall give the Architect written notice thereof within twenty days after the occurrence of the event giving rise to such claim. This notice shall be given by the Contractor before proceeding to execute the Work, except in an emergency endangering life or property in which case the Contractor shall proceed in accordance with Paragraph 10.3. No such claim shall be valid unless so made. If the Owner and the Contractor cannot agree on the amount of the adjustment in the Contract Sum, it shall be determined by the Architect. Any change in the Contract Sum resulting from such claim shall be authorized by Change Order.

12.3.2 If the Contractor claims that additional cost is involved because of, but not limited to, (1) any written interpretation pursuant to Subparagraph 2.2.8, (2) any order by the Owner to stop the Work pursuant to Paragraph 3.3 where the Contractor was not at fault, (3) any written order for a minor change in the Work issued pursuant to Paragraph 12.4, or (4) failure of payment by the Owner pursuant to Paragraph 9.7, the Contractor shall make such claim as provided in Subparagraph 12.3.1.

12.4 MINOR CHANGES IN THE WORK

12.4.1 The Architect will have authority to order minor changes in the Work not involving an adjustment in the Contract Sum or an extension of the Contract Time and not inconsistent with the intent of the Contract Documents. Such changes shall be effected by written order, and shall be binding on the Owner and the Contractor. The Contractor shall carry out such written orders promptly.

ARTICLE 13

UNCOVERING AND CORRECTION OF WORK

13.1 UNCOVERING OF WORK

13.1.1 If any portion of the Work should be covered contrary to the request of the Architect or to requirements specifically expressed in the Contract Documents, it must, if required in writing by the Architect, be uncovered for his observation and shall be replaced at the Contractor's expense.

13.1.2 If any other portion of the Work has been covered which the Architect has not specifically requested to observe prior to being covered, the Architect may request to see such Work and it shall be uncovered by the Contractor. If such Work be found in accordance with the Contract Documents, the cost of uncovering and replacement shall, by appropriate Change Order, be charged to the Owner. If such Work be found not in accordance with the Contract Documents, the Contractor shall pay such costs unless it be found that this condition was caused by the Owner or a separate contractor as provided in Article 6, in which event the Owner shall be responsible for the payment of such costs.

13.2 CORRECTION OF WORK

13.2.1 The Contractor shall promptly correct all Work rejected by the Architect as defective or as failing to conform to the Contract Documents whether observed before or after Substantial Completion and whether or not fabricated, installed or completed. The Contractor shall bear all costs of correcting such rejected Work, including compensation for the Architect's additional services made necessary thereby.

13.2.2 If, within one year after the Date of Substantial Completion of the Work or designated portion thereof or within one year after acceptance by the Owner of designated equipment or within such longer period of time as may be prescribed by law or by the terms of any applicable special warranty required by the Contract Documents, any of the Work is found to be defective or not in accordance with the Contract Documents, the Contractor shall correct it promptly after receipt of a written notice from the Owner to do so unless the Owner has previously given the Contractor a written acceptance of such condition. This obligation shall survive termination of the Contract. The Owner shall give such notice promptly after discovery of the condition.

13.2.3 The Contractor shall remove from the site all portions of the Work which are defective or non-conforming and which have not been corrected under Subparagraphs 4.5.1, 13.2.1 and 13.2.2, unless removal is waived by the Owner.

13.2.4 If the Contractor fails to correct defective or nonconforming Work as provided in Subparagraphs 4.5.1, 13.2.1 and 13.2.2, the Owner may correct it in accordance with Paragraph 3.4.

13.2.5 If the Contractor does not proceed with the correction of such defective or non-conforming Work within a reasonable time fixed by written notice from the Architect, the Owner may remove it and may store the materials or equipment at the expense of the Contractor. If the Contractor does not pay the cost of such removal and storage within ten days thereafter, the Owner may upon ten additional days' written notice sell such Work at auction or at private sale and shall account for the net proceeds thereof, after deducting all the costs that should have been borne by the Contractor, including compensation for the Architect's additional services made necessary thereby. If such proceeds of sale do not cover all costs which the Contractor should have borne, the difference shall be charged to the Contractor and an appropriate Change Order shall be issued. If the payments then or thereafter due the Contractor are not sufficient to cover such amount, the Contractor shall pay the difference to the Owner.

13.2.6 The Contractor shall bear the cost of making good all work of the Owner or separate contractors destroyed or damaged by such correction or removal.

13.2.7 Nothing contained in this Paragraph 13.2 shall be construed to establish a period of limitation with respect to any other obligation which the Contractor might have under the Contract Documents, including Paragraph 4.5 hereof. The establishment of the time period of one year after the Date of Substantial Completion or such longer period of time as may be prescribed by law or by the terms of any warranty required by the Contract Documents relates only to the specific obligation of the Contractor to correct the Work, and has no relationship to the time within which his obligation to comply with the Contract Documents may be sought to be enforced, nor to the time within which proceedings may be commenced to establish the Contractor's liability with respect to his obligations other than specifically to correct the Work.

13.3 ACCEPTANCE OF DEFECTIVE OR NON-CONFORMING WORK

13.3.1 If the Owner prefers to accept defective or non-conforming Work, he may do so instead of requiring its removal and correction, in which case a Change Order will be issued to reflect a reduction in the Contract Sum where appropriate and equitable. Such adjustment shall be effected whether or not final payment has been made.

ARTICLE 14

TERMINATION OF THE CONTRACT

14.1 TERMINATION BY THE CONTRACTOR

14.1.1 If the Work is stopped for a period of thirty days under an order of any court or other public authority having jurisdiction, or as a result of an act of government, such as a declaration of a national emergency making materials unavailable, through no act or fault of the Contractor or a Subcontractor or their agents or employees or any other persons performing any of the Work under a contract with the Contractor, or if the Work should be stopped for a period of thirty days by the Contractor because the Architect has not issued a Certificate for Payment as provided in Paragraph 9.7 or because the Owner has not made payment thereon as provided in Paragraph 9.7, then the Contractor may, upon seven additional days' written notice to the Owner and the Architect, terminate the Contract and recover from the Owner payment for all Work executed and for any proven loss sustained upon any materials, equipment, tools, construction equipment and machinery, including reasonable profit and damages.

14.2 TERMINATION BY THE OWNER

14.2.1 If the Contractor is adjudged a bankrupt, or if he makes a general assignment for the benefit of his creditors, or if a receiver is appointed on account of his insolvency, or if he persistently or repeatedly refuses or fails, except in cases for which extension of time is provided, to supply enough properly skilled workmen or proper materials, or if he fails to make prompt payment to Subcontractors or for materials or labor, or persistently disregards laws, ordinances, rules, regulations or orders of any public authority having jurisdiction, or otherwise is guilty of a substantial violation of a provision of the Contract Documents, then the Owner, upon certification by the Architect that sufficient cause exists to justify such action, may, without prejudice to any right or remedy and after giving the Contractor and his surety, if any, seven days' written notice, terminate the employment of the Contractor and take possession of the site and of all materials, equipment, tools, construction equipment and machinery thereon owned by the Contractor and may finish the Work by whatever method he may deem expedient. In such case the Contractor shall not be entitled to receive any further payment until the Work is finished.

14.2.2 If the unpaid balance of the Contract Sum exceeds the costs of finishing the Work, including compensation for the Architect's additional services made necessary thereby, such excess shall be paid to the Contractor. If such costs exceed the unpaid balance, the Contractor shall pay the difference to the Owner. The amount to be paid to the Contractor or to the Owner, as the case may be, shall be certified by the Architect, upon application, in the manner provided in Paragraph 9.4, and this obligation for payment shall survive the termination of the Contract.

APPENDIX B

Typical Considerations Effecting the Decision to Bid

TYPICAL CONSIDERATIONS: THE DECISION TO BID (OR NOT)*

A. Goals and Present Capabilities of Your Company (Plans for Growth; Type of Work; Market Conditions)

1. It is quite reasonable to actually want to stay where you are if you are satisfied with a situation of making a good living and staying active in work.
   • If so, is this job the kind you like doing? Does it have a good profit potential?
2. If you wish to grow larger, how fast do you wish to grow? Do you have the people and capital to do so?
   • Will the project to be bid help you in your growth?
   • Or will you have to bid it low just to keep your present men and equipment working, thus tying them up and postponing growth? (If you prefer type 1 goals, this latter strategy may be O.K.)
3. *Type of work.* Which type of work do you presently have the capability and experience to do? What types of work do you want to do in the future? Can you handle this particular project now? Will it give you good experience for the type of work you want to do in the future?
4. Consider the present and future competitive market conditions in this type of work.
   • Is it possible to earn a fair and reasonable profit? Or is the competition heavy?
   • Think of the job as an investment of your time, your talent and your money.
   It should earn a good return—in money, in satisfaction and pride; or provide some other return.

B. Location Of The Work

1. Is the project located in an area in which you normally like to operate?

*Based on material prepared by Prof. Boyd C. Paulson, Jr., Stanford University.

2. If not, would too large a portion of your time be consumed traveling to and from this job?
3. Do you have an associate or assistant who you believe can do a good job of supervising the job if you cannot often visit the site yourself?
4. Do you plan to expand your area of operations anyway, and if so, is this job in an area in which you want to expand?

C. Time And Place For Bid

1. When is the bid due (day and hour)? Will you have time to prepare an accurate and careful estimate? (For example, if you need two weeks to prepare a good bid and only 4 days remain, don't bid the job.)
2. Where is the bid to be submitted? How will you get it there? Do you have to allow two or three days for the mail?
3. Are there special rules for late delivery? For telegraphing last-minute changes?

D. How To Obtain Plans And Specifications

1. If you are a prime, you must find out who will provide the plans and specifications.
   - Is there a fee? How much?
   - Is there a deposit? How much? Is it refundable?
   - Is a plans room open and available? Where? What hours?
2. If you are a subcontractor, you want to know which prime contractors have plans and specifications.
   - Will they give you a copy of those that apply to your work?
   - Do they have a plans room for subcontractors? Where? What hours?
   - Can you get your plans and specifications directly from the owner? Fees? Deposits? How much? Refunds?

E. Legal And Other Official Requirements

1. *Licensing.* Some states, counties, cities, and towns require that a contractor have a license to work in their area.
   - If required, it is a legal necessity.
   - In some cases, unlicensed contractors can be fined without it.
   - Unlicensed contractors may not have recourse to the courts, even if wronged.
   - Especially note this when working on local government funded projects.
2. *Prequalification* may be required. If so, documents such as a financial statement, a statement of work in progress and experience, as well as a past litigation and performance history will be required.

3. *Bonding.*
   - Does project require (a) bid bond? (b) performance bond? (c) payment bond?
   - What is your bonding limit?
   - Can you qualify for bonds on this project.?

F. Scope of Work

1. What is the approximate size of the project (or subcontract):
   (a) In dollars—is it within your financial and bonding limits?
   (b) In major units of work (e.g. of earthmoving, cubic yards of concrete, pounds of steel, etc.) is it within the capacity of your available man power and equipment resources?)
2. What are the major types of work on the project or subcontract?
   (a) Are they the kind your company prefers to do?
   (b) Are they the kind your company is qualified to do?
3. How much time is available to complete the work?
   (a) When does it start; when does it finish?
   (b) How much other work do you plan to have going at that time? Can you handle this job as well?

G. Comparison of Resources

Compare the resources available to you to those that will be needed (order of magnitude only) on the job to be bid.

1. *Men*: Do you have a supervisor or foreman for the job? Can you get the laborers and craftsmen that will be needed?
2. *Equipment*: What major items of equipment (truck, crane, loader, etc.) will be needed? Do you own it already? Will it be available? Can you purchase new equipment? Can you rent or lease the equipment you will need?
3. *Money*: Will loans or credit be needed? How much? Can you get the financing needed?

H. Summary

All of these items should be considered in making the Decision To Bid or not bid on a particular job.

- This is an *executive decision*.
- It is a decision *you* as the contractor must make.

APPENDIX C

Performance and Payment Bonds

THE AMERICAN INSTITUTE OF ARCHITECTS

AIA Document A311

Performance Bond

KNOW ALL MEN BY THESE PRESENTS: that

(Here insert full name and address or legal title of Contractor)

as Principal, hereinafter called Contractor, and,

(Here insert full name and address or legal title of Surety)

as Surety, hereinafter called Surety, are held and firmly bound unto

(Here insert full name and address or legal title of Owner)

as Obligee, hereinafter called Owner, in the amount of

Dollars ($),

for the payment whereof Contractor and Surety bind themselves, their heirs, executors, administrators, successors and assigns, jointly and severally, firmly by these presents.

WHEREAS,

Contractor has by written agreement dated 19 , entered into a contract with Owner for

(Here insert full name, address and description of project)

in accordance with Drawings and Specifications prepared by

(Here insert full name and address or legal title of Architect)

which contract is by reference made a part hereof, and is hereinafter referred to as the Contract.

PERFORMANCE BOND

NOW, THEREFORE, THE CONDITION OF THIS OBLIGATION is such that, if Contractor shall promptly and faithfully perform said Contract, then this obligation shall be null and void; otherwise it shall remain in full force and effect.

The Surety hereby waives notice of any alteration or extension of time made by the Owner.

Whenever Contractor shall be, and declared by Owner to be in default under the Contract, the Owner having performed Owner's obligations thereunder, the Surety may promptly remedy the default, or shall promptly

1) Complete the Contract in accordance with its terms and conditions, or

2) Obtain a bid or bids for completing the Contract in accordance with its terms and conditions, and upon determination by Surety of the lowest responsible bidder, or, if the Owner elects, upon determination by the Owner and the Surety jointly of the lowest responsible bidder, arrange for a contract between such bidder and Owner, and make available as Work progresses (even though there should be a default or a succession of defaults under the contract or contracts of completion arranged under this paragraph) sufficient funds to pay the cost of completion less the balance of the contract price; but not exceeding, including other costs and damages for which the Surety may be liable hereunder, the amount set forth in the first paragraph hereof. The term "balance of the contract price," as used in this paragraph, shall mean the total amount payable by Owner to Contractor under the Contract and any amendments thereto, less the amount properly paid by Owner to Contractor.

Any suit under this bond must be instituted before the expiration of two (2) years from the date on which final payment under the Contract falls due.

No right of action shall accrue on this bond to or for the use of any person or corporation other than the Owner named herein or the heirs, executors, administrators or successors of the Owner.

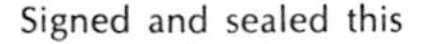

Signed and sealed this day of 19

	(Principal) (Seal)
(Witness)	
	(Title)
	(Surety) (Seal)
(Witness)	
	(Title)

THE AMERICAN INSTITUTE OF ARCHITECTS

AIA Document A311

Labor and Material Payment Bond

THIS BOND IS ISSUED SIMULTANEOUSLY WITH PERFORMANCE BOND IN FAVOR OF THE OWNER CONDITIONED ON THE FULL AND FAITHFUL PERFORMANCE OF THE CONTRACT

KNOW ALL MEN BY THESE PRESENTS: that

(Here insert full name and address or legal title of Contractor)

as Principal, hereinafter called Principal, and,

(Here insert full name and address or legal title of Surety)

as Surety, hereinafter called Surety, are held and firmly bound unto

(Here insert full name and address or legal title of Owner)

as Obligee, hereinafter called Owner, for the use and benefit of claimants as hereinbelow defined, in the amount of

(Here insert a sum equal to at least one-half of the contract price) Dollars ($),

for the payment whereof Principal and Surety bind themselves, their heirs, executors, administrators, successors and assigns, jointly and severally, firmly by these presents.

WHEREAS,

Principal has by written agreement dated 19 , entered into a contract with Owner for

(Here insert full name, address and description of project)

in accordance with Drawings and Specifications prepared by

(Here insert full name and address or legal title of Architect)

which contract is by reference made a part hereof, and is hereinafter referred to as the Contract.

LABOR AND MATERIAL PAYMENT BOND

NOW, THEREFORE, THE CONDITION OF THIS OBLIGATION is such that, if Principal shall promptly make payment to all claimants as hereinafter defined, for all labor and material used or reasonably required for use in the performance of the Contract, then this obligation shall be void; otherwise it shall remain in full force and effect, subject, however, to the following conditions:

1. A claimant is defined as one having a direct contract with the Principal or with a Subcontractor of the Principal for labor, material, or both, used or reasonably required for use in the performance of the Contract, labor and material being construed to include that part of water, gas, power, light, heat, oil, gasoline, telephone service or rental of equipment directly applicable to the Contract.

2. The above named Principal and Surety hereby jointly and severally agree with the Owner that every claimant as herein defined, who has not been paid in full before the expiration of a period of ninety (90) days after the date on which the last of such claimant's work or labor was done or performed, or materials were furnished by such claimant, may sue on this bond for the use of such claimant, prosecute the suit to final judgment for such sum or sums as may be justly due claimant, and have execution thereon. The Owner shall not be liable for the payment of any costs or expenses of any such suit.

3. No suit or action shall be commenced hereunder by any claimant:

a) Unless claimant, other than one having a direct contract with the Principal, shall have given written notice to any two of the following: the Principal, the Owner, or the Surety above named, within ninety (90) days after such claimant did or performed the last of the work or labor, or furnished the last of the materials for which said claim is made, stating with substantial accuracy the amount claimed and the name of the party to whom the materials were furnished, or for whom the work or labor was done or performed. Such notice shall be served by mailing the same by registered mail or certified mail, postage prepaid, in an envelope addressed to the Principal, Owner or Surety, at any place where an office is regularly maintained for the transaction of business, or served in any manner in which legal process may be served in the state in which the aforesaid project is located, save that such service need not be made by a public officer.

b) After the expiration of one (1) year following the date on which Principal ceased Work on said Contract, it being understood, however, that if any limitation embodied in this bond is prohibited by any law controlling the construction hereof such limitation shall be deemed to be amended so as to be equal to the minimum period of limitation permitted by such law.

c) Other than in a state court of competent jurisdiction in and for the county or other political subdivision of the state in which the Project, or any part thereof, is situated, or in the United States District Court for the district in which the Project, or any part thereof, is situated, and not elsewhere.

4. The amount of this bond shall be reduced by and to the extent of any payment or payments made in good faith hereunder, inclusive of the payment by Surety of mechanics' liens which may be filed of record against said improvement, whether or not claim for the amount of such lien be presented under and against this bond.

Signed and sealed this day of 19

(Principal) (Seal)

(Witness)

(Title)

(Surety) (Seal)

(Witness)

(Title)

APPENDIX D

AIA Document A101 Standard Form of Agreement between Owner and Contractor–Stipulated Sum

THE AMERICAN INSTITUTE OF ARCHITECTS

AIA Document A101

Standard Form of Agreement Between Owner and Contractor

where the basis of payment is a

STIPULATED SUM

1977 EDITION

THIS DOCUMENT HAS IMPORTANT LEGAL CONSEQUENCES; CONSULTATION WITH AN ATTORNEY IS ENCOURAGED WITH RESPECT TO ITS COMPLETION OR MODIFICATION

Use only with the 1976 Edition of AIA Document A201, General Conditions of the Contract for Construction.

This document has been approved and endorsed by The Associated General Contractors of America.

AGREEMENT

made as of the day of in the year of Nineteen Hundred and

BETWEEN the Owner:

and the Contractor:

The Project:

The Architect:

The Owner and the Contractor agree as set forth below.

ARTICLE 1

THE CONTRACT DOCUMENTS

The Contract Documents consist of this Agreement, the Conditions of the Contract (General, Supplementary and other Conditions), the Drawings, the Specifications, all Addenda issued prior to and all Modifications issued after execution of this Agreement. These form the Contract, and all are as fully a part of the Contract as if attached to this Agreement or repeated herein. An enumeration of the Contract Documents appears in Article 7.

ARTICLE 2

THE WORK

The Contractor shall perform all the Work required by the Contract Documents for

(Here insert the caption descriptive of the Work as used on other Contract Documents.)

ARTICLE 3

TIME OF COMMENCEMENT AND SUBSTANTIAL COMPLETION

The Work to be performed under this Contract shall be commenced

and, subject to authorized adjustments, Substantial Completion shall be achieved not later than

(Here Insert any special provisions for liquidated damages relating to failure to complete on time.)

ARTICLE 4

CONTRACT SUM

The Owner shall pay the Contractor in current funds for the performance of the Work, subject to additions and deductions by Change Order as provided in the Contract Documents, the Contract Sum of

The Contract Sum is determined as follows:
(State here the base bid or other lump sum amount, accepted alternates, and unit prices, as applicable.)

ARTICLE 5

PROGRESS PAYMENTS

Based upon Applications for Payment submitted to the Architect by the Contractor and Certificates for Payment issued by the Architect, the Owner shall make progress payments on account of the Contract Sum to the Contractor as provided in the Contract Documents for the period ending the day of the month as follows:

Not later than days following the end of the period covered by the Application for Payment percent (%) of the portion of the Contract Sum properly allocable to labor, materials and equipment incorporated in the Work and percent (%) of the portion of the Contract Sum properly allocable to materials and equipment suitably stored at the site or at some other location agreed upon in writing, for the period covered by the Application for Payment, less the aggregate of previous payments made by the Owner; and upon Substantial Completion of the entire Work, a sum sufficient to increase the total payments to percent (%) of the Contract Sum, less such amounts as the Architect shall determine for all incomplete Work and unsettled claims as provided in the Contract Documents.

(If not covered elsewhere in the Contract Documents, here insert any provision for limiting or reducing the amount retained after the Work reaches a certain stage of completion.)

Payments due and unpaid under the Contract Documents shall bear interest from the date payment is due at the rate entered below, or in the absence thereof, at the legal rate prevailing at the place of the Project.
(Here insert any rate of interest agreed upon.)

(Usury laws and requirements under the Federal Truth in Lending Act, similar state and local consumer credit laws and other regulations at the Owner's and Contractor's principal places of business, the location of the Project and elsewhere may affect the validity of this provision. Specific legal advice should be obtained with respect to deletion, modification, or other requirements such as written disclosures or waivers.)

ARTICLE 6

FINAL PAYMENT

Final payment, constituting the entire unpaid balance of the Contract Sum, shall be paid by the Owner to the Contractor when the Work has been completed, the Contract fully performed, and a final Certificate for Payment has been issued by the Architect.

ARTICLE 7

MISCELLANEOUS PROVISIONS

7.1 Terms used in this Agreement which are defined in the Conditions of the Contract shall have the meanings designated in those Conditions.

7.2 The Contract Documents, which constitute the entire agreement between the Owner and the Contractor, are listed in Article 1 and, except for Modifications issued after execution of this Agreement, are enumerated as follows:

(List below the Agreement, the Conditions of the Contract (General, Supplementary, and other Conditions), the Drawings, the Specifications, and any Addenda and accepted alternates, showing page or sheet numbers in all cases and dates where applicable.)

This Agreement entered into as of the day and year first written above.

OWNER CONTRACTOR

APPENDIX E

AIA Document A111 Standard Form of Agreement between Owner and Contractor–Cost of Work Plus a Fee

THE AMERICAN INSTITUTE OF ARCHITECTS

AIA Document A111

Standard Form of Agreement Between Owner and Contractor

where the basis of payment is the

COST OF THE WORK PLUS A FEE

1978 EDITION

THIS DOCUMENT HAS IMPORTANT LEGAL CONSEQUENCES; CONSULTATION WITH AN ATTORNEY IS ENCOURAGED WITH RESPECT TO ITS COMPLETION OR MODIFICATION

Use only with the 1976 Edition of AIA Document A201, General Conditions of the Contract for Construction.

This document has been approved and endorsed by The Associated General Contactors of America

AGREEMENT

made as of the day of in the year of Nineteen Hundred and

BETWEEN the Owner:

and the Contractor:

the Project:

the Architect:

The Owner and the Contractor agree as set forth below.

ARTICLE 1

THE CONTRACT DOCUMENTS

1.1 The Contract Documents consist of this Agreement, the Conditions of the Contract (General, Supplementary and other Conditions), the Drawings, the Specifications, all Addenda issued prior to and all Modifications issued after execution of this Agreement. These form the Contract, and all are as fully a part of the Contract as if attached to this Agreement or repeated herein. An enumeration of the Contract Documents appears in Article 16. If anything in the Contract Documents is inconsistent with this Agreement, the Agreement shall govern.

ARTICLE 2

THE WORK

2.1 The Contractor shall perform all the Work required by the Contract Documents for

(Here insert the caption descriptive of the Work as used on other Contract Documents.)

ARTICLE 3

THE CONTRACTOR'S DUTIES AND STATUS

3.1 The Contractor accepts the relationship of trust and confidence established between him and the Owner by this Agreement. He covenants with the Owner to furnish his best skill and judgment and to cooperate with the Architect in furthering the interests of the Owner. He agrees to furnish efficient business administration and superintendence and to use his best efforts to furnish at all times an adequate supply of workmen and materials, and to perform the Work in the best way and in the most expeditious and economical manner consistent with the interests of the Owner.

ARTICLE 4

TIME OF COMMENCEMENT AND SUBSTANTIAL COMPLETION

4.1 The Work to be performed under this Contract shall be commenced and, subject to authorized adjustments, Substantial Completion shall be achieved not later than

(Here insert any special provisions for liquidated damages relating to failure to complete on time.)

ARTICLE 5

COST OF THE WORK AND GUARANTEED MAXIMUM COST

5.1 The Owner agrees to reimburse the Contractor for the Cost of the Work as defined in Article 8. Such reimbursement shall be in addition to the Contractor's Fee stipulated in Article 6.

5.2 The maximum cost to the Owner, including the Cost of the Work and the Contractor's Fee, is guaranteed not to exceed the sum of dollars ($); such Guaranteed Maximum Cost shall be increased or decreased for Changes in the Work as provided in Article 7.

(Here insert any provision for distribution of any savings. Delete Paragraph 5.2 if there is no Guaranteed Maximum Cost.)

ARTICLE 6

CONTRACTOR'S FEE

6.1 In consideration of the performance of the Contract, the Owner agrees to pay the Contractor in current funds as compensation for his services a Contractor's Fee as follows:

6.2 For Changes in the Work, the Contractor's Fee shall be adjusted as follows:

6.3 The Contractor shall be paid percent (%) of the proportional amount of his Fee with each progress payment, and the balance of his Fee shall be paid at the time of final payment.

ARTICLE 7

CHANGES IN THE WORK

7.1 The Owner may make Changes in the Work as provided in the Contract Documents. The Contractor shall be reimbursed for Changes in the Work on the basis of Cost of the Work as defined in Article 8.

7.2 The Contractor's Fee for Changes in the Work shall be as set forth in Paragraph 6.2, or in the absence of specific provisions therein, shall be adjusted by negotiation on the basis of the Fee established for the original Work.

ARTICLE 8

COSTS TO BE REIMBURSED

8.1 The term Cost of the Work shall mean costs necessarily incurred in the proper performance of the Work and paid by the Contractor. Such costs shall be at rates not higher than the standard paid in the locality of the Work except with prior consent of the Owner, and shall include the items set forth below in this Article 8.

8.1.1 Wages paid for labor in the direct employ of the Contractor in the performance of the Work under applicable collective bargaining agreements, or under a salary or wage schedule agreed upon by the Owner and Contractor, and including such welfare or other benefits, if any, as may be payable with respect thereto.

8.1.2 Salaries of Contractor's personnel when stationed at the field office, in whatever capacity employed. Personnel engaged, at shops or on the road, in expediting the production or transportation of materials or equipment, shall be considered as stationed at the field office and their salaries paid for that portion of their time spent on this Work.

8.1.3 Cost of contributions, assessments or taxes incurred during the performance of the Work for such items as unemployment compensation and social security, insofar as such cost is based on wages, salaries, or other remuneration paid to employees of the Contractor and included in the Cost of the Work under Subparagraphs 8.1.1 and 8.1.2.

8.1.4 The portion of reasonable travel and subsistence expenses of the Contractor or of his officers or employees incurred while traveling in discharge of duties connected with the Work.

8.1.5 Cost of all materials, supplies and equipment incorporated in the Work, including costs of transportation thereof.

8.1.6 Payments made by the Contractor to Subcontractors for Work performed pursuant to subcontracts under this Agreement.

8.1.7 Cost, including transportation and maintenance, of all materials, supplies, equipment, temporary facilities and hand tools not owned by the workers, which are consumed in the performance of the Work, and cost less salvage value on such items used but not consumed which remain the property of the Contractor.

8.1.8 Rental charges of all necessary machinery and equipment, exclusive of hand tools, used at the site of the Work, whether rented from the Contractor or others, including installation, minor repairs and replacements, dismantling, removal, transportation and delivery costs thereof, at rental changes consistent with those prevailing in the area.

8.1.9 Cost of premiums for all bonds and insurance which the Contractor is required by the Contract Documents to purchase and maintain.

8.1.10 Sales, use or similar taxes related to the Work and for which the Contractor is liable imposed by any governmental authority.

8.1.11 Permit fees, royalties, damages for infringement of patents and costs of defending suits therefor, and deposits lost for causes other than the Contractor's negligence.

8.1.12 Losses and expenses, not compensated by insurance or otherwise, sustained by the Contractor in connection with the Work, provided they have resulted from causes other than the fault or neglect of the Contractor. Such losses shall include settlements made with the written consent and approval of the Owner. No such losses and expenses shall be included in the Cost of the Work for the purpose of determining the Contractor's Fee. If, however, such loss requires reconstruction and the Contractor is placed in charge thereof, he shall be paid for his services a Fee proportionate to that stated in Paragraph 6.1.

8.1.13 Minor expenses such as telegrams, long distance telephone calls, telephone service at the site, expressage, and similar petty cash items in connection with the Work.

8.1.14 Cost of removal of all debris.

8.1.15 Costs incurred due to an emergency affecting the safety of persons and property.

8.1.16 Other costs incurred in the performance of the Work if and to the extent approved in advance in writing by the Owner.

(Here insert modifications or limitations to any of the above Subparagraphs, such as equipment rental charges and small tool charges applicable to the Work.)

ARTICLE 9

COSTS NOT TO BE REIMBURSED

9.1 The term Cost of the Work shall not include any of the items set forth below in this Article 9.

9.1.1 Salaries or other compensation of the Contractor's personnel at the Contractor's principal office and branch offices.

9.1.2 Expenses of the Contractor's principal and branch offices other than the field office.

9.1.3 Any part of the Contractor's capital expenses, including interest on the Contractor's capital employed for the Work.

9.1.4 Except as specifically provided for in Subparagraph 8.1.8 or in modifications thereto, rental costs of machinery and equipment.

9.1.5 Overhead or general expenses of any kind, except as may be expressly included in Article 8.

9.1.6 Costs due to the negligence of the Contractor, any Subcontractor, anyone directly or indirectly employed by any of them, or for whose acts any of them may be liable, including but not limited to the correction of defective or nonconforming Work, disposal of materials and equipment wrongly supplied, or making good any damage to property.

9.1.7 The cost of any item not specifically and expressly included in the items described in Article 8.

9.1.8 Costs in excess of the Guaranteed Maximum Cost, if any, as set forth in Article 5 and adjusted pursuant to Article 7.

ARTICLE 10

DISCOUNTS, REBATES AND REFUNDS

10.1 All cash discounts shall accrue to the Contractor unless the Owner deposits funds with the Contractor with which to make payments, in which case the cash discounts shall accrue to the Owner. All trade discounts, rebates and refunds, and all returns from sale of surplus materials and equipment shall accrue to the Owner, and the Contractor shall make provisions so that they can be secured.

(Here insert any provisions relating to deposits by the Owner to permit the Contractor to obtain cash discounts.)

ARTICLE 11

SUBCONTRACTS AND OTHER AGREEMENTS

11.1 All portions of the Work that the Contractor's organization does not perform shall be performed under Subcontracts or by other appropriate agreement with the Contractor. The Contractor shall request bids from Subcontractors and shall deliver such bids to the Architect. The Owner will then determine, with the advice of the Contractor and subject to the reasonable objection of the Architect, which bids will be accepted.

11.2 All Subcontracts shall conform to the requirements of the Contract Documents. Subcontracts awarded on the basis of the cost of such work plus a fee shall also be subject to the provisions of this Agreement insofar as applicable.

ARTICLE 12

ACCOUNTING RECORDS

12.1 The Contractor shall check all materials, equipment and labor entering into the Work and shall keep such full and detailed accounts as may be necessary for proper financial management under this Agreement, and the system shall be satisfactory to the Owner. The Owner shall be afforded access to all the Contractor's records, books, correspondence, instructions, drawings, receipts, vouchers, memoranda and similar data relating to this Contract, and the Contractor shall preserve all such records for a period of three years, or for such longer period as may be required by law, after the final payment.

ARTICLE 13

APPLICATIONS FOR PAYMENT

13.1 The Contractor shall, at least ten days before each payment falls due, deliver to the Architect an itemized statement, notarized if required, showing in complete detail all moneys paid out or costs incurred by him on account of the Cost of the Work during the previous month for which he is to be reimbursed under Article 5 and the amount of the Contractor's Fee due as provided in Article 6, together with payrolls for all labor and such other data supporting the Contractor's right to payment for Subcontracts or materials as the Owner or the Architect may require.

ARTICLE 14

PAYMENTS TO THE CONTRACTOR

14.1 The Architect will review the Contractor's Applications for Payment and will promptly take appropriate action thereon as provided in the Contract Documents. Such amount as he may recommend for payment shall be payable by the Owner not later than the day of the month.

14.1.1 In taking action on the Contractor's Applications for Payment, the Architect shall be entitled to rely on the accuracy and completeness of the information furnished by the Contractor and shall not be deemed to represent that he has made audits of the supporting data, exhaustive or continuous on-site inspections or that he has made any examination to ascertain how or for what purposes the Contractor has used the moneys previously paid on account of the Contract.

14.2 Final payment, constituting the entire unpaid balance of the Cost of the Work and of the Contractor's Fee, shall be paid by the Owner to the Contractor days after Substantial Completion of the Work unless otherwise stipulated in the Certificate of Substantial Completion, provided the Work has been completed, the Contract fully performed, and final payment has been recommended by the Architect.

14.3 Payments due and unpaid under the Contract Documents shall bear interest from the date payment is due at the rate entered below, or in the absence thereof, at the legal rate prevailing at the place of the Project.

(Here insert any rate of interest agreed upon.)

(Usury laws and requirements under the Federal Truth in Lending Act, similar state and local consumer credit laws and other regulations at the Owner's and Contractor's principal places of business, the location of the Project and elsewhere may affect the validity of this provision. Specific legal advice should be obtained with respect to deletion, modification, or other requirements such as written disclosures or waivers.)

ARTICLE 15

TERMINATION OF CONTRACT

15.1 The Contract may be terminated by the Contractor as provided in the Contract Documents.

15.2 If the Owner terminates the Contract as provided in the Contract Documents, he shall reimburse the Contractor for any unpaid Cost of the Work due him under Article 5, plus (1) the unpaid balance of the Fee computed upon the Cost of the Work to the date of termination at the rate of the percentage named in Article 6, or (2) if the Contractor's Fee be stated as a fixed sum, such an amount as will increase the payments on account of his Fee to a sum which bears the same ratio to the said fixed sum as the Cost of the Work at the time of termination bears to the adjusted Guaranteed Maximum Cost, if any, otherwise to a reasonable estimated Cost of the Work when completed. The Owner shall also pay to the Contractor fair compensation, either by purchase or rental at the election of the Owner, for any equipment retained. In case of such termination of the Contract the Owner shall further assume and become liable for obligations, commitments and unsettled claims that the Contractor has previously undertaken or incurred in good faith in connection with said Work. The Contractor shall, as a condition of receiving the payments referred to in this Article 15, execute and deliver all such papers and take all such steps, including the legal assignment of his contractual rights, as the Owner may require for the purpose of fully vesting in himself the rights and benefits of the Contractor under such obligations or commitments.

ARTICLE 16

MISCELLANEOUS PROVISIONS

16.1 Terms used in this Agreement which are defined in the Contract Documents shall have the meanings designated in those Contract Documents.

16.2 The Contract Documents, which constitute the entire agreement between the Owner and the Contractor, are listed in Article 1 and, except for Modifications issued after execution of this Agreement, are enumerated as follows:

(List below the Agreement, the Conditions of the Contract, [General, Supplementary, and other Conditions], the Drawings, the Specifications, and any Addenda and accepted Alternates, showing page or sheet numbers in all cases and dates where applicable.)

This Agreement entered into as of the day and year first written above.

OWNER CONTRACTOR

APPENDIX F

AGC Sample Specification Implementation and Use of Network Scheduling Methods

A SAMPLE SPECIFICATION

PROJECT PLANNING AND SCHEDULING

SCOPE

The work specified in this section includes planning, scheduling and reporting required by the contractor.

METHOD

The project management tool commonly called CPM shall be employed by the contractor for planning, scheduling and reporting all work required by the Contract Documents.

QUALIFICATIONS

A statement of CPM capability shall be submitted in writing prior to the award of the contract and will verify that either the contractor's organization has "in-house capability" qualified to use the technique or that the contractor employs a consultant (firm) which is so qualified.

"Capability" shall be verified by description of construction projects. to which the contractor or his consultant has successfully applied CPM and shall include at least two (2) projects valued at at least half the expected value of this project, and at least one project which was controlled throughout the duration of the project by means of periodic systematic review of the CPM Schedule.

SUBMITTAL PROCEDURES

Within . . . . calendar days of notice to proceed, the contractor shall submit for approval of the Contracting Officer an arrow diagram describing the activities to be accomplished in the project and their dependency relationships as well as a tabulated schedule (as defined below). The schedule produced and submitted shall indicate a project completion date on or before the Contract Completion Date. The Contracting Officer, within . . . . calendar days of receipt of the arrow diagram, shall meet with a representative of the contractor so as to familiarize himself with the proposed plan and schedule.

Within . . . . calendar days of the conclusion of the familiarization period, the contractor shall revise the arrow diagram as required and resubmit the arrow diagram and a tabulated schedule produced therefrom. The revised arrow diagram and tabulated schedule shall be revised and approved or rejected by the Contracting Officer within . . . . calendar days of receipt. The arrow diagram and tabulated schedule when approved by the Contracting Officer shall constitute the Project Work Schedule until a revised schedule is submitted due to delays beyond the control and without the fault or negligence of the contractor.

When the arrow diagram and tabulated schedule have been approved, the contractor shall submit to the Contracting Officer . . . . copies of the arrow diagram . . . . copies of a tabulated schedule in which the activities have been sequenced by Early Starting Date and . . . . copies of a tabulated schedule in which the activities have been sequenced by Float.*

Additional Work Schedules. The Contracting Officer may require the contractor to provide an additional Work Schedule if, at any time, the Contracting Officer considers the completion date to be in jeopardy because of "activities behind schedule." The additional Work Schedule shall include a new arrow diagram and tabulated schedule designed to show how the contractor intends to accomplish the work to meet the completion date. The form and method employed by the contractor shall be same as for the original Work Schedule.

Schedule Revisions. The Contracting Officer may require the contractor to modify any portions of the work schedule that become infeasible because of "activities behind schedule" or for any other valid reason.

Activities Behind Schedule. An activity that cannot be completed by its original latest completion date shall be deemed to be behind schedule.

CHANGE ORDERS

On approval of a change order by the Contracting Officer, the approved change will be reflected in the next submission.

APPROVED STANDARDS

Definition. CPM as required by this section, shall be interpreted to be generally, as outlined in the Associated General Contractors publication, CPM in Construction - A Manual for General Contractors.

Work Schedules. Shall include a graphic network and tabulated schedules, as described below.

Networks, The CPM network, or arrow diagram, shall be of the customary activity-on-arrow type, and may be divided into a number of separate pages with suitable notation relating the interface points among the pages. Individual pages shall not exceed 36" x 60". Notation on each activity arrow shall include a brief work description and a duration estimate (see below)

*The sequences here requested are a chronological listing (Early Start Date), and a listing of activities by degree of criticalness (Float). The specifier should analyze his needs when requesting sequences of reports. He should be aware of the fact the the more sequences he requests, the more expense will be incurred. This generally does not apply to the number of copies requested.

Duration Estimates. The duration estimate indicated for each activity shall be in working days and shall represent the single best estimate considering the scope of the work and the resources planned for the activity. Except for certain none-labor activities, such as curing concrete, or delivering materials etc., activity durations will not exceed 10 days - or be less than 1 day unless otherwise approved by the Contracting Officer.*

Tabulated Schedules. The initial schedule shall include the following minimum data for each activity:

1. Activity Beginning and Ending Event Numbers
2. Estimated Duration
3. Activity Description
4. Early Start Date (Calendar Dated)
5. Early Finish Date (Calendar Dated)
6. Latest Allowable Start Date (Calendar Dated)
7. Latest Allowable Finish Date (Calendar Dated)
8. Status (Whether Critical)
9. Total Float

Project Information. Each Tabulation shall be prefaced with the following summary data:

1. Project Name
2. Contractor
3. Type of Tabulation (Initial or Updated)
4. Project Duration
5. Project Scheduled Completion Date
6. The Effective or Starting Date of the Schedule
7. If an updated (revised) schedule, the new project completion date and project status.

SCHEDULE MONITORING+

At ....(monthly, bi-monthly, weekly, etc.) intervals, and at the request of the contracting officer, the contractor shall submit to the contracting officer a revised schedule for those activities that remain to occur.

The revised schedule will be submitted in the sequences and copies requested for the initial schedule.

* This clause cannot be considered standard for all projects and each project should be examined carefully by the specifier. Heavy construction for example can involve activities that normally have durations substantially in excess of 10 days. This must be taken into consideration.

+ The specifier should be aware of the fact that schedule monitoring, while quite valuable, increases the cost of CPM's use. Likewise, the frequency of schedule monitoring relates directly to the cost involved. Schedule monitoring should be requested if CPM is to be effectively used, but care must be exercised in the determination of the frequency of such monitoring.

APPENDIX G

AIA Document B141
Standard Form of Agreement between Owner and Architect

THE AMERICAN INSTITUTE OF ARCHITECTS

AIA Document B141

Standard Form of Agreement Between Owner and Architect

1977 EDITION

THIS DOCUMENT HAS IMPORTANT LEGAL CONSEQUENCES; CONSULTATION WITH AN ATTORNEY IS ENCOURAGED WITH RESPECT TO ITS COMPLETION OR MODIFICATION

AGREEMENT

made as of the day of in the year of Nineteen Hundred and

BETWEEN the Owner:

and the Architect:

For the following Project:
(Include detailed description of Project location and scope.)

The Owner and the Architect agree as set forth below.

TERMS AND CONDITIONS OF AGREEMENT BETWEEN OWNER AND ARCHITECT

ARTICLE 1

ARCHITECT'S SERVICES AND RESPONSIBILITIES

BASIC SERVICES

The Architect's Basic Services consist of the five phases described in Paragraphs 1.1 through 1.5 and include normal structural, mechanical and electrical engineering services and any other services included in Article 15 as part of Basic Services.

1.1 SCHEMATIC DESIGN PHASE

1.1.1 The Architect shall review the program furnished by the Owner to ascertain the requirements of the Project and shall review the understanding of such requirements with the Owner.

1.1.2 The Architect shall provide a preliminary evaluation of the program and the Project budget requirements, each in terms of the other, subject to the limitations set forth in Subparagraph 3.2.1.

1.1.3 The Architect shall review with the Owner alternative approaches to design and construction of the Project.

1.1.4 Based on the mutually agreed upon program and Project budget requirements, the Architect shall prepare, for approval by the Owner, Schematic Design Documents consisting of drawings and other documents illustrating the scale and relationship of Project components.

1.1.5 The Architect shall submit to the Owner a Statement of Probable Construction Cost based on current area, volume or other unit costs.

1.2 DESIGN DEVELOPMENT PHASE

1.2.1 Based on the approved Schematic Design Documents and any adjustments authorized by the Owner in the program or Project budget, the Architect shall prepare, for approval by the Owner, Design Development Documents consisting of drawings and other documents to fix and describe the size and character of the entire Project as to architectural, structural, mechanical and electrical systems, materials and such other elements as may be appropriate.

1.2.2 The Architect shall submit to the Owner a further Statement of Probable Construction Cost.

1.3 CONSTRUCTION DOCUMENTS PHASE

1.3.1 Based on the approved Design Development Documents and any further adjustments in the scope or quality of the Project or in the Project budget authorized by the Owner, the Architect shall prepare, for approval by the Owner, Construction Documents consisting of Drawings and Specifications setting forth in detail the requirements for the construction of the Project.

1.3.2 The Architect shall assist the Owner in the preparation of the necessary bidding information, bidding forms, the Conditions of the Contract, and the form of Agreement between the Owner and the Contractor.

1.3.3 The Architect shall advise the Owner of any adjustments to previous Statements of Probable Construction Cost indicated by changes in requirements or general market conditions.

1.3.4 The Architect shall assist the Owner in connection with the Owner's responsibility for filing documents required for the approval of governmental authorities having jurisdiction over the Project.

1.4 BIDDING OR NEGOTIATION PHASE

1.4.1 The Architect, following the Owner's approval of the Construction Documents and of the latest Statement of Probable Construction Cost, shall assist the Owner in obtaining bids or negotiated proposals, and assist in awarding and preparing contracts for construction.

1.5 CONSTRUCTION PHASE—ADMINISTRATION OF THE CONSTRUCTION CONTRACT

1.5.1 The Construction Phase will commence with the award of the Contract for Construction and, together with the Architect's obligation to provide Basic Services under this Agreement, will terminate when final payment to the Contractor is due, or in the absence of a final Certificate for Payment or of such due date, sixty days after the Date of Substantial Completion of the Work, whichever occurs first.

1.5.2 Unless otherwise provided in this Agreement and incorporated in the Contract Documents, the Architect shall provide administration of the Contract for Construction as set forth below and in the edition of AIA Document A201, General Conditions of the Contract for Construction, current as of the date of this Agreement.

1.5.3 The Architect shall be a representative of the Owner during the Construction Phase, and shall advise and consult with the Owner. Instructions to the Contractor shall be forwarded through the Architect. The Architect shall have authority to act on behalf of the Owner only to the extent provided in the Contract Documents unless otherwise modified by written instrument in accordance with Subparagraph 1.5.16.

1.5.4 The Architect shall visit the site at intervals appropriate to the stage of construction or as otherwise agreed by the Architect in writing to become generally familiar with the progress and quality of the Work and to determine in general if the Work is proceeding in accordance with the Contract Documents. However, the Architect shall not be required to make exhaustive or continuous on-site inspections to check the quality or quantity of the Work. On the basis of such on-site observations as an architect, the Architect shall keep the Owner informed of the progress and quality of the Work, and shall endeavor to guard the Owner against defects and deficiencies in the Work of the Contractor.

1.5.5 The Architect shall not have control or charge of and shall not be responsible for construction means, methods, techniques, sequences or procedures, or for safety precautions and programs in connection with the Work, for the acts or omissions of the Contractor, Subcontractors or any other persons performing any of the Work, or for the failure of any of them to carry out the Work in accordance with the Contract Documents.

1.5.6 The Architect shall at all times have access to the Work wherever it is in preparation or progress.

1.5.7 The Architect shall determine the amounts owing to the Contractor based on observations at the site and on evaluations of the Contractor's Applications for Payment, and shall issue Certificates for Payment in such amounts, as provided in the Contract Documents.

1.5.8 The issuance of a Certificate for Payment shall constitute a representation by the Architect to the Owner, based on the Architect's observations at the site as provided in Subparagraph 1.5.4 and on the data comprising the Contractor's Application for Payment, that the Work has progressed to the point indicated; that, to the best of the Architect's knowledge, information and belief, the quality of the Work is in accordance with the Contract Documents (subject to an evaluation of the Work for conformance with the Contract Documents upon Substantial Completion, to the results of any subsequent tests required by or performed under the Contract Documents, to minor deviations from the Contract Documents correctable prior to completion, and to any specific qualifications stated in the Certificate for Payment); and that the Contractor is entitled to payment in the amount certified. However, the issuance of a Certificate for Payment shall not be a representation that the Architect has made any examination to ascertain how and for what purpose the Contractor has used the moneys paid on account of the Contract Sum.

1.5.9 The Architect shall be the interpreter of the requirements of the Contract Documents and the judge of the performance thereunder by both the Owner and Contractor. The Architect shall render interpretations necessary for the proper execution or progress of the Work with reasonable promptness on written request of either the Owner or the Contractor, and shall render written decisions, within a reasonable time, on all claims, disputes and other matters in question between the Owner and the Contractor relating to the execution or progress of the Work or the interpretation of the Contract Documents.

1.5.10 Interpretations and decisions of the Architect shall be consistent with the intent of and reasonably inferable from the Contract Documents and shall be in written or graphic form. In the capacity of interpreter and judge, the Architect shall endeavor to secure faithful performance by both the Owner and the Contractor, shall not show partiality to either, and shall not be liable for the result of any interpretation or decision rendered in good faith in such capacity.

1.5.11 The Architect's decisions in matters relating to artistic effect shall be final if consistent with the intent of the Contract Documents. The Architect's decisions on any other claims, disputes or other matters, including those in question between the Owner and the Contractor, shall be subject to arbitration as provided in this Agreement and in the Contract Documents.

1.5.12 The Architect shall have authority to reject Work which does not conform to the Contract Documents. Whenever, in the Architect's reasonable opinion, it is necessary or advisable for the implementation of the intent of the Contract Documents, the Architect will have authority to require special inspection or testing of the Work in accordance with the provisions of the Contract Documents, whether or not such Work be then fabricated, installed or completed.

1.5.13 The Architect shall review and approve or take other appropriate action upon the Contractor's submittals such as Shop Drawings, Product Data and Samples, but only for conformance with the design concept of the Work and with the information given in the Contract Documents. Such action shall be taken with reasonable promptness so as to cause no delay. The Architect's approval of a specific item shall not indicate approval of an assembly of which the item is a component.

1.5.14 The Architect shall prepare Change Orders for the Owner's approval and execution in accordance with the Contract Documents, and shall have authority to order minor changes in the Work not involving an adjustment in the Contract Sum or an extension of the Contract Time which are not inconsistent with the intent of the Contract Documents.

1.5.15 The Architect shall conduct inspections to determine the Dates of Substantial Completion and final completion, shall receive and forward to the Owner for the Owner's review written warranties and related documents required by the Contract Documents and assembled by the Contractor, and shall issue a final Certificate for Payment.

1.5.16 The extent of the duties, responsibilities and limitations of authority of the Architect as the Owner's representative during construction shall not be modified or extended without written consent of the Owner, the Contractor and the Architect.

1.6 PROJECT REPRESENTATION BEYOND BASIC SERVICES

1.6.1 If the Owner and Architect agree that more extensive representation at the site than is described in Paragraph 1.5 shall be provided, the Architect shall provide one or more Project Representatives to assist the Architect in carrying out such responsibilities at the site.

1.6.2 Such Project Representatives shall be selected, employed and directed by the Architect, and the Architect shall be compensated therefor as mutually agreed between the Owner and the Architect as set forth in an exhibit appended to this Agreement, which shall describe the duties, responsibilities and limitations of authority of such Project Representatives.

1.6.3 Through the observations by such Project Representatives, the Architect shall endeavor to provide further protection for the Owner against defects and deficiencies in the Work, but the furnishing of such project representation shall not modify the rights, responsibilities or obligations of the Architect as described in Paragraph 1.5.

1.7 ADDITIONAL SERVICES

The following Services are not included in Basic Services unless so identified in Article 15. They shall be provided if authorized or confirmed in writing by the Owner, and they shall be paid for by the Owner as provided in this Agreement, in addition to the compensation for Basic Services.

1.7.1 Providing analyses of the Owner's needs, and programming the requirements of the Project.

1.7.2 Providing financial feasibility or other special studies.

1.7.3 Providing planning surveys, site evaluations, environmental studies or comparative studies of prospective sites, and preparing special surveys, studies and submissions required for approvals of governmental authorities or others having jurisdiction over the Project.

1.7.4 Providing services relative to future facilities, systems and equipment which are not intended to be constructed during the Construction Phase.

1.7.5 Providing services to investigate existing conditions or facilities or to make measured drawings thereof, or to verify the accuracy of drawings or other information furnished by the Owner.

1.7.6 Preparing documents of alternate, separate or sequential bids or providing extra services in connection with bidding, negotiation or construction prior to the completion of the Construction Documents Phase, when requested by the Owner.

1.7.7 Providing coordination of Work performed by separate contractors or by the Owner's own forces.

1.7.8 Providing services in connection with the work of a construction manager or separate consultants retained by the Owner.

1.7.9 Providing Detailed Estimates of Construction Cost, analyses of owning and operating costs, or detailed quantity surveys or inventories of material, equipment and labor.

1.7.10 Providing interior design and other similar services required for or in connection with the selection, procurement or installation of furniture, furnishings and related equipment.

1.7.11 Providing services for planning tenant or rental spaces.

1.7.12 Making revisions in Drawings, Specifications or other documents when such revisions are inconsistent with written approvals or instructions previously given, are required by the enactment or revision of codes, laws or regulations subsequent to the preparation of such documents or are due to other causes not solely within the control of the Architect.

1.7.13 Preparing Drawings, Specifications and supporting data and providing other services in connection with Change Orders to the extent that the adjustment in the Basic Compensation resulting from the adjusted Construction Cost is not commensurate with the services required of the Architect, provided such Change Orders are required by causes not solely within the control of the Architect.

1.7.14 Making investigations, surveys, valuations, inventories or detailed appraisals of existing facilities, and services required in connection with construction performed by the Owner.

1.7.15 Providing consultation concerning replacement of any Work damaged by fire or other cause during construction, and furnishing services as may be required in connection with the replacement of such Work.

1.7.16 Providing services made necessary by the default of the Contractor, or by major defects or deficiencies in the Work of the Contractor, or by failure of performance of either the Owner or Contractor under the Contract for Construction.

1.7.17 Preparing a set of reproducible record drawings showing significant changes in the Work made during construction based on marked-up prints, drawings and other data furnished by the Contractor to the Architect.

1.7.18 Providing extensive assistance in the utilization of any equipment or system such as initial start-up or testing, adjusting and balancing, preparation of operation and maintenance manuals, training personnel for operation and maintenance, and consultation during operation.

1.7.19 Providing services after issuance to the Owner of the final Certificate for Payment, or in the absence of a final Certificate for Payment, more than sixty days after the Date of Substantial Completion of the Work.

1.7.20 Preparing to serve or serving as an expert witness in connection with any public hearing, arbitration proceeding or legal proceeding.

1.7.21 Providing services of consultants for other than the normal architectural, structural, mechanical and electrical engineering services for the Project.

1.7.22 Providing any other services not otherwise included in this Agreement or not customarily furnished in accordance with generally accepted architectural practice.

1.8 TIME

1.8.1 The Architect shall perform Basic and Additional Services as expeditiously as is consistent with professional skill and care and the orderly progress of the Work. Upon request of the Owner, the Architect shall submit for the Owner's approval, a schedule for the performance of the Architect's services which shall be adjusted as required as the Project proceeds, and shall include allowances for periods of time required for the Owner's review and approval of submissions and for approvals of authorities having jurisdiction over the Project. This schedule, when approved by the Owner, shall not, except for reasonable cause, be exceeded by the Architect.

ARTICLE 2

THE OWNER'S RESPONSIBILITIES

2.1 The Owner shall provide full information regarding requirements for the Project including a program, which shall set forth the Owner's design objectives, constraints and criteria, including space requirements and relationships, flexibility and expandability, special equipment and systems and site requirements.

2.2 If the Owner provides a budget for the Project it shall include contingencies for bidding, changes in the Work during construction, and other costs which are the responsibility of the Owner, including those described in this Article 2 and in Subparagraph 3.1.2. The Owner shall, at the request of the Architect, provide a statement of funds available for the Project, and their source.

2.3 The Owner shall designate, when necessary, a representative authorized to act in the Owner's behalf with respect to the Project. The Owner or such authorized representative shall examine the documents submitted by the Architect and shall render decisions pertaining thereto promptly, to avoid unreasonable delay in the progress of the Architect's services.

2.4 The Owner shall furnish a legal description and a certified land survey of the site, giving, as applicable, grades and lines of streets, alleys, pavements and adjoining property; rights-of-way, restrictions, easements, encroachments, zoning, deed restrictions, boundaries and contours of the site; locations, dimensions and complete data pertaining to existing buildings, other improvements and trees; and full information concerning available service and utility lines both public and private, above and below grade, including inverts and depths.

2.5 The Owner shall furnish the services of soil engineers or other consultants when such services are deemed necessary by the Architect. Such services shall include test borings, test pits, soil bearing values, percolation tests, air and water pollution tests, ground corrosion and resistivity tests, including necessary operations for determining subsoil, air and water conditions, with reports and appropriate professional recommendations.

2.6 The Owner shall furnish structural, mechanical, chemical and other laboratory tests, inspections and reports as required by law or the Contract Documents.

2.7 The Owner shall furnish all legal, accounting and insurance counseling services as may be necessary at any time for the Project, including such auditing services as the Owner may require to verify the Contractor's Applications for Payment or to ascertain how or for what purposes the Contractor uses the moneys paid by or on behalf of the Owner.

2.8 The services, information, surveys and reports required by Paragraphs 2.4 through 2.7 inclusive shall be furnished at the Owner's expense, and the Architect shall be entitled to rely upon the accuracy and completeness thereof.

2.9 If the Owner observes or otherwise becomes aware of any fault or defect in the Project or nonconformance with the Contract Documents, prompt written notice thereof shall be given by the Owner to the Architect.

2.10 The Owner shall furnish required information and services and shall render approvals and decisions as expeditiously as necessary for the orderly progress of the Architect's services and of the Work.

ARTICLE 3

CONSTRUCTION COST

3.1 DEFINITION

3.1.1 The Construction Cost shall be the total cost or estimated cost to the Owner of all elements of the Project designed or specified by the Architect.

3.1.2 The Construction Cost shall include at current market rates, including a reasonable allowance for overhead and profit, the cost of labor and materials furnished by the Owner and any equipment which has been designed, specified, selected or specially provided for by the Architect.

3.1.3 Construction Cost does not include the compensation of the Architect and the Architect's consultants, the cost of the land, rights-of-way, or other costs which are the responsibility of the Owner as provided in Article 2.

3.2 RESPONSIBILITY FOR CONSTRUCTION COST

3.2.1 Evaluations of the Owner's Project budget, Statements of Probable Construction Cost and Detailed Estimates of Construction Cost, if any, prepared by the Architect, represent the Architect's best judgment as a design professional familiar with the construction industry. It is recognized, however, that neither the Architect nor the Owner has control over the cost of labor, materials or equipment, over the Contractor's methods of determining bid prices, or over competitive bidding, market or negotiating conditions. Accordingly, the Architect cannot and does not warrant or represent that bids or negotiated prices will not vary from the Project budget proposed, established or approved by the Owner, if any, or from any Statement of Probable Construction Cost or other cost estimate or evaluation prepared by the Architect.

3.2.2 No fixed limit of Construction Cost shall be established as a condition of this Agreement by the furnishing, proposal or establishment of a Project budget under Subparagraph 1.1.2 or Paragraph 2.2 or otherwise, unless such fixed limit has been agreed upon in writing and signed by the parties hereto. If such a fixed limit has been established, the Architect shall be permitted to include contingencies for design, bidding and price escalation, to determine what materials, equipment, component systems and types of construction are to be included in the Contract Documents, to make reasonable adjustments in the scope of the Project and to include in the Contract Documents alternate bids to adjust the Construction Cost to the fixed limit. Any such fixed limit shall be increased in the amount of any increase in the Contract Sum occurring after execution of the Contract for Construction.

3.2.3 If the Bidding or Negotiation Phase has not commenced within three months after the Architect submits the Construction Documents to the Owner, any Project budget or fixed limit of Construction Cost shall be adjusted to reflect any change in the general level of prices in the construction industry between the date of submission of the Construction Documents to the Owner and the date on which proposals are sought.

3.2.4 If a Project budget or fixed limit of Construction Cost (adjusted as provided in Subparagraph 3.2.3) is exceeded by the lowest bona fide bid or negotiated proposal, the Owner shall (1) give written approval of an increase in such fixed limit, (2) authorize rebidding or renegotiating of the Project within a reasonable time, (3) if the Project is abandoned, terminate in accordance with Paragraph 10.2, or (4) cooperate in revising the Project scope and quality as required to reduce the Construction Cost. In the case of (4), provided a fixed limit of Construction Cost has been established as a condition of this Agreement, the Architect, without additional charge, shall modify the Drawings and Specifications as necessary to comply with the fixed limit. The providing of such service shall be the limit of the Architect's responsibility arising from the establishment of such fixed limit, and having done so, the Architect shall be entitled to compensation for all services performed, in accordance with this Agreement, whether or not the Construction Phase is commenced.

ARTICLE 4

DIRECT PERSONNEL EXPENSE

4.1 Direct Personnel Expense is defined as the direct salaries of all the Architect's personnel engaged on the Project, and the portion of the cost of their mandatory and customary contributions and benefits related thereto, such as employment taxes and other statutory employee benefits, insurance, sick leave, holidays, vacations, pensions and similar contributions and benefits.

ARTICLE 5

REIMBURSABLE EXPENSES

5.1 Reimbursable Expenses are in addition to the Compensation for Basic and Additional Services and include actual expenditures made by the Architect and the Architect's employees and consultants in the interest of the Project for the expenses listed in the following Subparagraphs:

5.1.1 Expense of transportation in connection with the Project; living expenses in connection with out-of-town travel; long distance communications, and fees paid for securing approval of authorities having jurisdiction over the Project.

5.1.2 Expense of reproductions, postage and handling of Drawings, Specifications and other documents, excluding reproductions for the office use of the Architect and the Architect's consultants.

5.1.3 Expense of data processing and photographic production techniques when used in connection with Additional Services.

5.1.4 If authorized in advance by the Owner, expense of overtime work requiring higher than regular rates.

5.1.5 Expense of renderings, models and mock-ups requested by the Owner.

5.1.6 Expense of any additional insurance coverage or limits, including professional liability insurance, requested by the Owner in excess of that normally carried by the Architect and the Architect's consultants.

ARTICLE 6

PAYMENTS TO THE ARCHITECT

6.1 PAYMENTS ON ACCOUNT OF BASIC SERVICES

6.1.1 An initial payment as set forth in Paragraph 14.1 is the minimum payment under this Agreement.

6.1.2 Subsequent payments for Basic Services shall be made monthly and shall be in proportion to services performed within each Phase of services, on the basis set forth in Article 14.

6.1.3 If and to the extent that the Contract Time initially established in the Contract for Construction is exceeded or extended through no fault of the Architect, compensation for any Basic Services required for such extended period of Administration of the Construction Contract shall be computed as set forth in Paragraph 14.4 for Additional Services.

6.1.4 When compensation is based on a percentage of Construction Cost, and any portions of the Project are deleted or otherwise not constructed, compensation for such portions of the Project shall be payable to the extent services are performed on such portions, in accordance with the schedule set forth in Subparagraph 14.2.2, based on (1) the lowest bona fide bid or negotiated proposal or, (2) if no such bid or proposal is received, the most recent Statement of Probable Construction Cost or Detailed Estimate of Construction Cost for such portions of the Project.

6.2 PAYMENTS ON ACCOUNT OF ADDITIONAL SERVICES

6.2.1 Payments on account of the Architect's Additional Services as defined in Paragraph 1.7 and for Reimbursable Expenses as defined in Article 5 shall be made monthly upon presentation of the Architect's statement of services rendered or expenses incurred.

6.3 PAYMENTS WITHHELD

6.3.1 No deductions shall be made from the Architect's compensation on account of penalty, liquidated damages or other sums withheld from payments to contractors, or on account of the cost of changes in the Work other than those for which the Architect is held legally liable.

6.4 PROJECT SUSPENSION OR TERMINATION

6.4.1 If the Project is suspended or abandoned in whole or in part for more than three months, the Architect shall be compensated for all services performed prior to receipt of written notice from the Owner of such suspension or abandonment, together with Reimbursable Expenses then due and all Termination Expenses as defined in Paragraph 10.4. If the Project is resumed after being suspended for more than three months, the Architect's compensation shall be equitably adjusted.

ARTICLE 7

ARCHITECT'S ACCOUNTING RECORDS

7.1 Records of Reimbursable Expenses and expenses pertaining to Additional Services and services performed on the basis of a Multiple of Direct Personnel Expense shall be kept on the basis of generally accepted accounting principles and shall be available to the Owner or the Owner's authorized representative at mutually convenient times.

ARTICLE 8

OWNERSHIP AND USE OF DOCUMENTS

8.1 Drawings and Specifications as instruments of service are and shall remain the property of the Architect whether the Project for which they are made is executed or not. The Owner shall be permitted to retain copies, including reproducible copies, of Drawings and Specifications for information and reference in connection with the Owner's use and occupancy of the Project. The Drawings and Specifications shall not be used by the Owner on other projects, for additions to this Project, or for completion of this Project by others provided the Architect is not in default under this Agreement, except by agreement in writing and with appropriate compensation to the Architect.

8.2 Submission or distribution to meet official regulatory requirements or for other purposes in connection with the Project is not to be construed as publication in derogation of the Architect's rights.

ARTICLE 9

ARBITRATION

9.1 All claims, disputes and other matters in question between the parties to this Agreement, arising out of or relating to this Agreement or the breach thereof, shall be decided by arbitration in accordance with the Construction Industry Arbitration Rules of the American Arbitration Association then obtaining unless the parties mutually agree otherwise. No arbitration, arising out of or relating to this Agreement, shall include, by consolidation, joinder or in any other manner, any additional person not a party to this Agreement except by written consent containing a specific reference to this Agreement and signed by the Architect, the Owner, and any other person sought to be joined. Any consent to arbitration involving an additional person or persons shall not constitute consent to arbitration of any dispute not described therein or with any person not named or described therein. This Agreement to arbitrate and any agreement to arbitrate with an additional person or persons duly consented to by the parties to this Agreement shall be specifically enforceable under the prevailing arbitration law.

9.2 Notice of the demand for arbitration shall be filed in writing with the other party to this Agreement and with the American Arbitration Association. The demand shall be made within a reasonable time after the claim, dispute or other matter in question has arisen. In no event shall the demand for arbitration be made after the date when institution of legal or equitable proceedings based on such claim, dispute or other matter in question would be barred by the applicable statute of limitations.

9.3 The award rendered by the arbitrators shall be final, and judgment may be entered upon it in accordance with applicable law in any court having jurisdiction thereof.

ARTICLE 10

TERMINATION OF AGREEMENT

10.1 This Agreement may be terminated by either party upon seven days' written notice should the other party fail substantially to perform in accordance with its terms through no fault of the party initiating the termination.

10.2 This Agreement may be terminated by the Owner upon at least seven days' written notice to the Architect in the event that the Project is permanently abandoned.

10.3 In the event of termination not the fault of the Architect, the Architect shall be compensated for all services performed to termination date, together with Reimbursable Expenses then due and all Termination Expenses as defined in Paragraph 10.4.

10.4 Termination Expenses include expenses directly attributable to termination for which the Architect is not otherwise compensated, plus an amount computed as a percentage of the total Basic and Additional Compensation earned to the time of termination, as follows:

.1 20 percent if termination occurs during the Schematic Design Phase; or

.2 10 percent if termination occurs during the Design Development Phase; or

.3 5 percent if termination occurs during any subsequent phase.

ARTICLE 11

MISCELLANEOUS PROVISIONS

11.1 Unless otherwise specified, this Agreement shall be governed by the law of the principal place of business of the Architect.

11.2 Terms in this Agreement shall have the same meaning as those in AIA Document A201, General Conditions of the Contract for Construction, current as of the date of this Agreement.

11.3 As between the parties to this Agreement: as to all acts or failures to act by either party to this Agreement, any applicable statute of limitations shall commence to run and any alleged cause of action shall be deemed to have accrued in any and all events not later than the relevant Date of Substantial Completion of the Work, and as to any acts or failures to act occurring after the relevant Date of Substantial Completion, not later than the date of issuance of the final Certificate for Payment.

11.4 The Owner and the Architect waive all rights against each other and against the contractors, consultants, agents and employees of the other for damages covered by any property insurance during construction as set forth in the edition of AIA Document A201, General Conditions, current as of the date of this Agreement. The Owner and the Architect each shall require appropriate similar waivers from their contractors, consultants and agents.

ARTICLE 12

SUCCESSORS AND ASSIGNS

12.1 The Owner and the Architect, respectively, bind themselves, their partners, successors, assigns and legal representatives to the other party to this Agreement and to the partners, successors, assigns and legal representatives of such other party with respect to all covenants of this Agreement. Neither the Owner nor the Architect shall assign, sublet or transfer any interest in this Agreement without the written consent of the other.

ARTICLE 13

EXTENT OF AGREEMENT

13.1 This Agreement represents the entire and integrated agreement between the Owner and the Architect and supersedes all prior negotiations, representations or agreements, either written or oral. This Agreement may be amended only by written instrument signed by both Owner and Architect.

ARTICLE 14

BASIS OF COMPENSATION

The Owner shall compensate the Architect for the Scope of Services provided, in accordance with Article 6, Payments to the Architect, and the other Terms and Conditions of this Agreement, as follows:

14.1 AN INITIAL PAYMENT of dollars ($)

shall be made upon execution of this Agreement and credited to the Owner's account as follows:

14.2 BASIC COMPENSATION

14.2.1 FOR BASIC SERVICES, as described in Paragraphs 1.1 through 1.5, and any other services included in Article 15 as part of Basic Services, Basic Compensation shall be computed as follows:

(Here insert basis of compensation, including fixed amounts, multiples or percentages, and identify Phases to which particular methods of compensation apply, if necessary.)

14.2.2 Where compensation is based on a Stipulated Sum or Percentage of Construction Cost, payments for Basic Services shall be made as provided in Subparagraph 6.1.2, so that Basic Compensation for each Phase shall equal the following percentages of the total Basic Compensation payable:

(Include any additional Phases as appropriate.)

Schematic Design Phase: percent (%)
Design Development Phase: percent (%)
Construction Documents Phase: percent (%)
Bidding or Negotiation Phase: percent (%)
Construction Phase: percent (%)

14.3 FOR PROJECT REPRESENTATION BEYOND BASIC SERVICES, as described in Paragraph 1.6, Compensation shall be computed separately in accordance with Subparagraph 1.6.2.

14.4 COMPENSATION FOR ADDITIONAL SERVICES

14.4.1 FOR ADDITIONAL SERVICES OF THE ARCHITECT, as described in Paragraph 1.7, and any other services included in Article 15 as part of Additional Services, but excluding Additional Services of consultants, Compensation shall be computed as follows:

(Here insert basis of compensation, including rates and/or multiples of Direct Personnel Expense for Principals and employees, and identify Principals and classify employees, if required. Identify specific services to which particular methods of compensation apply, if necessary.)

14.4.2 FOR ADDITIONAL SERVICES OF CONSULTANTS, including additional structural, mechanical and electrical engineering services and those provided under Subparagraph 1.7.21 or identified in Article 15 as part of Additional Services, a multiple of ( ) times the amounts billed to the Architect for such services.

(Identify specific types of consultants in Article 15, if required.)

14.5 FOR REIMBURSABLE EXPENSES, as described in Article 5, and any other items included in Article 15 as Reimbursable Expenses, a multiple of ( ) times the amounts expended by the Architect, the Architect's employees and consultants in the interest of the Project.

14.6 Payments due the Architect and unpaid under this Agreement shall bear interest from the date payment is due at the rate entered below, or in the absence thereof, at the legal rate prevailing at the principal place of business of the Architect.

(Here insert any rate of interest agreed upon.)

(Usury laws and requirements under the Federal Truth in Lending Act, similar state and local consumer credit laws and other regulations at the Owner's and Architect's principal places of business, the location of the Project and elsewhere may affect the validity of this provision. Specific legal advice should be obtained with respect to deletion, modification, or other requirements such as written disclosures or waivers.)

14.7 The Owner and the Architect agree in accordance with the Terms and Conditions of this Agreement that:

14.7.1 IF THE SCOPE of the Project or of the Architect's Services is changed materially, the amounts of compensation shall be equitably adjusted.

14.7.2 IF THE SERVICES covered by this Agreement have not been completed within

( ) months of the date hereof, through no fault of the Architect, the amounts of compensation, rates and multiples set forth herein shall be equitably adjusted.

ARTICLE 15

OTHER CONDITIONS OR SERVICES

This Agreement entered into as of the day and year first written above.

OWNER	ARCHITECT
______________________	______________________
______________________	______________________
______________________	______________________
BY ______________________	BY ______________________

APPENDIX H

AGC Builders Association of Chicago: Typical Agent Job Descriptions

PROJECT MANAGER

A. General Functions

The project manager in the construction industry is usually an "inside" and "outside" man. The position may vary considerably from company to company. The project manager in some companies may be an estimator, and expeditor, and even handle some duties normally done by the job superintendent, while with other companies he may merely supervise superintendents.

B. Detailed Functions

1. May procure the invitation to bid on jobs.
2. May, when working as an estimator, prepare bids.
3. May handle the legal requirements for a contract.
4. May negotiate the specialty contractor's arrangements and agreements.
5. Set up completion schedules by bar graph or critical path method.
6. Supervise subcontractors and coordinate their material deliveries.
7. Arrange for sufficient man power for the project.
8. Supervise superintendents on the job—"walk the job" each day to see progress being made and, during this time, review the work with a superintendent.
9. Control the movement of workers from one job to another.
10. Arrange for permits from the city, county, etc.
11. Hire and fire superintendents, foremen, engineers, and other personnel under his supervision.
12. Set up occupancy dates for buildings.
13. Act as public relations representative.
14. Coordinate with architect and owner requested revisions or errors found in drawings.

ESTIMATOR

A. General Functions

An estimator makes as close an estimate as possible of what the costs will be. In order to do so, he must itemize all of the building materials and calculate labor costs for the entire project—the cost estimate may also include a percentage for profit, though this may be done with or by top management.

B. Detailed Functions

1. Mail or telephone bid proposals to subcontractors.
2. Follow up with subcontractors on submission of their bids.
3. Review bid with subcontractors.
4. Prior to bid, inspect job site to determine access and that the land is the same as on the plans. Look for water conditions and other problems that might arise.
5. Analyze plans and specification, that is, "learn the job."
6. Make a take-off for each type of work to be done by general contractor forces.
7. Does take-off for subcontractors when necessary.
8. May sit in on owner, architect, and contractor conferences.
9. May check on other estimator's work or have his work checked by another estimator.
10. Price the quantity take-offs.
11. Read prints, noting discrepancies.
12. Make itemized lists of prices for materials.
13. Review and preview subcontractors' bids.
14. In some companies, purchase steel, lumber, and all other materials necessary for the job.
15. Compute a percentage for overhead and profit, which would be added to estimated cost.
16. Arrive at final bid price or cost price.
17. Prepare change order (estimates cost of changes) as needed or required—if major item and not handled by field personnel.
18. Expedite distribution of plans, including general and mechanical.
19. Serve in quality control capacity, because of position in purchasing, and plan review.
20. Make bar graphs, network, or CPM for scheduling.
21. Make cost breakdown of work performed by company forces for cost control purposes.

EXPEDITOR

A. General Functions

An expeditor may schedule or coordinate job material requirements. He serves as a trouble shooter when there is a breakdown in delivery schedule. He foresees problems by reviewing plans and specifications of the subcontractors and coordinating these with the plans and specifications of the architect.

B. Detailed Functions

1. In some companies receives the plans and specifications and breaks the specifications down by trade.
2. In some companies writes to all subcontractors advising them what is necessary to do on their plans.
3. Follows up on drawings, that is, shop drawings or the detailed drawings of project.
4. Submits drawings to the architect after having checked them to see if they match, that the job is correct, that the materials used are those specified, and analyzes the drawings.
5. Maintains constant follow-up on plans and drawings to ensure documents reach the proper place at the proper time.
6. Distributes approved plans to subs, or to anyone else who should get them. Has to order enough plans from subcontractors so that entrusted parties will have sufficient documentation for project schedule.
7. In some companies establishes delivery time for materials, equipment, or labor, based on when they will be required and when they can be acquired, and determines the lead time required for acquisition.
8. May follow a CPM printout, make out delivery schedules, use a bar graph method or the critical path method. Makes sure the shop items are on the critical path method or the bar graph.
9. In some companies does small buying such as purchasing mailboxes, signs, and finish items.
10. Maintains constant follow-up to ensure that schedule is accomplished.
11. Checks all incoming tests to ensure they meet specifications.
12. In some companies checks the concrete design, that is, the mix or fixed formula of the concrete used.
13. Plans material delivery and schedules with job superintendents.
14. Keeps in contact each day with subcontractors.
15. Writes memos as needed to architects, superintendents, subcontractors, etc.
16. Follows up daily on trouble areas, that is, those places where delivery of materials may be lagging.
17. In some companies accumulates change order information.
18. Generally troubleshoots, especially for delivery problems.

EQUIPMENT SUPERINTENDENT

A. General Functions

Maintains and repairs equipment owned by company. To do this, he supervises garage and yard personnel and coordinates delivery of equipment to the specific job sites and expedites repairs and deliveries.

B. Detailed Functions

1. Supervises, maintains, and repairs.
2. Purchases parts for maintenance and repairs.
3. Expedites repairs of equipment that cannot be done in the company garage.
4. Keeps detailed records of equipment, including maintenance costs for each piece of equipment.
5. Keeps track of equipment, that is, which job is using it at any given time.
6. Makes recommendations concerning purchase of new equipment.
7. Provides delivery of equipment to job sites, helps to plan the time, provides means of delivery to the site, and provides equipment setup at site.
8. Keeps weekly repair costs on his crew.
9. Prepares an annual budget for operation.

FIELD SUPERINTENDENT

A. General Functions

Builds the building. Manages men and materials on the job site so that the project is built for profit. Coordinates schedules so that men and materials are available to promote efficient erection of the building at a profit level.

B. Detailed Functions

1. "Learns the building." Studies plans and specifications so that he can plan the work to be accomplished.
2. Tries to anticipate problems.
3. Studies the costs.
4. Arranges scheduling and manufacture of building parts or components.
5. Coordinates building when the manufactured items become available for the building.
6. Does survey and layout work or supervises technical or field engineer who does this.
7. Keeps constant check on all trades, overseeing workmanship and materials.
8. Hires and fires workmen.
9. Supplies information to Accounting Department so that records of costs can be maintained.
10. Supervises men directly or indirectly (i.e., through the foreman).
11. May be responsible for deliveries.
12. Is responsible for drawings and seeing that drawings are made of changes or incomplete items.
13. Arranges for plan changes as needed.
14. May be responsible for written schedules or physical schedules.
15. Does on-the-spot estimating (material or labor).
16. May price out extra items or charges.
17. Does limited buying (supplies and items missed by the purchasing department).
18. Makes daily safety inspections.
19. May record daily field activities in a log.

MECHANICAL SUPERINTENDENT

A. General Functions

The mechanical superintendent coordinates subcontractor's work with that of the general contractor to ensure that project remains on schedule and quality is maintained.

B. Detailed Functions

1. Compiles listing of major mechanical electrical equipment required.
2. Expedites shop drawings and equipment deliveries.
3. Assists in preparation of project schedules.
4. Prepares weekly progress reports on electrical and mechanical work.
5. Coordinates subcontractors' work with general contractor.
6. Checks schedule to ensure project is on schedule.
7. Supervises general contractor's work done for subcontractors (equipment production, excavations, etc.).
8. Processes and distributes shop drawings.
9. Supervises, inspects, and evaluates work performed by subcontractors—ensures there is compliance with plans and specifications.
10. Supervises project closely to ensure that the owner is getting his money's worth on subcontractor work.

SCHEDULING ENGINEER (FIELD ENGINEER)

A. General Functions

Scheduling engineer schedules and coordinates. He serves as a troubleshooter when there is a breakdown in delivery schedule. He maintains a constant follow-up on the schedule to ensure progress as previously planned.

B. Detailed Functions

1. Receives plans and specifications and breaks them down by trade.
2. Writes to all subcontractors telling them when their work is necessary on the schedule.
3. Expedites follow-up for drawings, that is, shop drawings or detailed drawings. Checks with own staff for follow-up.
4. Keeps a close follow-up to ensure that plans and drawings reach the right people at the right time.
5. Establishes delivery times for materials, equipment, or labor, based on when they can be acquired, and determines the lead time required for acquisitions.
6. Makes out delivery schedules, using a bar graph method or critical path method. Makes sure that the shop items are on the bar graph or CPM.
7. Discusses material delivery and scheduling with job superintendent.
8. Keeps in touch with subcontractors as needed.
9. Writes memos as needed to superintendents, subcontractors, etc.
10. Follows up daily on trouble areas, where delivery of materials may be lagging.
11. Generally troubleshoots.

TIMEKEEPER

A. General Functions

A timekeeper is primarily concerned with maintaining cost control of the labor force on a project. He maintains payroll records and may also maintain records on material deliveries.

B. Detailed Functions

1. Ensures that the men are on the job, checks what specific tasks they are performing, and checks this against job sheets given to him daily by the foreman.
2. Checks with the foreman to determine exact job and classification of work each man is doing so that the work can be coded and entered against the correct account.
3. Walks-the-job a few times each day.
4. Computes previous day's work sheets to obtain costs.
5. Projects daily costs to determine if work was completed within the allocated budget.
6. Talks over costs with the superintendent.
7. Posts workers' hours to the payroll on a daily basis.
8. Types a cost report each week. In some companies this may be done by central office staff.
9. Types payroll each week. In some companies this may be done by central office staff.
10. Types paychecks each week. In some companies this may be done by central office staff.
11. Types all back charges and time tickets.
12. Estimates costs of requests from subcontractors for sheds, shanties, carpenters, and concrete work performed for them.
13. May compile subcontractors' invoices for payment and discuss these with architect to determine accuracy.
14. Codes all delivery tickets to maintain costs on all building parts.
15. Keeps records of all reinforcing steel deliveries.
16. Records all concrete pours.
17. May assist superintendent by ordering labor, lumber, and other materials.
18. On certain big load days may call the union halls for extra men; will sign these men for the day and pay them by check at night.

19. Signs up all new workmen (W-4 forms, applications, etc.) and submits originals to central office.
20. Enters new employees' names and proper wage rate for the particular trade on payroll.
21. Types monthly report on welfare and pension. In some companies this may be done at the central office.
22. Balances the payroll and types it each Monday. Submits it to the main office so that checks can be made out and returned to the job site by Wednesday. In some companies this may be done at central office.
23. May travel to various job sites and perform same duties for each of the projects.
24. On projects involving federal funds, he collects payroll data from subcontractors for submission to the government in compliance with their regulations.
25. Maintains time record on company truck drivers when material deliveries are made.
26. May supervise "time checkers" on larger project.

APPENDIX I
AGC Standard Subcontract Agreement

STANDARD SUBCONTRACT AGREEMENT

THIS AGREEMENT made this day of in the year Nineteen

Hundred and by and between

hereinafter

called the Subcontractor and

hereinafter called the Contractor.

WITNESSETH, That the Subcontractor and Contractor for the consideration hereinafter named agree as follows:

(Developed as a guide by The Associated General Contractors of America, The National Electrical Contractors Association, The Mechanical Contractors Association of America, The Sheet Metal and Air Conditioning Contractors National Association and the National Association of Plumbing - Heating - Cooling Contractors

ARTICLE I

The Subcontractor agrees to furnish all material and perform all work as described in Article II hereof for

(Here name the project.)

for

(Here name the Contractor.)

at

(Here insert the location of the work and name of Owner.)

in accordance with this Agreement, the Agreement between the Owner and Contractor, and in accordance with the General Conditions of the Contract, Supplementary General Conditions, the Drawings and Specifications and addenda prepared by

hereinafter called the Architect or Owner's authorized agent, all of which documents, signed by the parties thereto or identified by the Architect or Owner's authorized agent, form a part of a Contract

between the Contractor and the Owner dated , 19 , and hereby become a part of this contract, and herein referred to as the Contract Documents, and shall be made available to the Subcontractor upon his request prior to and at anytime subsequent to signing this Subcontract.

ARTICLE II

The Subcontractor and the Contractor agree that the materials and equipment to be furnished and work to be done by the Subcontractor are:

(Here insert a precise description of the work, preferably by reference to the numbers of the drawings and the pages of the specifications including addenda and accepted alternates.)

ARTICLE III

Time is of the essence and the Subcontractor agrees to commence and to complete the work as described in Article II as follows:

(Here insert any information pertaining to the method of notification for commencement of work, starting and completion dates, or duration, and any liquidated damage requirements.)

(a) No extension of time of this contract will be recognized without the written consent of the Contractor which consent shall not be withheld unreasonably consistent with Article X-4 of this Contract, subject to the arbitration provisions herein provided.

ARTICLE IV

The Contractor agrees to pay the Subcontractor for the performance of this work the

sum of ($)
in current funds, subject to additions and deductions for changes as may be agreed upon in writing, and to make monthly payments on account thereof in accordance with Article X, Sections 20-23 inclusive.

(Here insert additional details—unit prices, etc., payment procedure including date of monthly applications for payment, payment procedure if other than on a monthly basis, consideration of materials safely and suitably stored at the site or at some other location agreed upon in writing by the parties—and any provisions made for limiting or reducing the amount retained after the work reaches a certain stage of completion which should be consistent with the Contract Documents.)

ARTICLE V

Final payment shall be due when the work described in this contract is fully completed and performed in accordance with the Contract Documents, and payment to be consistent with Article IV and Article X, Sections 18, 20-23 inclusive of this contract.

Before issuance of the final payment the Subcontractor if required shall submit evidence satisfactory to the Contractor that all payrolls, material bills, and all known indebtedness connected with the Subcontractor's work have been satisfied.

ARTICLE VI

Performance and Payment Bonds

(Here insert any requirement for the furnishing of performance and payment bonds.)

ARTICLE VII

Temporary Site Facilities

(Here insert any requirements and terms concerning temporary site facilities, i.e., storage, sheds, water, heat, light, powe toilets, hoists, elevators, scaffolding, cold weather protection, ventilating, pumps, watchman service, etc.)

ARTICLE VIII

Insurance

Unless otherwise provided herein, the Subcontractor shall have a direct liability for the acts of his employees and agents for which he is legally responsible, and the Subcontractor shall not be required to assume the liability for the acts of any others.

Prior to starting work the insurance required to be furnished shall be obtained from a responsible company or companies to provide proper and adequate coverage and satisfactory evidence will be furnished to the Contractor that the Subcontractor has complied with the requirements as stated in this Section.

(Here insert any insurance requirements and Subcontractor's responsibility for obtaining, maintaining and paying for necessary insurance, not less than limits as may be specified in the Contract Documents or required by laws. This to include fire insurance and extended coverage, consideration of public liability, property damage, employer's liability, and workmen's compensation insurance for the Subcontractor and his employees. The insertion should provide the agreement of the Contractor and the Subcontractor on subrogation waivers, provision for notice of cancellation, allocation of insurance proceeds, and other aspects of insurance.)

(It is recommended that the AGC Insurance and Bonds Checklist (AGC Form No. 29) be referred to as a guide for other insurance coverages.)

ARTICLE IX

Job Conditions

(Here insert any applicable arrangements and necessary cooperation concerning labor matters for the project.)

ARTICLE X

In addition to the foregoing provisions the parties also agree:
That the Subcontractor shall:

(1) Be bound to the Contractor by the terms of the Contractor Documents and this Agreement, and assume toward the Contractor all the obligations and responsibilities that the Contractor, by those documents, assumes toward the Owner, as applicable to this Subcontract. (a) Not discriminate against any employee or applicant for employment because of race, creed, color, or national origin.

(2) Submit to the Contractor applications for payment at such times as stipulated in Article IV so as to enable the Contractor to apply for payment.

If payments are made on valuations of work done, the Subcontractor shall, before the first application, submit to the Contractor a schedule of values of the various parts of the work, aggregating the total sum of the Contract, made out in such detail as the Subcontractor and Contractor may agree upon, or as required by the Owner, and, if required, supported by such evidence as to its correctness as the Contractor may direct. This schedule, when approved by the Contractor, shall be used as a basis for Certificates for Payment, unless it be found to be in error. In applying for payment, the Subcontractor shall submit a statement based upon this schedule.

If payments are made on account of materials not incorporated in the work but delivered and suitably stored at the site, or at some other location agreed upon in writing, such payments shall be in accordance with the terms and conditions of the Contract Documents.

(3) Pay for all materials and labor used in, or in connection with, the performance of this contract, through the period covered by previous payments received from the Contractor, and furnish satisfactory evidence when requested by the Contractor, to verify compliance with the above requirements.

(4) Make all claims for extras, for extensions of time and for damage for delays or otherwise, promptly to the Contractor consistent with the Contract Documents.

(5) Take necessary precaution to properly protect the finished work of other trades.

(6) Keep the building and premises clean at all times of debris arising out of the operation of this subcontract. The Subcontractor shall not be held responsible for unclean conditions caused by other contractors or subcontractors, unless otherwise provided for.

(7) Comply with all statutory and/or contractual safety requirements applying to his work and/or initiated by the Contractor, and shall report within 3 days to the Contractor any injury to the Subcontractor's employees at the site of the project.

(8) (a) Not assign this subcontract or any amounts due or to become due thereunder without the written consent of the contractor. (b) Nor subcontract the whole of this subcontract without the written consent of the contractor. (c) Nor further subcontract portions of this subcontract without written notification to the contractor when such notification is requested by the contractor.

(9) Guarantee his work against all defects of materials and/or workmanship as called for in the plans, specifications and addenda, or if no guarantee is called for, then for a period of one year from the dates of partial or total acceptance of the Subcontractor's work by the Owner.

(10) And does hereby agree that if the Subcontractor should neglect to prosecute the work diligently and properly or fail to perform any provision of this contract, the Contractor, after three days written notice to the Subcontractor, may, without prejudice to any other remedy he may have, make good such deficiencies and may deduct the cost thereof from the payment then or thereafter due the Subcontractor, provided, however, that if such action is based upon faulty workmanship the Architect or Owner's authorized agent, shall first have determined that the workmanship and/or materials is defective.

(11) And does hereby agree that the Contractor's equipment will be available to the Subcontractor only at the Contractor's discretion and on mutually satisfactory terms.

(12) Furnish periodic progress reports of the work as mutually agreed including the progress of materials or equipment under this Agreement that may be in the course of preparation or manufacture.

(13) Make any and all changes or deviations from the original plans and specifications without nullifying the original contract when specifically ordered to do so in writing by the Contractor. The Subcontractor prior to the commencement of this revised work, shall submit promptly to the Contractor written copies of the cost or credit proposal for such revised work in a manner consistent with the Contract Documents.

(14) Cooperate with the Contractor and other Subcontractors whose work might interfere with the Subcontractor's work and to participate in the preparation of coordinated drawings in areas of congestion as required by the Contract Documents, specifically noting and advising the Contractor of any such interference.

(15) Cooperate with the Contractor in scheduling his work so as not conflict or interfere with the work of others. To promptly submit shop drawings, drawings, and samples, as required in order to carry on said work efficiently and at speed that will not cause delay in the progress of the Contractor's work or other branches of the work carried on by other Subcontractors.

(16) Comply with all Federal, State and local laws and ordinances applying to the building or structure and to comply and give adequate notices relating to the work to proper authorities and to secure and pay for all necessary licenses or permits to carry on the work as described in the Contract Documents as applicable to this Subcontract.

(17) Comply with Federal, State and local tax laws, Social Security laws and Unemployment Compensation laws and Workmen's Compensation Laws insofar as applicable to the performance of this subcontract.

(18) And does hereby agree that all work shall be done subject to the final approval of the Architect or Owner's authorized agent, and his decision in matters relating to artistic effect shall be final, if within the terms of the Contract Documents.

That the Contractor shall—

(19) Be bound to the Subcontractor by all the obligations that the Owner assumes to the Contractor under the Contract Documents and by all the provisions thereof affording remedies and redress to the Contractor from the Owner insofar as applicable to this Subcontract.

(20) Pay the Subcontractor within seven days, unless otherwise provided in the Contract Documents, upon the payment of certificates issued under the Contractor's schedule of values, or as described in Article IV herein. The amount of the payment shall be equal to the percentage of completion certified by the Owner or his authorized agent for the work of this Subcontractor applied to the amount set forth under Article IV and allowed to the Contractor on account of the Subcontractor's work to the extent of the Subcontractor's interest therein.

(21) Permit the Subcontractor to obtain direct from the Architect or Owner's authorized agent, evidence of percentages of completion certified on his account.

(22) Pay the Subcontractor on demand for his work and/or materials as far as executed and fixed in place, less the retained percentage, at the time the payment should be made to the Subcontractor if the Architect or Owner's authorized agent fails to issue the certificate for any fault of the Contractor and not the fault of the Subcontractor or as otherwise provided herein.

(23) And does hereby agree that the failure to make payments to the Subcontractor as herein provided for any cause not the fault of the Subcontractor, within 7 days from the Contractor's receipt of payment or from time payment should be made as

provided in Article X, Section 22, or maturity, then the Subcontractor may upon 7 days written notice to the Contractor stop work without prejudice to any other remedy he may have.

(24) Not issue or give any instructions, order or directions directly to employees or workmen of the Subcontractor other than to the persons designated as the authorized representative(s) of the Subcontractor.

(25) Make no demand for liquidated damages in any sum in excess of such amount as may be specifically named in the subcontract, provided, however, no liquidated damages shall be assessed for delays or causes attributable to other Subcontractors or arising outside the scope of this Subcontract.

(26) And does hereby agree that no claim for services rendered or materials furnished by the Contractor to the Subcontractor shall be valid unless written notice thereof is given by the Contractor to the Subcontractor during the first ten days of the calendar month following that in which the claim originated.

(27) Give the Subcontractor an opportunity to be present and to submit evidence in any arbitration involving his rights.

(28) Name as arbitor under arbitration proceedings as provided in the General Conditions the person nominated by the Subcontractor, if the sole cause of dispute is the work, materials, rights or responsibilities of the Subcontractor; or if, of the Subcontractor and any other Subcontractor jointly, to name as such arbitrator the person upon whom they agree.

That the Contractor and the Subcontractor agree—

(29) That in the matter of arbitration, their rights and obligations and all procedure shall be analogous to those set forth in the Contract Documents provided, however, that a decision by the Architect or Owner's authorized agent, shall not be a condition precedent to arbitration.

(30) This subcontract is solely for the benefit of the signatories hereto.

ARTICLE XI

IN WITNESS WHEREOF the parties hereto have executed this Agreement under seal, the day and year first above written.

Attest: ______________________ Subcontractor

______________________ (Seal) By ______________________ (Title)

Attest: ______________________ Contractor

______________________ (Seal) By ______________________ (Title)

APPENDIX J
Interest Tables

5% FACTORS INTEREST

	Single Payment		Uniform Series				
n	Compound Amount Factor CAF	Present Worth Factor PWSP	Compound Amount Factor USCA	Sinking Fund Factor SFF	Present Worth Factor PWUS	Capital Recovery Factor CRF	n
	Given P to Find F $(1+i)^n$	Given F to Find P $\frac{1}{(1+i)^n}$	Given A to Find F $\frac{(1+i)^n-1}{i}$	Given F to Find A $\frac{i}{(1+i)^n-1}$	Given A to Find P $\frac{(1+i)^n-1}{i(1+i)^n}$	Given P to Find A $\frac{i(1+i)^n}{(1+i)^n-1}$	
1	1.050	0.9524	1.000	1.00001	0.952	1.05001	1
2	1.102	0.9070	2.050	0.48781	1.859	0.53781	2
3	1.158	0.8638	3.152	0.31722	2.723	0.36722	3
4	1.216	0.8227	4.310	0.23202	3.546	0.28202	4
5	1.276	0.7835	5.526	0.18098	4.329	0.23098	5
6	1.340	0.7462	6.802	0.14702	5.076	0.19702	6
7	1.407	0.7107	8.142	0.12282	5.786	0.17282	7
8	1.477	0.6768	9.549	0.10472	6.463	0.15472	8
9	1.551	0.6446	11.026	0.09069	7.108	0.14069	9
10	1.629	0.6139	12.578	0.07951	7.722	0.12951	10
11	1.710	0.5847	14.206	0.07039	8.306	0.12039	11
12	1.796	0.5568	15.917	0.06283	8.863	0.11283	12
13	1.886	0.5303	17.712	0.05646	9.393	0.10646	13
14	1.980	0.5051	19.598	0.05103	9.899	0.10103	14
15	2.079	0.4810	21.578	0.04634	10.380	0.09634	15
16	2.183	0.4581	23.657	0.04227	10.838	0.09227	16
17	2.292	0.4363	25.840	0.03870	11.274	0.08870	17
18	2.407	0.4155	28.132	0.03555	11.689	0.08555	18
19	2.527	0.3957	30.538	0.03275	12.085	0.08275	19
20	2.653	0.3769	33.065	0.03024	12.462	0.08024	20
21	2.786	0.3589	35.718	0.02800	12.821	0.07800	21
22	2.925	0.3419	38.504	0.02597	13.163	0.07597	22
23	3.071	0.3256	41.429	0.02414	13.488	0.07414	23
24	3.225	0.3101	44.500	0.02247	13.798	0.07247	24
25	3.386	0.2953	47.725	0.02095	14.094	0.07095	25
26	3.556	0.2812	51.112	0.01957	14.375	0.06956	26
27	3.733	0.2679	54.667	0.01829	14.643	0.06829	27
28	3.920	0.2551	58.400	0.01712	14.898	0.06712	28
29	4.116	0.2430	62.320	0.01605	15.141	0.06605	29
30	4.322	0.2314	66.436	0.01505	15.372	0.06505	30
35	5.516	0.1813	90.316	0.01107	16.374	0.06107	35
40	7.040	0.1421	120.794	0.00828	17.159	0.05828	40
50	11.467	0.0872	209.336	0.00478	18.256	0.05478	50
75	38.830	0.0258	756.594	0.00132	19.485	0.05132	75
100	131.488	0.0076	2609.761	0.00038	19.848	0.05038	100

6% FACTORS INTEREST

	Single Payment		Uniform Series				
	Compound Amount Factor CAF	Present Worth Factor PWSP	Compound Amount Factor USCA	Sinking Fund Factor SFF	Present Worth Factor PWUS	Capital Recovery Factor CRF	
n	Given P to Find F $(1+i)^n$	Given F to Find P $\frac{1}{(1+i)^n}$	Given A to Find F $\frac{(1+i)^n-1}{i}$	Given F to Find A $\frac{i}{(1+i)^n-1}$	Given A to Find P $\frac{(1+i)^n-1}{i(1+i)^n}$	Given P to Find A $\frac{i(1+i)^n}{(1+i)^n-1}$	n
1	1.060	0.9434	1.000	1.00001	0.943	1.06001	1
2	1.124	0.8900	2.060	0.48544	1.833	0.54544	2
3	1.191	0.8396	3.184	0.31411	2.673	0.37411	3
4	1.262	0.7921	4.375	0.22860	3.465	0.28860	4
5	1.338	0.7473	5.637	0.17740	4.212	0.23740	5
6	1.419	0.7050	6.975	0.14337	4.917	0.20337	6
7	1.504	0.6651	8.394	0.11914	5.582	0.17914	7
8	1.594	0.6274	9.897	0.10104	6.210	0.16104	8
9	1.689	0.5919	11.491	0.08702	6.802	0.14702	9
10	1.791	0.5584	13.181	0.07587	7.360	0.13587	10
11	1.898	0.5268	14.971	0.06679	7.887	0.12679	11
12	2.012	0.4970	16.870	0.05928	8.384	0.11928	12
13	2.133	0.4688	18.882	0.05296	8.853	0.11296	13
14	2.261	0.4423	21.015	0.04759	9.295	0.10759	14
15	2.397	0.4173	23.275	0.04296	9.712	0.10296	15
16	2.540	0.3937	25.672	0.03895	10.106	0.09895	16
17	2.693	0.3714	28.212	0.03545	10.477	0.09545	17
18	2.854	0.3503	30.905	0.03236	10.828	0.09236	18
19	3.026	0.3305	33.759	0.02962	11.158	0.08962	19
20	3.207	0.3118	36.785	0.02719	11.470	0.08719	20
21	3.399	0.2942	39.992	0.02501	11.764	0.08501	21
22	3.603	0.2775	43.391	0.02305	12.041	0.08305	22
23	3.820	0.2618	46.994	0.02128	12.303	0.08128	23
24	4.049	0.2470	50.814	0.01968	12.550	0.07968	24
25	4.292	0.2330	54.863	0.01823	12.783	0.07823	25
26	4.549	0.2198	59.154	0.01690	13.003	0.07690	26
27	4.822	0.2074	63.704	0.01570	13.210	0.07570	27
28	5.112	0.1956	68.526	0.01459	13.406	0.07459	28
29	5.418	0.1846	73.637	0.01358	13.591	0.07358	29
30	5.743	0.1741	79.055	0.01265	13.765	0.07265	30
35	7.686	0.1301	111.430	0.00897	14.498	0.06897	35
40	10.285	0.0972	154.755	0.00646	15.046	0.06646	40
50	18.419	0.0543	290.321	0.00344	15.762	0.06344	50
75	79.051	0.0127	1300.852	0.00077	16.456	0.06077	75
100	339.269	0.0029	5637.809	0.00018	16.618	0.06018	100

7% FACTORS INTEREST

n	Single Payment		Uniform Series				n
	Compound Amount Factor CAF	Present Worth Factor PWSP	Compound Amount Factor USCA	Sinking Fund Factor SFF	Present Worth Factor PWUS	Capital Recovery Factor CRF	
	Given P to Find F $(1+i)^n$	Given F to Find P $\frac{1}{(1+i)^n}$	Given A to Find F $\frac{(1+i)^n-1}{i}$	Given F to Find A $\frac{i}{(1+i)^n-1}$	Given A to Find P $\frac{(1+i)^n-1}{i(1+i)^n}$	Given P to Find A $\frac{i(1+i)^n}{(1+i)^n-1}$	
1	1.070	0.9346	1.000	1.00000	0.935	1.07000	1
2	1.145	0.8734	2.070	0.48310	1.808	0.55310	2
3	1.225	0.8163	3.215	0.31106	2.624	0.38105	3
4	1.311	0.7629	4.440	0.22523	3.387	0.29523	4
5	1.403	0.7130	5.751	0.17389	4.100	0.24389	5
6	1.501	0.6663	7.153	0.13980	4.766	0.20980	6
7	1.606	0.6228	8.654	0.11555	5.389	0.18555	7
8	1.718	0.5820	10.260	0.09747	5.971	0.16747	8
9	1.838	0.5439	11.978	0.08349	6.515	0.15349	9
10	1.967	0.5084	13.816	0.07238	7.024	0.14238	10
11	2.105	0.4751	15.783	0.06336	7.499	0.13336	11
12	2.252	0.4440	17.888	0.05590	7.943	0.12590	12
13	2.410	0.4150	20.140	0.04965	8.358	0.11965	13
14	2.579	0.3878	22.550	0.04435	8.745	0.11435	14
15	2.759	0.3625	25.129	0.03980	9.108	0.10980	15
16	2.952	0.3387	27.887	0.03586	9.447	0.10586	16
17	3.159	0.3166	30.840	0.03243	9.763	0.10243	17
18	3.380	0.2959	33.998	0.02941	10.059	0.09941	18
19	3.616	0.2765	37.378	0.02675	10.336	0.09675	19
20	3.870	0.2584	40.995	0.02439	10.594	0.09439	20
21	4.140	0.2415	44.864	0.02229	10.835	0.09229	21
22	4.430	0.2257	49.005	0.02041	11.061	0.09041	22
23	4.740	0.2110	53.435	0.01871	11.272	0.08871	23
24	5.072	0.1972	58.175	0.01719	11.469	0.08719	24
25	5.427	0.1843	63.247	0.01581	11.654	0.08581	25
26	5.807	0.1722	68.675	0.01456	11.826	0.08456	26
27	6.214	0.1609	74.482	0.01343	11.987	0.08343	27
28	6.649	0.1504	80.695	0.01239	12.137	0.08239	28
29	7.114	0.1406	87.344	0.01145	12.278	0.08145	29
30	7.612	0.1314	94.458	0.01059	12.409	0.08059	30
35	10.676	0.0937	138.233	0.00723	12.948	0.07723	35
40	14.974	0.0668	199.628	0.00501	13.332	0.07501	40
50	29.456	0.0339	406.511	0.00246	13.801	0.07246	50
75	159.866	0.0063	2269.516	0.00044	14.196	0.07044	75
100	867.644	0.0012	12380.633	0.00008	14.269	0.07008	100

8% FACTORS INTEREST

n	Single Payment: Compound Amount Factor CAF, Given P to Find F, $(1+i)^n$	Single Payment: Present Worth Factor PWSP, Given F to Find P, $\frac{1}{(1+i)^n}$	Uniform Series: Compound Amount Factor USCA, Given A to Find F, $\frac{(1+i)^n-1}{i}$	Uniform Series: Sinking Fund Factor SFF, Given F to Find A, $\frac{i}{(1+i)^n-1}$	Uniform Series: Present Worth Factor PWUS, Given A to Find P, $\frac{(1+i)^n-1}{i(1+i)^n}$	Uniform Series: Capital Recovery Factor CRF, Given P to Find A, $\frac{i(1+i)^n}{(1+i)^n-1}$	n
1	1.080	0.9259	1.000	1.00000	0.926	1.08000	1
2	1.166	0.8573	2.080	0.48077	1.783	0.56077	2
3	1.260	0.7938	3.246	0.30804	2.577	0.38804	3
4	1.360	0.7350	4.506	0.22192	3.312	0.30192	4
5	1.469	0.6806	5.867	0.17046	3.993	0.25046	5
6	1.587	0.6302	7.336	0.13632	4.623	0.21632	6
7	1.714	0.5835	8.923	0.11207	5.206	0.19207	7
8	1.851	0.5403	10.637	0.09402	5.747	0.17402	8
9	1.999	0.5003	12.487	0.08008	6.247	0.16008	9
10	2.159	0.4632	14.486	0.06903	6.710	0.14903	10
11	2.332	0.4289	16.645	0.06008	7.139	0.14008	11
12	2.518	0.3971	18.977	0.05270	7.536	0.13270	12
13	2.720	0.3677	21.495	0.04652	7.904	0.12652	13
14	2.937	0.3405	24.215	0.04130	8.244	0.12130	14
15	3.172	0.3152	27.152	0.03683	8.559	0.11683	15
16	3.426	0.2919	30.324	0.03298	8.851	0.11298	16
17	3.700	0.2703	33.750	0.02963	9.122	0.10963	17
18	3.996	0.2503	37.450	0.02670	9.372	0.10670	18
19	4.316	0.2317	41.446	0.02413	9.604	0.10413	19
20	4.661	0.2146	45.761	0.02185	9.818	0.10185	20
21	5.034	0.1987	50.422	0.01983	10.017	0.09983	21
22	5.436	0.1839	55.456	0.01803	10.201	0.09803	22
23	5.871	0.1703	60.892	0.01642	10.371	0.09642	23
24	6.341	0.1577	66.764	0.01498	10.529	0.09498	24
25	6.848	0.1460	73.105	0.01368	10.675	0.09368	25
26	7.396	0.1352	79.953	0.01251	10.810	0.09251	26
27	7.988	0.1252	87.349	0.01145	10.935	0.09145	27
28	8.627	0.1159	95.337	0.01049	11.051	0.09049	28
29	9.317	0.1073	103.964	0.00962	11.158	0.08962	29
30	10.062	0.0994	113.281	0.00883	11.258	0.08883	30
35	14.785	0.0676	172.313	0.00580	11.655	0.08580	35
40	21.724	0.0460	259.050	0.00386	11.925	0.08386	40
50	46.900	0.0213	573.753	0.00174	11.233	0.08174	50
75	321.190	0.0031	4002.378	0.00025	11.461	0.08025	75
100	2199.630	0.0005	27482.879	0.00004	11.494	0.08004	100

APPENDIX K
AGC-CSI List of Cost Accounts

CONDITIONS OF THE CONTRACT

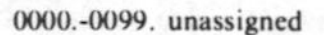

0000.-0099. unassigned

1

GENERAL REQUIREMENTS

0100. ALTERNATES OF PROJECT SCOPE
0101.-0109. unassigned
0110. SCHEDULES & REPORTS
0111.-0119. unassigned
0120. SAMPLES & SHOP DRAWINGS
0121.-0129. unassigned
0130. TEMPORARY FACILITIES
0131.-0139. unassigned
0140. CLEANING UP
0141.-0149. unassigned
0150. PROJECT CLOSEOUT
0151.-0159. unassigned
0160. ALLOWANCES
0161.-0199. unassigned

SITE WORK

0200. ALTERNATES
0201.-0209. unassigned
0210. CLEARING OF SITE
0211. Demolition
0212. Structures Moving
0213. Clearing & Grubbing
0214.-0219. unassigned
0220. EARTHWORK
0221. Site Grading
0222. Excavating & Backfilling
0223. Dewatering
0224. Subdrainage
0225. Soil Poisoning
0226. Soil Compaction Control
0227. Soil Stabilization
0228.-0229. unassigned
0230. PILING
0231.-0234. unassigned
0235. CAISSONS
0236.-0239. unassigned
0240. SHORING & BRACING
0241. Sheeting
0242. Underpinning
0243.-0249. unassigned
0250. SITE DRAINAGE
0251.-0254. unassigned
0255. SITE UTILITIES
0256.-0259. unassigned
0260. ROADS & WALKS
0261. Paving
0262. Curbs & Gutters
0263. Walks
0264. Road & Parking Appurtenances
0265.-0269. unassigned
0270. SITE IMPROVEMENTS
0271. Fences
0272. Playing Fields
0273. Fountains
0274. Irrigation System
0275. Yard Improvements
0276.-0279. unassigned
0280. LAWNS & PLANTING
0281. Soil Preparation
0282. Lawns
0283. Ground Covers & Other Plants
0284. Trees & Shrubs
0285.-0289. unassigned
0290. RAILROAD WORK
0291.-0294. unassigned
0295. MARINE WORK
0296. Boat Facilities
0297. Protective Marine Structures
0298. Dredging
0299. unassigned

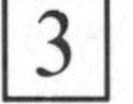

CONCRETE

0300. ALTERNATES
0301.-0309. unassigned
0310. CONCRETE FORMWORK
0311.-0319. unassigned
0320. CONCRETE REINFORCEMENT
0321.-0329. unassigned
0330. CAST-IN-PLACE CONCRETE
0331. Heavyweight Aggregate Concrete
0332. Lightweight Aggregate Concrete
0333. Post-Tensioned Concrete
0334. Nailable Concrete
0335. Specially Finished Concrete
0336. Specially Placed Concrete
0337.-0339. unassigned
0340. PRECAST CONCRETE
0341. Precast Concrete Panels
0342. Precast Structural Concrete
0343. Precast Prestressed Concrete
0344.-0349. unassigned
0350. CEMENTITIOUS DECKS
0351. Poured Gypsum Deck
0352. Insulating Concrete Roof Decks
0353. Cementitious Unit Decking
0354.-0399. unassigned

4

MASONRY

0400. ALTERNATES
0401.-0409. unassigned
0410. MORTAR
0411.-0419. unassigned
0420. UNIT MASONRY
0421. Brick Masonry
0422. Concrete Unit Masonry
0423. Clay Backing Tile
0424. Clay Facing Tile
0425. Creamic Veneer
0426. Pavers
0427. Glass Unit Masonry
0428. Gypsum Unit Masonry
0429. Reinforced Masonry
0430.-0439. unassigned
0440. STONE
0441. Rough Stone
0442. Cut Stone
0443. Simulated Stone
0444. Flagstone
0445.-0449. unassigned
0450. MASONRY RESTORATION
0451.-0499. unassigned

5 METALS

0500. ALTERNATES
0501.-0509. unassigned
0510. STRUCTURAL METAL
0511.-0519. unassigned
0520. OPEN-WEB JOISTS
0521.-0529. unassigned
0530. METAL DECKING
0531.-0539. unassigned
0540. LIGHTGAGE FRAMING
0541.-0549. unassigned
0550. MISCELLANEOUS METAL
0551. Metal Stairs
0552. Floor Gratings
0553. Construction Castings
0554.-0569. unassigned
0570. ORNAMENTAL METAL
0571.-0579. unassigned
0580. SPECIAL FORMED METAL
0581.-0599. unassigned

6 CARPENTRY

0600. ALTERNATES
0601.-0609. unassigned
0610. ROUGH CARPENTRY
0611. Framing & Sheathing
0612. Heavy Timber Work
0613.-0619. unassigned
0620. FINISH CARPENTRY
0621. Wood Trim
0622. Millwork
0623. Wood Siding
0624.-0629. unassigned
0630. GLUE-LAMINATED WOOD
0631.-0639. unassigned
0640. CUSTOM WOODWORK
0641. Custom Cabinetwork
0642. Custom Panelwork
0643.-0699. unassigned

7 MOISTURE PROTECTION

0700. ALTERNATES
0701.-0709. unassigned
0710. WATERPROOFING
0711. Membrane Waterproofing
0712. Hydrolithic Waterproofing
0713. Liquid Waterproofing
0714. Metallic Oxide Waterproofing
0715. DAMPPROOFING
0716. Bituminous Dampproofing
0717. Silicone Dampproofing
0718. Cementitious Dampproofing
0719. Preformed Vapor Barrier
0720. BUILDING INSULATION
0721.-0729. unassigned
0730. SHINGLES & ROOFING TILES
0731. Asphalt Shingles
0732. Asbestos-Cement Shingles
0733. Wood Shingles
0734. Slate Shingles
0735. Clay Roofing Tiles
0736. Concrete Roofing Tiles
0737. Porcelain Enamel Shingles
0738. Metal Shingles
0739. unassigned
0740. PREFORMED ROOFING & SIDING
0741. Preformed Metal Roofing
0742. Preformed Metal Siding
0743. Asbestos-Cement Panels
0744. Preformed Plastic Panels
0745. Custom Panel Roofing
0746.-0749. unassigned
0750. MEMBRANE ROOFING
0751. Builtup Bituminous Roofing
0752. Prepared Roll Roofing
0753. Elastic Sheet Roofing
0754. Elastic Liquid Roofing
0755.-0759. unassigned
0760. SHEET METAL WORK
0761. Sheet Metal Roofing
0762. Metal Roof Flashing & Trim
0763. Gutters & Downspouts
0764. Grilles & Louvers
0765. Decorative Sheet Metal Work
0766.-0769. unassigned
0770. WALL FLASHING
0771.-0779. unassigned
0780. ROOF ACCESSORIES
0781. Plastic Skylights
0782. Metal-Framed Skylights
0783. Roof Hatches
0784. Gravity Ventilators
0785.-0789. unassigned
0790. CALKING & SEALANTS
0791.-0799. unassigned

8 DOORS, WINDOWS, & GLASS

0800. ALTERNATES
0801.-0809. unassigned
0810. METAL DOORS & FRAMES
0811. Hollow Metal Doors & Frames
0812. Aluminum Doors & Frames
0813. Stainless Steel Doors & Frames
0814. Bronze Doors & Frames
0815. Metal Storm & Screen Doors
0816.-0819. unassigned
0820. WOOD DOORS
0821.-0829. unassigned
0830. SPECIAL DOORS
0831. Sliding Metal Firedoors
0832. Metal-Covered Doors
0833. Coiling Doors & Grilles
0834. Plastic-Faced Doors
0835. Folding Doors
0836. Overhead Doors
0837. Sliding Glass Doors
0838. Tempered Glass Doors
0839. Revolving Doors
0840. Flexible Doors
0841. Hangar Doors
0842.-0849. unassigned
0850. METAL WINDOWS
0851. Steel Windows
0852. Aluminum Windows
0853. Stainless Steel Windows
0854. Bronze Windows
0855.-0859. unassigned
0860. WOOD WINDOWS
0861.-0869. unassigned
0870. FINISH HARDWARE
0871.-0874. unassigned
0875. OPERATORS
0876.-0879. unassigned
0880. WEATHERSTRIPPING
0881.-0884. unassigned
0885. GLASS & GLAZING
0886.-0889. unassigned
0890. CURTAINWALL SYSTEM
0891.-0894. unassigned
0895. STOREFRONT SYSTEM
0896.-0899. unassigned

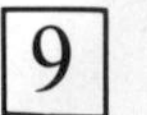

FINISHES

0900. ALTERNATES
0901.-0909. unassigned
0910. LATH & PLASTER
0911. Metal Furring
0912. Metal Lath
0913. Gypsum Lath
0914. Plaster Partition Systems
0915. Plastering Accessories
0916. Plaster
0917. Stucco
0918. Acoustical Plaster
0919. Plaster Moldings & Ornaments
0920.-0924. unassigned
0925. GYPSUM DRYWALL
0926. Gypsum Drywall Systems
0927. Gypsum Drywall Finishing
0928.-0929. unassigned
0930. TILE WORK
0931. Ceramic Tile
0932. Ceramic Mosaics
0933. Quarry Tile
0934. Glass Mosaics
0935. Conductive Ceramic Tile
0936.-0939. unassigned
0940. TERRAZZO
0941. Cast-In-Place Terrazzo
0942. Precast Terrazzo
0943. Conductive Terrazzo
0944. unassigned
0945. VENEER STONE
0946.-0949. unassigned
0950. ACOUSTICAL TREATMENT
0951.-0954. unassigned
0955. WOOD FLOORING
0956. Wood Strip Flooring
0957. Wood Parquet Flooring
0958. Plywood Block Flooring
0959. Resilient Wood Floor Systems
0960. Wood Block Industrial Floor
0961.-0964. unassigned
0965. RESILIENT FLOORING
0966. Resilient Tile Flooring
0967. Resilient Sheet Flooring
0968. Conductive Resilient Floors

9 **FINISHES, continued**

0969. unassigned
0970. SPECIAL FLOORING
0971. Magnesium Oxychloride Floors
0972. Epoxy-Marble-Chip Flooring
0973. Elastomeric Liquid Floors
0974. Heavy-Duty Concrete Toppings
0975.-0979. unassigned
0980. SPECIAL COATINGS
0981. Cementitious Coatings
0982. Elastomeric Coatings
0983. Fire-Resistant Coatings
0984.-0989. unassigned
0990. PAINTING
0991.-0994. unassigned
0995. WALL COVERING
0996.-0999. unassigned

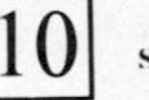

SPECIALTIES

1000. ALTERNATES
1001.-1009. unassigned
1010. CHALKBOARD & TACKBOARD
1011.-1012. unassigned
1013. CHUTES
1014. unassigned
1015. COMPARTMENTS & CUBICLES
1016. Hospital Cubicles
1017. Office Cubicles
1018. Toilet & Shower Compartments
1019. unassigned
1020. DEMOUNTABLE PARTITIONS
1021.-1022. unassigned
1023. DISAPPEARING STAIRS
1024. unassigned

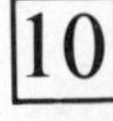

SPECIALTIES, continued

1025. FIREFIGHTING DEVICES
1026.-1029. unassigned
1030. FIREPLACE EQUIPMENT
1031. Fireplace Accessories
1032. Fireplace Dampers
1033. Prefabricated Fireplaces
1034. unassigned
1035. FLAGPOLES
1036. unassigned
1037. FOLDING GATES
1038.-1039. unassigned
1040. IDENTIFYING DEVICES
1041. Directory & Bulletin Boards
1042. Painted Signs
1043. Plaques
1044. Three-Dimensional Signs
1045.-1049. unassigned
1050. LOCKERS
1051.-1052. unassigned
1053. MESH PARTITIONS
1054. unassigned
1055. POSTAL SPECIALTIES
1056.-1059. unassigned
1060. RETRACTABLE PARTITIONS
1061. Coiling Partitions
1062. Folding Partitions
1063.-1064. unassigned
1065. SCALES
1066. unassigned
1067. STORAGE SHELVING
1068.-1069. unassigned
1070. SUN CONTROL DEVICES
1071.-1074. unassigned
1075. TELEPHONE BOOTHS
1076.-1079. unassigned
1080. TOILET & BATH ACCESSORIES
1081.-1084. unassigned
1085. VENDING MACHINES
1086.-1089. unassigned
1090. WARDROBE SPECIALTIES
1091.-1094. unassigned
1095. WASTE DISPOSAL UNITS
1096. Packaged Incinerators
1097. Waste Compactors
1098.-1099. unassigned

11 EQUIPMENT

1100. ALTERNATES
1101.-1109. unassigned
1110. BANK EQUIPMENT
1111. Depository Units
1112. Outdoor Tellers' Windows
1113. Safes
1114. Tellers' Counters
1115. COMMERCIAL EQUIPMENT
1116.-1117. unassigned
1118. DARKROOM EQUIPMENT
1119. unassigned
1120. ECCLESIASTICAL EQUIPMENT
1121. Baptismal Tanks
1122. Bells
1123. Carillons
1124. Chancel Fittings
1125. Organs
1126. Pews
1127.-1129. unassigned
1130. EDUCATIONAL EQUIPMENT
1131. Art & Draft Equipment
1132. Audio-Visual Aids
1133. Language Laboratories
1134. Prefabricated Astro-Observatories
1135. Vocational Shop Equipment
1136.-1139. unassigned
1140. FOOD SERVICE EQUIPMENT
1141. Bar Units
1142. Cooking Equipment
1143. Dishwashing Equipment
1144. Food Preparation Machines
1145. Food Preparation Tables
1146. Food Serving Units
1147. Refrigerated Cases
1148. Sinks & Drainboards
1149. Soda Fountains
1150. GYMNASIUM EQUIPMENT
1151.-1154. unassigned
1155. INDUSTRIAL EQUIPMENT
1156.-1159. unassigned
1160. Laboratory Equipment
1161.-1162. unassigned
1163. LAUNDRY EQUIPMENT
1164. unassigned
1165. LIBRARY EQUIPMENT
1166. Bookshelving
1167. Bookstacks
1168. Charging Counters
1169. unassigned
1170. MEDICAL EQUIPMENT
1171. Dental Equipment
1172. Examination Room Equipment

11 EQUIPMENT, continued

1173. Hospital Casework
1174. Incubators
1175. Patient Care Equipment
1176. Radiology Equipment
1177. Sterilizers
1178. Surgery Equipment
1179. Therapy Equipment
1180. MORTUARY EQUIPMENT
1181.-1184. unassigned
1185. PARKING EQUIPMENT
1186.-1187. unassigned
1188. PRISON EQUIPMENT
1189. unassigned
1190. RESIDENTIAL EQUIPMENT
1191. Central Vacuum Cleaner
1192. Kitchen & Lavatory Cabinets
1193. Residential Kitchen Equipment
1194. Residential Laundry Equipment
1195. Unit Kitchens
1196. unassigned
1197. STAGE EQUIPMENT
1198.-1199. unassigned

12 FURNISHINGS

1200. ALTERNATES
1201.-1209. unassigned
1210. ARTWORK
1211.-1219. unassigned
1220. BLINDS & SHADES
1221.-1229. unassigned
1230. CABINETS & FIXTURES
1231. Classroom Cabinets
1232. Dormitory Units
1233.-1239. unassigned
1240. CARPETS & MATS
1241.-1249. unassigned
1250. DRAPERY & CURTAINS
1251. Drapery Tracks
1252. Fabrics
1253.-1259. unassigned
1260. FURNITURE
1261.-1269. unassigned
1270. SEATING
1271. Auditorium Seating
1272. Classroom Seating
1273. Stadium Seating
1274.-1299. unassigned

13 SPECIAL CONSTRUCTION

1300. ALTERNATES
1301.-1309. unassigned
1310. AUDIOMETRIC ROOMS
1311.-1314. unassigned
1315. BOWLING ALLEYS
1316.-1319. unassigned
1320. BROADCASTING STUDIOS
1321.-1324. unassigned
1325. CLEAN ROOMS
1326.-1329. unassigned
1330. CONSERVATORIES
1331.-1334. unassigned
1335. HYPERBARIC ROOMS
1336.-1339. unassigned
1340. INCINERATORS
1341.-1344. unassigned
1345. INSULATED ROOMS
1346.-1349. unassigned
1350. INTEGRATED CEILINGS
1351.-1354. unassigned
1355. OBSERVATORIES
1356.-1359. unassigned
1360. PEDESTAL FLOORS
1361.-1364. unassigned
1365. PREFABRICATED STRUCTURES
1366.-1369. unassigned
1370. RADIATION PROTECTION
1371.-1374. unassigned
1375. SPECIAL CHIMNEY CONSTRUCTION
1376.-1379. unassigned
1380. STORAGE VAULTS
1381.-1384. unassigned
1385. SWIMMING POOLS
1386.-1389. unassigned
1390. ZOO STRUCTURES
1391.-1399. unassigned

14 CONVEYING SYSTEM

1400. ALTERNATES
1401.-1409. unassigned
1410. DUMBWAITERS
1411.-1419. unassigned
1420. ELEVATORS
1421.-1429. unassigned
1430. HOISTS & CRANES
1431.-1439. unassigned
1440. LIFTS
1441.-1449. unassigned

MECHANICAL

1500. ALTERNATES
1501.-1509. unassigned
1510. BASIC MATERIALS & METHODS
1511. Pipe & Pipefittings
1512. Valves
1513. Piping Specialties
1514. Mechanical Supporting Devices
1515. Vibration Isolation
1516. Mechanical Systems Insulation
1517.-1519. unassigned
1520. WATER SUPPLY SYSTEM
1521. Water Supply Piping
1522. Domestic Hot Water System
1523. Domestic Iced Water System
1524. Water Well & Wellpump
1525. SOIL & WASTE SYSTEM
1526. Soil & Waste Piping
1527. Waste Treatment Equipment
1528. Sanitary Sewers
1529. ROOF DRAINAGE SYSTEM
1530. PLUMBING FIXTURES & TRIM
1531.-1534. unassigned
1535. GAS PIPING SYSTEM
1536.-1539. unassigned
1540. SPECIAL PIPING SYSTEMS
1541. Compressed Air System
1542. Vacuum Piping System
1543. Oxygen Piping System
1544. Nitrous Oxide Piping System
1545. Process Piping System
1546.-1549. unassigned
1550. FIRE EXTINGUISHING SYSTEM
1551. Automatic Sprinkler System
1552. Carbon Dioxide System
1553. Elevated Water Reservoir
1554. Standpipe & Firehose Stations
1555. Underground Fire Lines
1556. unassigned
1557. FUEL HANDLING SYSTEM
1558.-1559. unassigned
1560. STEAM HEATING SYSTEM
1561. Steam Boiler & Equipment
1562. Steam Circulating System
1563. Steam Terminal Units
1564. unassigned
1565. HOT WATER HEATING SYSTEM
1566. Hot Water Boiler & Equipment

15 MECHANICAL, continued

1567. Hot Water Circulating System
1568. Hot Water Terminal Units
1569. Hot Water Snow-Melting System
1570.-1574. unassigned
1575. CHILLED WATER SYSTEM
1576. unassigned
1577. DUAL-TEMPERATURE SYSTEM
1578. HEAT PUMPS (see also 1684).
1579. unassigned
1580. AIR-TEMPERING SYSTEM
1581. Warm Air Furnaces
1582. Air-Handling Equipment
1583. Air Filtration Equipment
1584. Humidity Control Equipment
1585. Packaged Air-Tempering Units
1586. Air Distribution Duct System
1587. Tempered Air Terminal Units
1588. Air Curtain
1589. unassigned
1590. REFRIGERATION
1591. Water Chillers
1592. Commercial Refrigeration Units
1593. Cooling Towers
1594. unassigned
1595. HVC CONTROLS & INSTRUMENTS
1596.-1599. unassigned

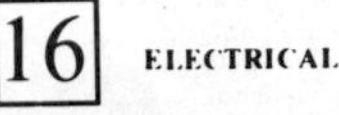

ELECTRICAL

1600. ALTERNATES
1601.-1609. unassigned
1610. BASIC MATERIALS & METHODS
1611. Raceways & Fittings
1612. Busways
1613. Conductors
1614. Electrical Supporting Devices
1615.-1619. unassigned
1620. ELECTRICAL SERVICE SYSTEM
1621. Overhead Electrical Service

ELECTRICAL, continued

1622. Underground Electrical Service
1623. Electrical Substations
1624. Electrical Entrance Equipment
1625. Grounding System
1626. Standby Electrical System
1627.-1629. unassigned
1630. ELECTRICAL DISTRIBUTION SYSTEM
1631. Feeder Circuits
1632. Branch Circuits
1633. Panelboards
1634. Wiring Devices
1635. Underfloor Electrical System
1636.-1639. unassigned
1640. LIGHTING FIXTURES
1641. Indoor Lighting Fixtures
1642. Outdoor Lighting Fixtures
1643.-1649. unassigned
1650. COMMUNICATION SYSTEM
1651. Telephone Equipment
1652. Intercommunication System
1653. Public Address System
1654. Paging System
1655. Nurses' Call System
1656. Alarm & Detection System
1657. Clock & Program System
1658. Audio-Video Reproducers
1659. Closed-Circuit Television
1660. Radiotelephone System
1661. Commercial Projection System
1662.-1669. unassigned
1670. ELECTRICAL POWER EQUIPMENT
1671. Motors & Motor Controls
1672. Special Transformers
1673. Frequency Converters
1674. Rectifiers
1675.-1679. unassigned
1680. ELECTRICAL COMFORT SYSTEM
1681. Electrical Heating System
1682. Packaged Air-Tempering Units
1683. Electrical Snow-Melting System
1684. Heat Pumps (see also 1578.)
1685.-1689. unassigned
1690. ELECTRICAL SYSTEM CONTROLS & INSTRUMENTS
1691.-1694. unassigned
1695. LIGHTNING PROTECTION SYSTEM
1696.-1699. unassigned

APPENDIX L
Plans for a Small Gas Station

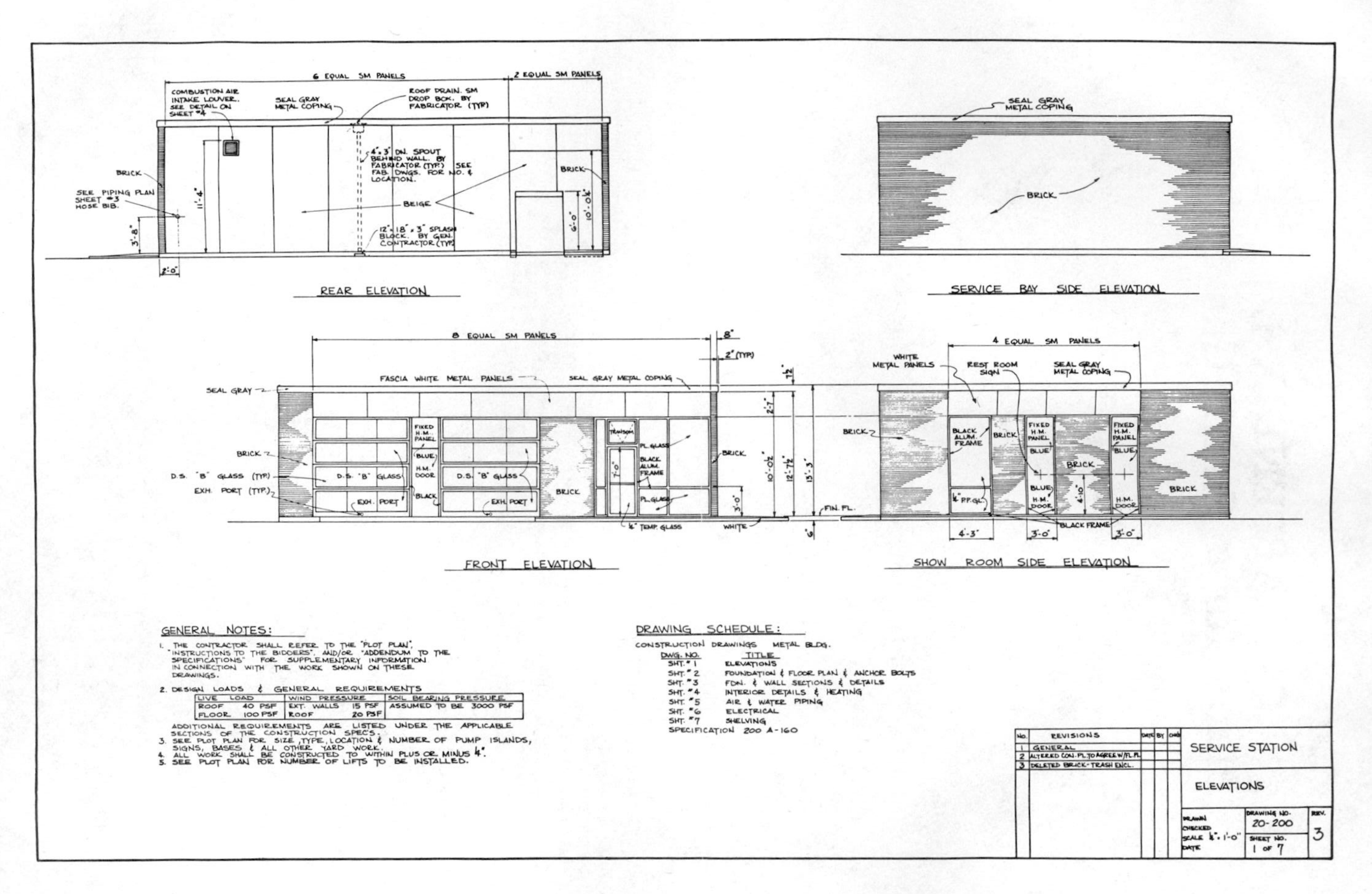

6 EQUAL SM PANELS
2 EQUAL SM PANELS
COMBUSTION AIR INTAKE LOUVER. SEE DETAIL ON SHEET #4
SEAL GRAY METAL COPING
ROOF DRAIN. SM DROP BOX. BY FABRICATOR (TYP)
4" x 3" DN. SPOUT BEHIND WALL. BY FABRICATOR (TYP.) SEE FAB. DWGS. FOR NO. & LOCATION.
BRICK
SEE PIPING PLAN SHEET #3 HOSE BIB.
BEIGE
12" x 18" x 3" SPLASH BLOCK. BY GEN. CONTRACTOR (TYP)
11'-4"
3'-8"
2'-0"
6'-0"
10'-0 1/4"
REAR ELEVATION
SEAL GRAY METAL COPING
BRICK
SERVICE BAY SIDE ELEVATION
8 EQUAL SM PANELS
8"
2" (TYP)
FASCIA WHITE METAL PANELS
SEAL GRAY METAL COPING
SEAL GRAY
BRICK
D.S. "B" GLASS (TYP)
EXH. PORT (TYP.)
D.S. "B" GLASS
EXH. PORT
FIXED H.M. PANEL
BLUE
H.M. DOOR
BLACK
TRANSOM
PL. GLASS
BLACK ALUM. FRAME
7'-0"
3'-0"
1/4" TEMP. GLASS
WHITE
2'-7"
10'-0 1/2"
12'-7 1/2"
13'-3"
FIN. FL.
6"
FRONT ELEVATION
4 EQUAL SM PANELS
WHITE METAL PANELS
REST ROOM SIGN
BLACK ALUM. FRAME
1/4" P.P. GL.
H.M. DOOR
BLACK FRAME
4'-3"
3'-0"
6'-10"
SHOW ROOM SIDE ELEVATION
GENERAL NOTES:
1. THE CONTRACTOR SHALL REFER TO THE "PLOT PLAN", "INSTRUCTIONS TO THE BIDDERS", AND/OR "ADDENDUM" TO THE SPECIFICATIONS" FOR SUPPLEMENTARY INFORMATION IN CONNECTION WITH THE WORK SHOWN ON THESE DRAWINGS.
2. DESIGN LOADS & GENERAL REQUIREMENTS
LIVE LOAD | WIND PRESSURE | SOIL BEARING PRESSURE
ROOF 40 PSF | EXT. WALLS 15 PSF | ASSUMED TO BE 3000 PSF
FLOOR 100 PSF | ROOF 20 PSF
ADDITIONAL REQUIREMENTS ARE LISTED UNDER THE APPLICABLE SECTIONS OF THE CONSTRUCTION SPECS.
3. SEE PLOT PLAN FOR SIZE, TYPE, LOCATION & NUMBER OF PUMP ISLANDS, SIGNS, BASES & ALL OTHER YARD WORK.
4. ALL WORK SHALL BE CONSTRUCTED TO WITHIN PLUS OR MINUS 1/4".
5. SEE PLOT PLAN FOR NUMBER OF LIFTS TO BE INSTALLED.
DRAWING SCHEDULE:
CONSTRUCTION DRAWINGS METAL BLDG.
DWG. NO. TITLE
SHT. #1 ELEVATIONS
SHT. #2 FOUNDATION & FLOOR PLAN & ANCHOR BOLTS
SHT. #3 FDN. & WALL SECTIONS & DETAILS
SHT. #4 INTERIOR DETAILS & HEATING
SHT. #5 AIR & WATER PIPING
SHT. #6 ELECTRICAL
SHT. #7 SHELVING
SPECIFICATION 200 A-160
NO. REVISIONS DATE BY CHK
1 GENERAL
2 ALTERED CON. FL. TO AGREE W/ FL. PL.
3 DELETED BRICK-TRASH ENCL.
SERVICE STATION
ELEVATIONS
DRAWN
CHECKED
SCALE 1/4" = 1'-0"
DATE
DRAWING NO. 20-200
SHEET NO. 1 OF 7
REV. 3

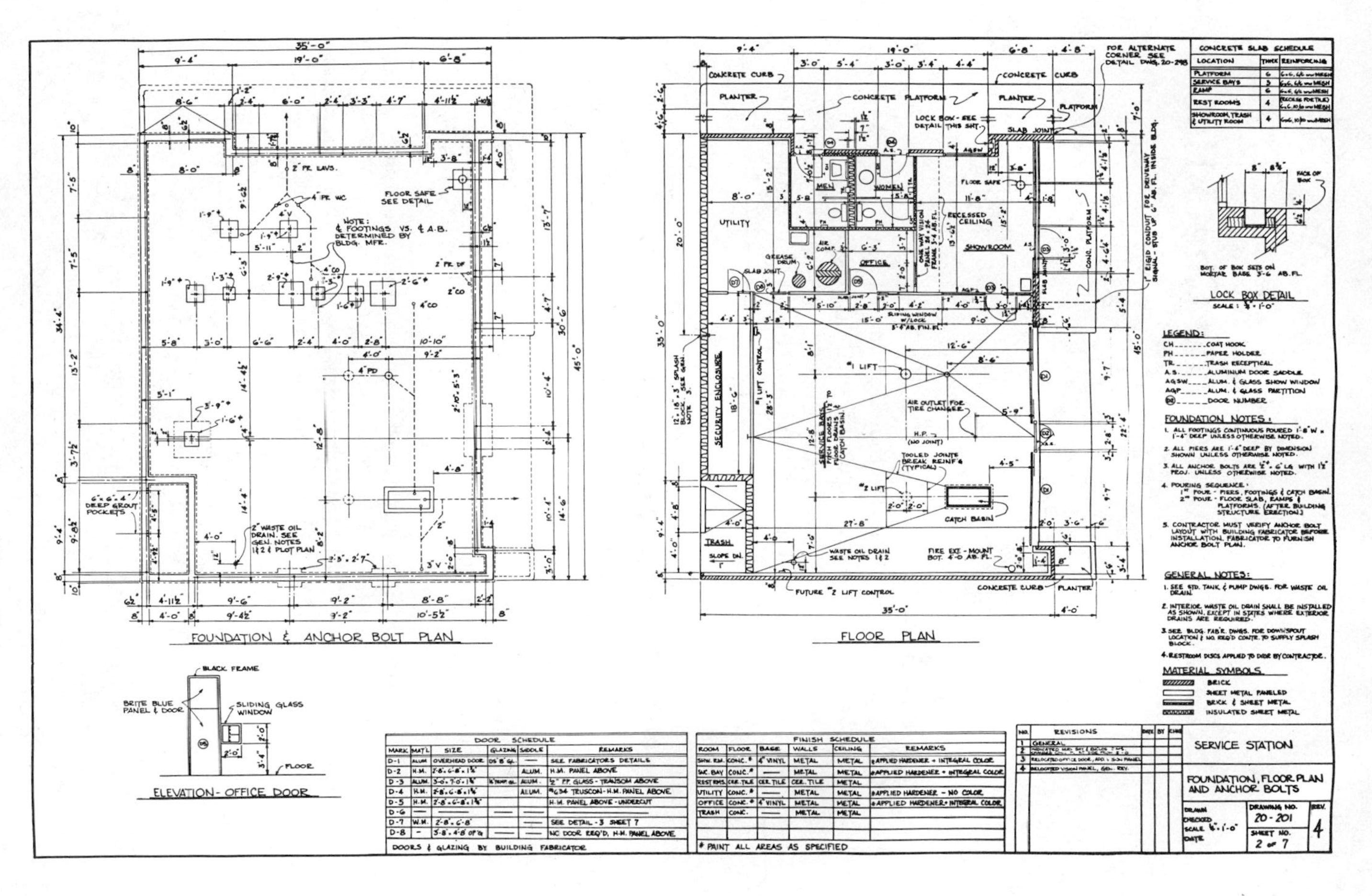

CONCRETE SLAB SCHEDULE		
LOCATION	THICK	REINFORCING
PLATFORM	6	6x6, 6/6 ww MESH
SERVICE BAYS	5	6x6, 6/6 ww MESH
RAMP	6	6x6, 6/6 ww MESH
REST ROOMS	4	(RECESS FOR TILE) 6x6, 10/10 ww MESH
SHOWROOM, TRASH & UTILITY ROOM	4	6x6, 10/10 ww MESH

DOOR SCHEDULE					
MARK	MAT'L	SIZE	GLAZING	SADDLE	REMARKS
D-1	ALUM	OVERHEAD DOOR	DS'B' GL.	—	SEE FABRICATORS DETAILS
D-2	H.M.	2'-8" x 6'-8" x 1 3/4"		ALUM.	H.M. PANEL ABOVE
D-3	ALUM.	3'-0" x 7'-0" x 1 3/4"	1/4" TEMP. GL.	ALUM.	1/2" P.P. GLASS - TRANSOM ABOVE
D-4	H.M.	2'-8" x 6'-8" x 1 3/4"		ALUM.	#634 TRUSCON - H.M. PANEL ABOVE
D-5	H.M.	2'-8" x 6'-8" x 1 3/4"			H.M. PANEL ABOVE - UNDERCUT
D-6	—	—	—	—	—
D-7	W.M.	2'-8" x 6'-8"		—	SEE DETAIL - 3 SHEET 7
D-8	-	3'-8" x 4'-8" OP'G	—	—	NO DOOR REQ'D, H.M. PANEL ABOVE
DOORS & GLAZING BY BUILDING FABRICATOR					

FINISH SCHEDULE					
ROOM	FLOOR	BASE	WALLS	CEILING	REMARKS
SHW. RM.	CONC.*	4" VINYL	METAL	METAL	*APPLIED HARDENER + INTEGRAL COLOR
SVC. BAY	CONC.*	—	METAL	METAL	*APPLIED HARDENER + INTEGRAL COLOR
REST RMS.	CER. TILE	CER. TILE	CER. TILE	METAL	
UTILITY	CONC.*	—	METAL	METAL	*APPLIED HARDENER - NO COLOR
OFFICE	CONC.*	4" VINYL	METAL	METAL	*APPLIED HARDENER + INTEGRAL COLOR
TRASH	CONC.	—	METAL	METAL	
* PAINT ALL AREAS AS SPECIFIED					

APPENDIX M
Site Reconnaissance Checklist

Site Reconnaissance Checklist

General Considerations:

I. What features are native to topology and climate?
II. What is required for construction method selected?
III. What features are needed to support construction force?
IV. What features might encroach on local society or environment?

I. Features native to topology and climate
- A. actual topology (excessive grades, etc.)
- B. elevation
- C. geology (soil characteristics, rock, etc.)
- D. ground cover
- E. excessive seasonal effects
- F. wind direction
- G. natural defenses
- H. drainage
- I. subsurface water conditions
- J. seismic zones

II. Features required that contribute to construction method
- A. accessibility to site (rail, road, water)
- B. labor availability (skill, cost, attitude)
- C. material availability (salvage, cost, attitude)
- D. locate borrow pits (gravel, sand, base, fill)
- E. locate storage areas, plant sites
- F. alternate building, campsites
- G. general working room about site
- H. location of existing structures and utilities
- I. conflicts with existing structures and utilities
- J. overhead
- K. disposal areas
- L. land usage
- M. local building practices

III. Features to support construction force
  A. billeting/shelter
  B. food (also on-job meals)
  C. special equipment
  D. clothing
  E. communications
  F. local hazards
  G. fire/security protection available
  H. local customs/culture
  I. potable H_2O
  J. sanitary facilities (also for job)
  K. entertainment
  L. small stores
  M. medical
  N. banking, currency
  O. transportation
  P. local maintenance available

IV. Features that might encroach on local society or environment
  A. noise
  B. dust
  C. blasting
  D. hauling over roads
  E. use of water
  F. burning (smoke)
  G. drainage (create problems)
  H. flight operations
  I. disposal areas
  J. utility disruption
  K. relocation problems
  L. work hours
  M. economy impact
  N. community attitude
  O. security
  P. political

Bibliography

Allen, C. R., "The Construction Management Concept," *AACE Bulletin*, vol. 15, no. 6, December 1973, pp. 169-173.

Antill, James M., and Ronald W. Woodhead, *Critical Path Methods in Construction Practice*, 2nd ed., John Wiley & Sons, Inc., New York, 1970.

Barrie, Donald S., "CM as Seen by an Engineer-Contractor," *Plant Engineering*, July 13, 1972, p. 85.

——, and Boyd C. Paulson, Jr., "Professional Construction Management," *Journal of the Construction Division*, ASCE, vol. 102, no. C03, proc. paper 12394, September 1976, pp. 425-436.

Barrie, Donald S., and Boyd C. Paulson, Jr., *Professional Construction Management*, McGraw-Hill Book Co., New York, 1978.

Battersby, A., *Network Analysis*, 3rd ed., The MacMillan Co., New York, 1970.

Bonny, John B., and Joseph P. Frein, *Handbook of Construction Management and Organization*, Von Nostrand Reinhold Co., New York, 1973.

Building Construction Cost Data, Robert Snow Means Co., Inc., Kingston, Mass. Published Annually.

Building Estimator's Reference Book, 18th ed., The Frank R. Walker Co., Chicago, 1973.

Burman, Peter J., *Precedence Networks for Project Planning and Control*, McGraw-Hill Book Co., London, 1972.

Bush, Vincent G., *Construction Management*, Reston Publishing Co., Inc., Reston, Va., 1973.

Bush, Vincent G., *Safety in the Construction Industry*, Reston Publishing Co., Inc., Reston, Va., 1975.

California Construction Safety Orders, Dept. of Industrial Relations, Division of Industrial Safety, San Francisco.

Caterpillar Performance Handbook, 9th ed., Caterpillar Tractor Company, Peoria, Ill., 1979.

Clough, Richard H., *Construction Contracting*, 3d ed., John Wiley & Sons, Inc., New York, 1975.

CM for the General Contractor: A Guide Manual for Construction Management, The Associated General Contractors of America, Washington, D.C., 1975.

Collier, Keith, *Construction Contracts*, Reston Publishing Co., Inc., Reston, Va., 1979.

Collins, Carroll J., "Impact—The Real Cost of Change Orders," *Transactions of the American Association of Cost Engineers*, Morgantown, W. Va., June 1970, pp. 188-191.

Construction Contracting Systems: A Report on the Systems Used by PBS and Other Organizations, General Services Administration, Public Buildings Service, Washington, D.C., March 1970.

Construction Industry: OSHA Safety and Health Standards Digest, OSHA 2202, U. S. Government Printing Office, Superintendent of Documents, Washington, D.C. Revised June 1975.

"Construction Management Guidelines for Use by AGC Members," Attachment to Special Contracting Methods Committee Report, The Associated General Contractors of America, Washington, D.C., February 15, 1972.

"Construction Management: Putting Professionalism into Contracting," *Construction Methods and Equipment*, vol. 54, no. 3, March 1972, pp. 69-75.

"Construction Management—Whirling in Evolution and in Ferment," *Engineering News-Record*, vol. 188, no. 18, May 4, 1972, pp. 14-19.

Construction Safety and Health Regulations: Part 1926, OSHA 2207, U.S. Government Printing Office, Superintendent of Documents, Washington, D.C., June 1974.

Construction Safety and Health Training, General Services Administration, National Audiovisual Center, Washington, D.C. (Manuals and slides for 30-hour course.)

Contractors Equipment Ownership Expense, Associated General Contractors of America, Washington, D.C., 1974.

Cost Control and CPM in Construction, The Associated General Contractors of America, Washington, D.C., 1968.

CPM in Construction, A Manual for General Contractors, The Associated General Contractors of America, Washington, D.C., 1965.

Credit Reports (individual subscription), Building Construction Division, Dun and Bradstreet, Inc., New York, N.Y.

Dallavia, Louis, *Estimating General Construction Costs*, 2nd ed., F. W. Dodge Co., New York, 1957.

Dell'Isola, Alphonse J., *Value Engineering in the Construction Industry*, Construction Publishing Company, Inc., New York, 1974 (now Van Nostrand Reinhold Co.).

deStwolinski, Lance W., *Occupational Health in the Construction Industry*, Technical Report No. 105, Stanford University, Dept. of Civil Engineering, The Construction Institute, May 1969.

______. *A Survey of the Safety Environment of the Construction Industry*, Technical Report No. 114, Stanford University, Dept. of Civil Engineering, The Construction Institute, October 1969.

Dodge Bulletin (daily publication), F.W. Dodge Corp., (Division of McGraw-Hill, Inc.) New York, N.Y.

Dodge Manual for Construction Pricing and Scheduling, McGraw-Hill Information Systems Co., New York. Published annually.

Douglas, James, *Construction Equipment Policy*, McGraw-Hill Book Co., New York, 1975.

Dunham, Clarence W., Robert D. Young, and Joseph T. Bockrath, *Contracts Specifications and Law for Engineers*, 3d ed., McGraw-Hill Book Co., New York, 1979.

Engineering News-Record. Published weekly by McGraw-Hill Inc., Inc.

Estey, Martin, *The Unions*, Harcourt, Brace Jovanovich, New York, 1967.

Fisk, Edward R., *Construction Project Administration*, John Wiley & Sons, Inc., New York, 1978.

Fondahl, John W., *A Non-Computer Approach to the Critical Path Method for the Construction Industry*, Technical Report No. 9, Stanford University, Dept. of Civil Engineering, The Construction Institute, 1964.

Fondahl, John W., and Ricardo R. Bacarreza, *Construction Contract Mark-up Related to Forecasted Cash Flow*, Technical Report No. 161, Stanford University, Department of Civil Engineering, The Construction Institute, November 1972.

Foxall, William B., *Professional Construction Management and Project Administration*, Architectural Record and The American Institute of Architects, New York, 1972.

Fundamentals of Earthmoving, Caterpillar Tractor Co., Peoria, Ill., April 1968.

General Safety Requirements, Manual EM 385-1-1, and Supplements 1 and 2, U.S. Army Corps of Engineers, Washington, D.C.,

Gibb, Thomas Wilson, Jr., *Building Construction in the Southeastern United States*. Report presented to the School of Civil Engineering, Georgia Institute of Technology, Atlanta, 1975.

Goldhaber, Stanley, Chandra K. Jha, and Manuel C. Macedu, Jr., *Construction Management Principles and Practices*, John Wiley & Sons, Inc., New York, 1977.

Gorman, James E., *Simplified Guide to Construction Management for Architects and Engineers*, Cahners Books International, Inc., Boston, 1976.

Grant, Eugene L., W. Grant Ireson, and Richard S. Leavenworth, *Principles of Engineering Economy*, 6th ed., The Ronald Press Company, New York, 1976.

The GSA System for Construction Management, General Services Administration, Public Buildings Service, Washington, D.C., April 1975.

Guide for Supplementary Conditions, AIA Document A 511, The American Institute of Architects, Washington, D.C., 1973.

Halperin, Don A. *Construction Funding*, John Wiley & Sons, Inc., New York, 1974.

Halpin, Daniel W., "CYCLONE"—Method for Modeling Job Site Processes." *Journal of the Construction Division*, American Society of Civil Engineers, vol. 103, No. C03, proc. paper 13234, September 1977, pp. 489-499.

Halpin, D. W., and Neathammer, R. D., *Construction Time Overruns*, Technical Report P-16, Construction Engineering Research Laboratory, Champaign, Ill., 1973.

Halpin, Daniel W., and Ronald W. Woodhead, *Constructo—A Heuristic Game for Construction Management*, University of Illinois Press, Urbana, Ill., 1973.

Halpin, Daniel W., and Ronald W. Woodhead, *Design of Construction and Process Operations*, John Wiley & Sons, Inc., New York, 1976.

Harris, Robert B., *Precedence and Arrow Networking Techniques for Construction*, John Wiley & Sons, Inc., New York, 1978.

Hauf, Harold D., *Building Contracts for Design and Construction*, John Wiley & sons, Inc., New York, 1968.

Herry, George T., "Construction Management Defined," *The Military Engineer*, No. 430, March-April 1974, p. 85.

Herry, George T., *Time, Cost and Architecture*, McGraw-Hill Book Co., New York, 1975.

Hinze, Jimmie, *The Effect of Middle Management on Safety in Construction*, Technical Report No. 209, Stanford University, Dept. of Civil Engineering, The Construction Institute, October 1969.

Jackson, John Howard, *Contract Law in Modern Society*, West Publishing Co., St. Paul, 1973.

Jordan, Mark H., and Robert I. Carr, "Education for the Professional Construction Manager," *Journal of the Construction Division*, ASCE, vol. 102, no. C03, proc. paper 12392, September 1976, pp. 511-519.

Kettle, Kenath A., "Project Delivery Systems for Construction Projects," *Journal of the Construction Division*, ASCE, vol. 102, no. C04, proc. paper 12594, December 1976, pp. 575-585.

Knox, H., "Construction Safety as it Relates to Insurance Costs," *AACE Bulletin*, vol. 16, no. 3, June 1974, pp. 71-73.

Levitt, Raymond E., *The Effect of Top Management on Safety in Construction*, Technical Report No. 196, Stanford University, Dept. of Civil Engineering, The Construction Institute, July 1975.

______, and Henry W. Parker, "Reducing Construction Accidents—Top Management's Role," *Journal of the Construction Division*, ASCE, vol. 102, no. C03, proc. paper 12384, September 1976, pp. 465-478.

Manual of Accident Prevention in Construction,, 6th ed., The Associated General Contractors of America, Washington, D.C., 1971.

Meredith, D. et al., *Design and Planning of Engineering Systems*, Prentice-Hall, Inc., Englewood Cliffs, N.J., 1973.

Mills, Daniel Quinn, *Industrial Relations and Manpower in Construction*, MIT Press, Cambridge, Mass., 1972.

Moder, Joseph J., and Cecil R. Phillips, *Project Management with CPM and PERT*, 2nd ed., van Nostrand Reinhold Co., New York, 1970.

Morgan, W. C., and L. Pearson, "Determining Shovel-Truck Productivity," *Mining Engineering*, December 1968.

Naaman, Antoine E., "Networking Methods for Project Planning and Control," *Journal of the Construction Division*, ASCE, Vol. 100, no. C03, proc. paper 10814, September 1974, pp. 357-372.

O'Brien, James J. (ed.), *CPM in Construction Management*, 2d ed., McGraw-Hill Book Company, New York, 1971.

______, *Scheduling Handbook*, McGraw-Hill Book Company, New York, 1969.

O'Brien, James J., and Robert G. Zilly (eds.), *Contractor's Management Handbook*, McGraw-Hill Book Company, New York, 1971.

"The Occupational Safety and Health Act of 1970," P.L. 91-596 (OSHA 2001), U.S. Government Printing Office, Superintendent of Documents, Washington, D.C., December 1970.

OSHA Safety and Health Training Guidelines for Construction, (PB-239 312/AS), U.S. Dept. of Commerce, National Technical Information Service, Springfield, Va.

Parker, Henry W., and Clarkson H. Oglesby, *Methods Improvement for Construction Managers*, McGraw-Hill book Company, New York, 1972.

Peurifoy, Robert L., *Estimating Construction Costs*, 3rd ed., McGraw-Hill Book Co., New York, 1975.

Pulver, Harry E., *Construction Estimates and Costs*, 4th ed., McGraw-Hill Book Co., New York, 1969.

Reiner, Lawrence E., *Handbook of Construction Management*, Prentice-Hall, Inc., Englewood Cliffs, N.J., 1972.

Rubey, Harry, John A. Logan, and Walker W. Milner, *The Engineer and Professional Management*, 3rd ed., The Iowa State Press, Ames, Iowa, 1970.

Safety Requirements for Construction by Contract, U.S. Dept. of the Interior, Bureau of Reclamation, Washington, D.C.

Samelson, Nancy Morse, *The Effect of Foremen on Safety in Construction*, Technical Report No. 219, Stanford University, Dept. of Civil Engineering, The Construction Institute, June, 1977.

Uniform Construction Index, the Construction Specifications Institute, Washington, D.C., 1972.

Value Engineering (Handbook), PBS P 8000.1 (Jan. 12, 1972) and Change 0.1 (March 2, 1973), U.S. General Services Administration, Washington, D.C.

Zehner, John R., *Builder's Guide to Contracting*, McGraw-Hill Book Company, New York, 1975.

Index